Human Anatomy Atlas

**Featuring Art from
Human Anatomy, First Edition**

By

Michael McKinley

Valerie Dean O'Loughlin

Higher Education

Boston   Burr Ridge, IL   Dubuque, IA   Madison, WI   New York   San Francisco   St. Louis
Bangkok   Bogotá   Caracas   Kuala Lumpur   Lisbon   London   Madrid   Mexico City
Milan   Montreal   New Delhi   Santiago   Seoul   Singapore   Sydney   Taipei   Toronto

The **McGraw·Hill** Companies

Human Anatomy Atlas featuring art from
HUMAN ANATOMY, FIRST EDITION
MICHAEL MCKINLEY AND VALERIE DEAN O'LOUGHLIN

Published by McGraw-Hill Higher Education, an imprint of The McGraw-Hill Companies, Inc., 1221 Avenue of the Americas, New York, NY 10020. Copyright © 2006 by The McGraw-Hill Companies, Inc. All rights reserved.

This book is printed on acid-free paper.

5 6 7 8 9 0 DOW/DOW 0 9 8

ISBN-13: 978-0-07-302841-5
ISBN-10: 0-07-302841-X

www.mhhe.com

Directory of Atlas Figures
Featuring Art from Human Anatomy, First Edition
Michael P. McKinley and Valerie Dean O'Loughlin

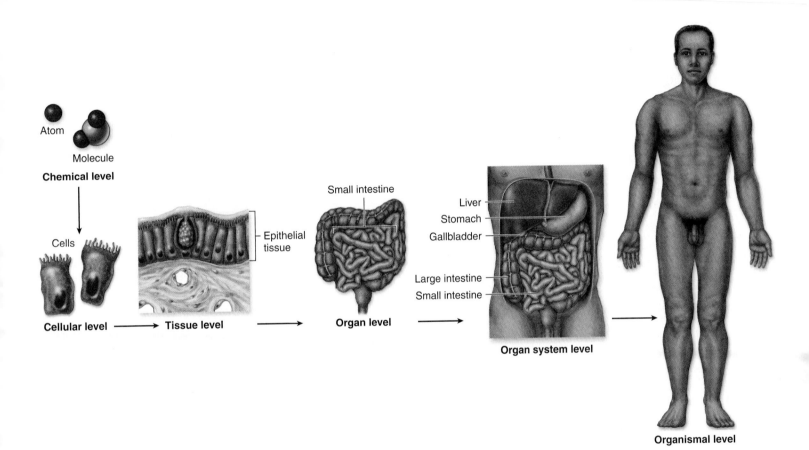

Atom

Molecule

Chemical level

Cells

Cellular level → **Tissue level** →

Epithelial tissue

Small intestine

Organ level →

Liver
Stomach
Gallbladder

Large intestine
Small intestine

Organ system level

Organismal level

Levels of Organization in the Human Body
Figure 1.2

1

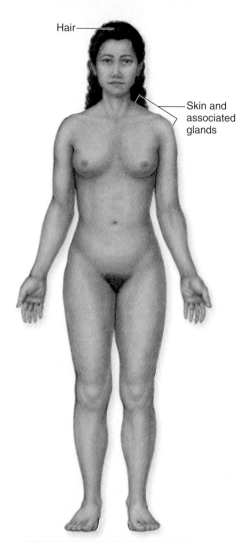

Hair

Skin and associated glands

Integumentary System (Chapter 5)

Provides protection, regulates body temperature, site of cutaneous receptors, synthesizes vitamin D, prevents water loss.

Integumentary System
Figure 1.3a

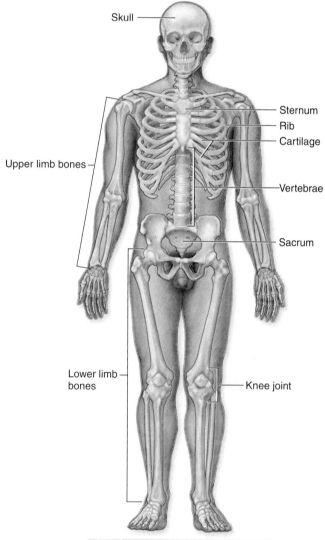

Skull

Upper limb bones

Sternum

Rib

Cartilage

Vertebrae

Sacrum

Lower limb bones

Knee joint

Skeletal System (Chapters 6–9)

Provides support and protection, site of hemopoiesis (blood cell production), stores calcium and phosphorus, allows for body movement.

Skeletal System
Figure 1.3b

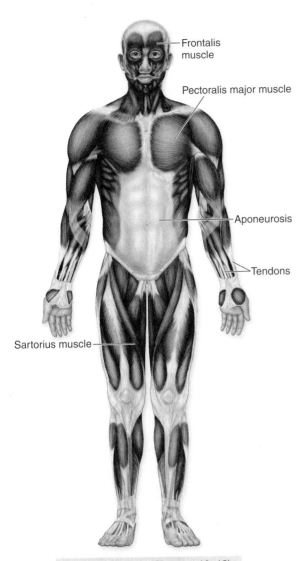

Frontalis muscle

Pectoralis major muscle

Aponeurosis

Tendons

Sartorius muscle

Muscular System (Chapters 10–12)
Produces body movement, generates heat when muscles contract.

Muscular System
Figure 1.3c

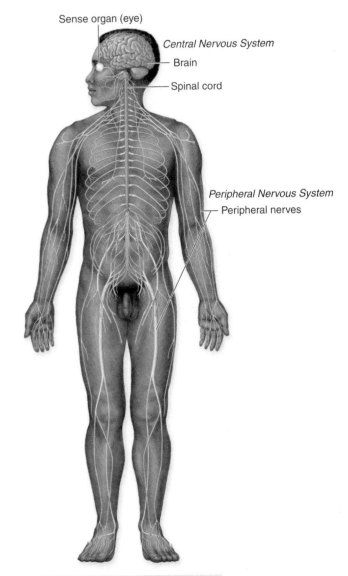

Sense organ (eye)

Central Nervous System

Brain

Spinal cord

Peripheral Nervous System

Peripheral nerves

Nervous System (Chapters 14–19)
A regulatory system that controls body movement, responds to sensory stimuli, and helps control all other systems of the body. Also responsible for consciousness, intelligence, memory.

Nervous System
Figure 1.3d

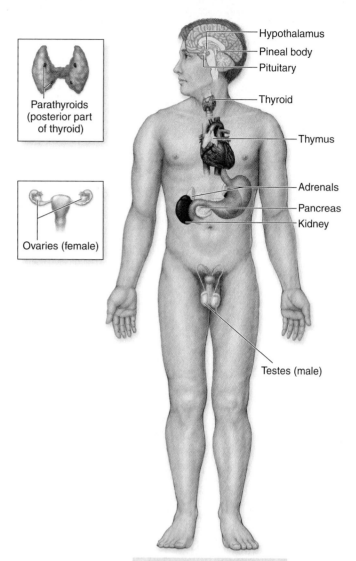

Parathyroids
(posterior part
of thyroid)

Ovaries (female)

Hypothalamus
Pineal body
Pituitary
Thyroid
Thymus
Adrenals
Pancreas
Kidney
Testes (male)

Endocrine System (Chapter 20)
Consists of glands and cell clusters that secrete hormones, some of which regulate body and cellular growth, chemical levels in the body, and reproductive functions.

Endocrine System
Figure 1.3e

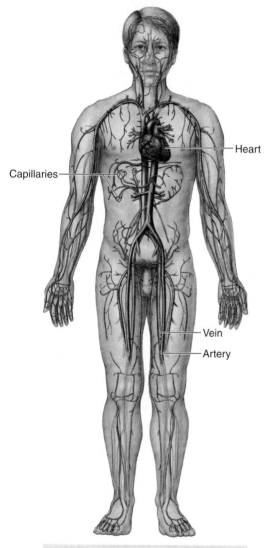

Heart
Capillaries
Vein
Artery

Cardiovascular System (Chapters 21–23)
Consists of a pump (the heart) that moves blood through blood vessels in order to distribute hormones, nutrients, and gases, and pick up waste products.

Cardiovascular System
Figure 1.3f

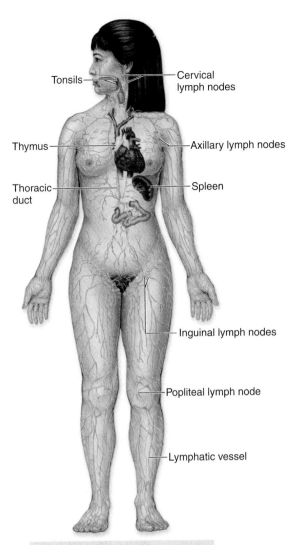

Tonsils

Cervical
lymph nodes

Thymus

Axillary lymph nodes

Thoracic
duct

Spleen

Inguinal lymph nodes

Popliteal lymph node

Lymphatic vessel

Lymphatic System (Chapter 24)

Transports and filters lymph (interstitial
fluid) and initiates an immune response
when necessary.

Lymphatic System
Figure 1.3g

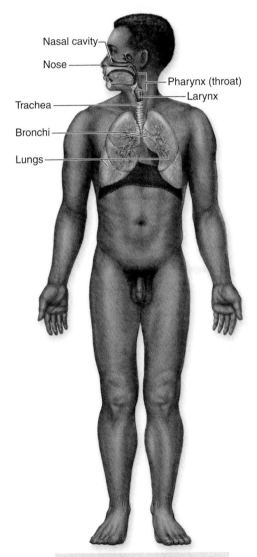

Nasal cavity

Nose

Pharynx (throat)

Larynx

Trachea

Bronchi

Lungs

Respiratory System (Chapter 25)

Responsible for exchange of gases
(oxygen and carbon dioxide) between
blood and the air in the lungs.

Respiratory System
Figure 1.3h

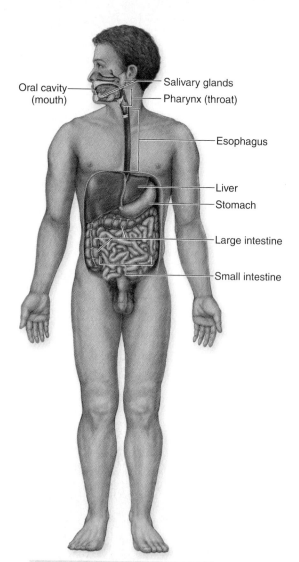

Oral cavity (mouth)

Salivary glands

Pharynx (throat)

Esophagus

Liver

Stomach

Large intestine

Small intestine

Digestive System (Chapter 26)

Mechanically and chemically digests food materials, absorbs nutrients, and expels waste products.

Digestive System
Figure 1.3i

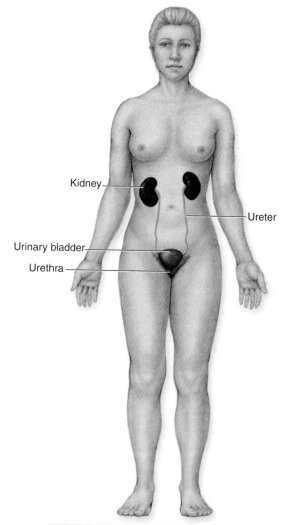

Kidney

Ureter

Urinary bladder

Urethra

Urinary System (Chapter 27)

Filters the blood and removes waste products from the blood, concentrates waste products in the form of urine, and expels urine from the body.

Urinary System
Figure 1.3j

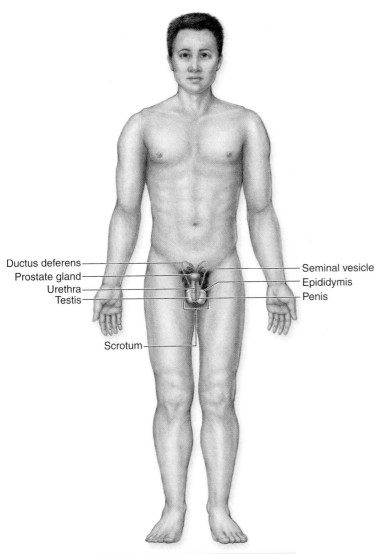

Ductus deferens
Prostate gland
Urethra
Testis

Seminal vesicle
Epididymis
Penis

Scrotum

Male Reproductive System (Chapter 28)

Produces male sex cells (sperm) and male
hormones (e.g., testosterone), transfers
sperm to the female.

Male Reproductive System
Figure 1.3k

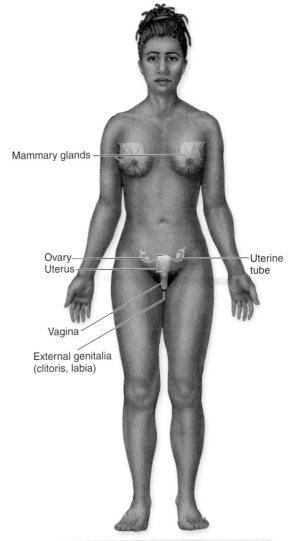

Mammary glands

Ovary
Uterus

Uterine
tube

Vagina

External genitalia
(clitoris, labia)

Female Reproductive System (Chapter 28)

Produces female sex cells (oocytes) and
female hormones (e.g., estrogen and
progesterone), receives sperm from male,
site of fertilization of oocyte, site of growth
and development of embryo and fetus.

Female Reproductive System
Figure 1.3l

7

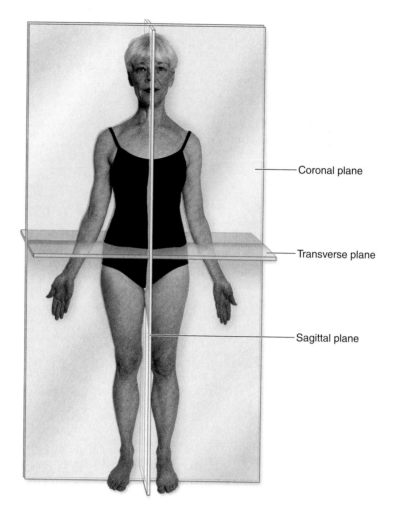

Coronal plane

Transverse plane

Sagittal plane

Anatomic Position and Planes of the Body
Figure 1.4

Figure 1.4: © The McGraw-Hill Companies, Inc./Photo by Jw Ramsey

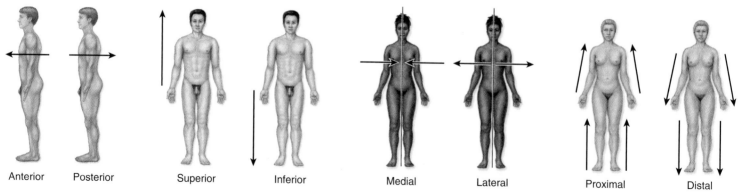

| Anterior | Posterior | Superior | Inferior | Medial | Lateral | Proximal | Distal |

Directional Terms in Anatomy
Figure 1.6

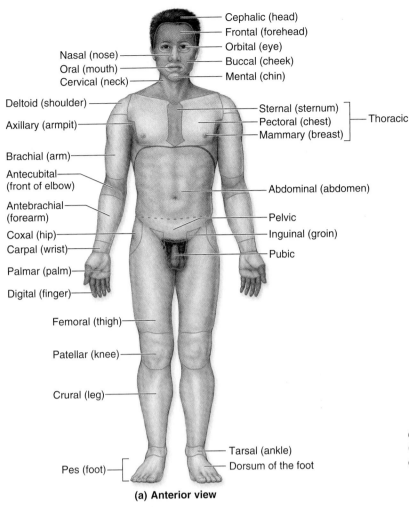

Cephalic (head)
Frontal (forehead)
Orbital (eye)
Nasal (nose)
Buccal (cheek)
Oral (mouth)
Mental (chin)
Cervical (neck)
Deltoid (shoulder)
Sternal (sternum)
Pectoral (chest)
Axillary (armpit)
Mammary (breast)
Thoracic
Brachial (arm)
Antecubital (front of elbow)
Abdominal (abdomen)
Antebrachial (forearm)
Coxal (hip)
Pelvic
Inguinal (groin)
Carpal (wrist)
Pubic
Palmar (palm)
Digital (finger)
Femoral (thigh)
Patellar (knee)
Crural (leg)
Tarsal (ankle)
Pes (foot)
Dorsum of the foot

(a) Anterior view

Regional Terms—Anterior View
Figure 1.7a

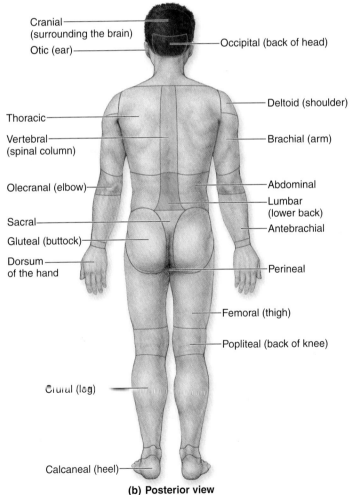

Cranial (surrounding the brain)
Otic (ear)
Occipital (back of head)
Deltoid (shoulder)
Thoracic
Vertebral (spinal column)
Brachial (arm)
Olecranal (elbow)
Abdominal
Lumbar (lower back)
Sacral
Antebrachial
Gluteal (buttock)
Dorsum of the hand
Perineal
Femoral (thigh)
Popliteal (back of knee)
Crural (leg)
Calcaneal (heel)

(b) Posterior view

Regional Terms—Posterior View
Figure 1.7b

9

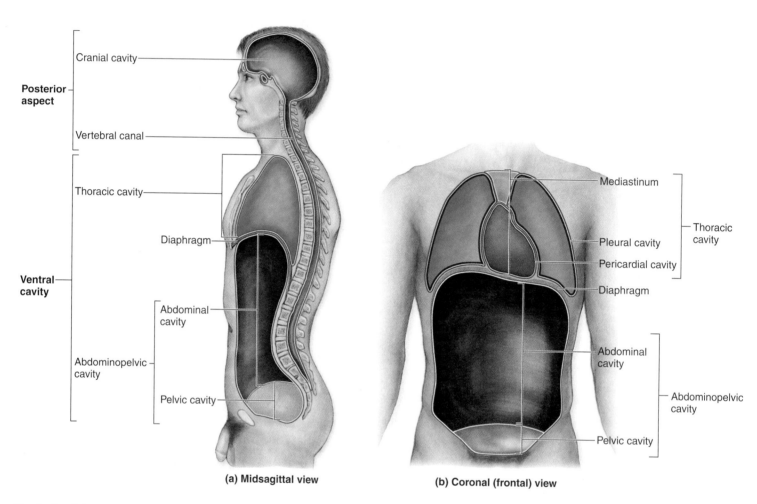

Posterior
aspect

Cranial cavity

Vertebral canal

Ventral
cavity

Thoracic cavity

Diaphragm

Abdominal
cavity

Abdominopelvic
cavity

Pelvic cavity

(a) Midsagittal view

Mediastinum

Thoracic
cavity

Pleural cavity

Pericardial cavity

Diaphragm

Abdominal
cavity

Abdominopelvic
cavity

Pelvic cavity

(b) Coronal (frontal) view

Body Cavities
Figure 1.8

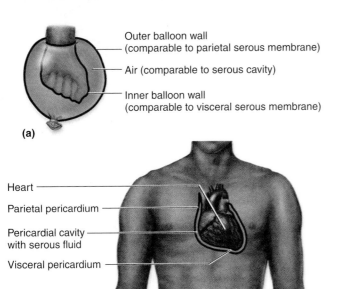

Outer balloon wall
(comparable to parietal serous membrane)

Air (comparable to serous cavity)

Inner balloon wall
(comparable to visceral serous membrane)

(a)

Heart

Parietal pericardium

Pericardial cavity
with serous fluid

Visceral pericardium

(b) Pericardium

Parietal pleura

Visceral pleura

Pleural cavity (between parietal
and visceral pleura)

Diaphragm

(c) Pleura

**Serous Membranes in the Ventral Body Cavities—
Pericardium, Pleura**
Figure 1.9a,b,c

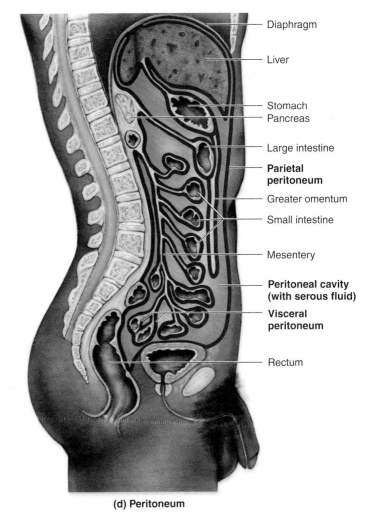

Diaphragm

Liver

Stomach
Pancreas

Large intestine

**Parietal
peritoneum**

Greater omentum

Small intestine

Mesentery

**Peritoneal cavity
(with serous fluid)**

**Visceral
peritoneum**

Rectum

(d) Peritoneum

**Serous Membranes in the Ventral Body Cavities—
Peritoneum**
Figure 1.9d

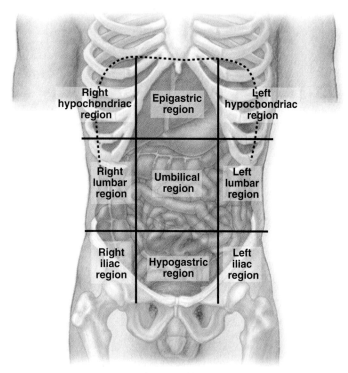

(a) Abdominopelvic regions

Right hypochondriac region

Epigastric region

Left hypochondriac region

Right lumbar region

Umbilical region

Left lumbar region

Right iliac region

Hypogastric region

Left iliac region

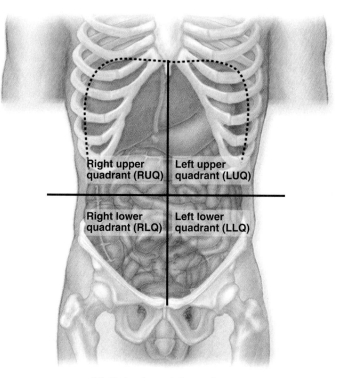

(b) Abdominopelvic quadrants

Right upper quadrant (RUQ)

Left upper quadrant (LUQ)

Right lower quadrant (RLQ)

Left lower quadrant (LLQ)

Abdominopelvic Regions and Quadrants
Figure 1.10

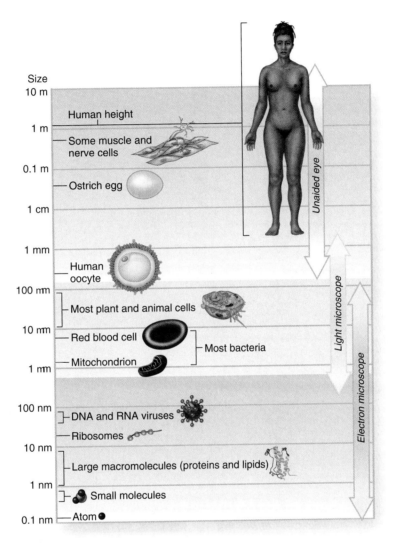

The Range of Cell Sizes
Figure 2.1

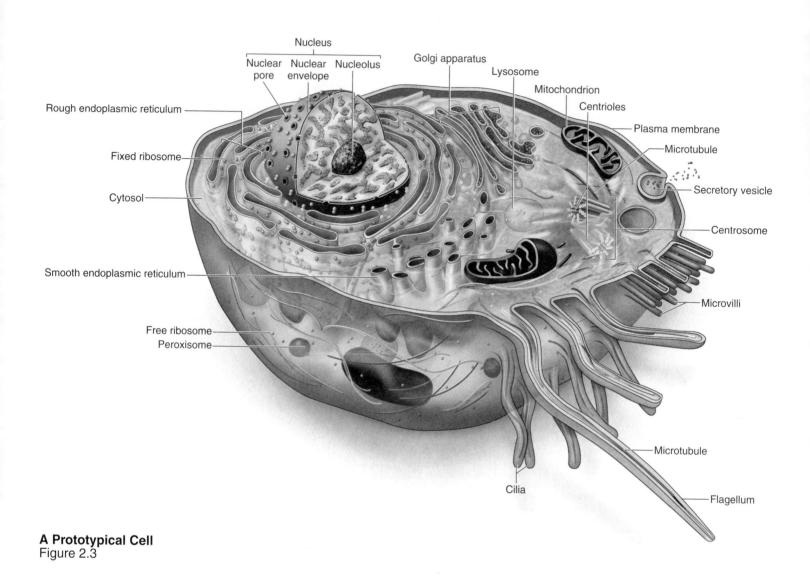

Nucleus

Nuclear pore

Nuclear envelope

Nucleolus

Golgi apparatus

Lysosome

Mitochondrion

Centrioles

Plasma membrane

Microtubule

Secretory vesicle

Centrosome

Microvilli

Microtubule

Flagellum

Cilia

Rough endoplasmic reticulum

Fixed ribosome

Cytosol

Smooth endoplasmic reticulum

Free ribosome

Peroxisome

A Prototypical Cell
Figure 2.3

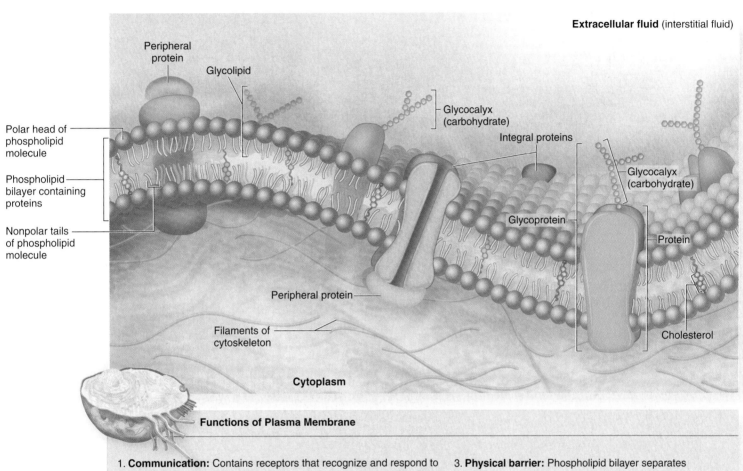

Extracellular fluid (interstitial fluid)

Peripheral protein

Glycolipid

Glycocalyx (carbohydrate)

Integral proteins

Polar head of phospholipid molecule

Phospholipid bilayer containing proteins

Nonpolar tails of phospholipid molecule

Glycocalyx (carbohydrate)

Glycoprotein

Protein

Peripheral protein

Filaments of cytoskeleton

Cholesterol

Cytoplasm

Functions of Plasma Membrane

1. **Communication:** Contains receptors that recognize and respond to molecular signals
2. **Intercellular connection:** Establishes a flexible boundary, protects cellular contents, and supports cell structure
3. **Physical barrier:** Phospholipid bilayer separates substances inside and outside the cell
4. **Selective permeability:** Regulates entry and exit of ions, nutrients, and waste molecules through the membrane

Structure of the Plasma Membrane
Figure 2.4

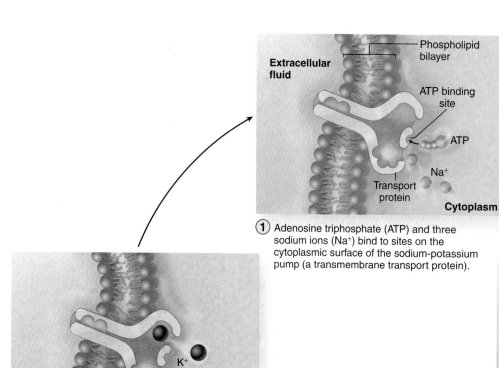

(1) Adenosine triphosphate (ATP) and three sodium ions (Na⁺) bind to sites on the cytoplasmic surface of the sodium-potassium pump (a transmembrane transport protein).

(2) ATP breaks down into adenosine diphosphate (ADP) and phosphate (P), resulting in a release of energy that causes the sodium-potassium pump to change conformation (shape) and release the Na⁺ ions to the extracellular fluid.

(3) As the three Na⁺ ions diffuse away from the sodium-potassium pump into the extracellular fluid, two K⁺ ions from the extracellular fluid bind to sites on the extracellular surface of the sodium-potassium pump. At the same time, the phosphate produced earlier by ATP hydrolysis is released into the cytoplasm.

(4) This transport protein reverts back to its original shape, resulting in the release of the K⁺ ions into the cytoplasm. After the K⁺ ions diffuse away from the sodium-potassium pump, it is ready to begin the process again.

Sodium-Potassium Pump
Figure 2.5

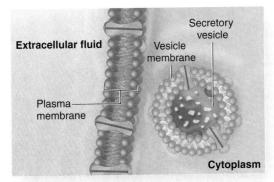

Extracellular fluid

Plasma membrane

Vesicle membrane

Secretory vesicle

Cytoplasm

① Vesicle nears plasma membrane

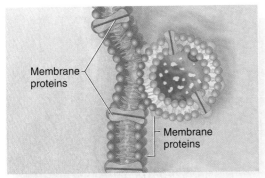

Membrane proteins

Membrane proteins

② Fusion of vesicle with membrane

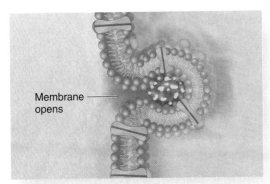

Membrane opens

③ Exocytosis as membrane opens externally

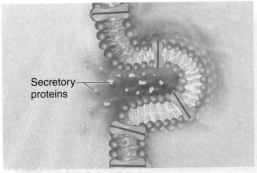

Secretory proteins

④ Release of vesicle components into the extracellular fluid and integration of vesicle membrane components into the plasma membrane

Exocytosis
Figure 2.6

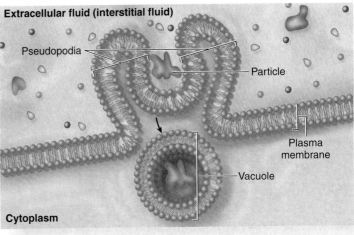

Extracellular fluid (interstitial fluid)

Pseudopodia

Particle

Plasma membrane

Vacuole

Cytoplasm

(a) Phagocytosis

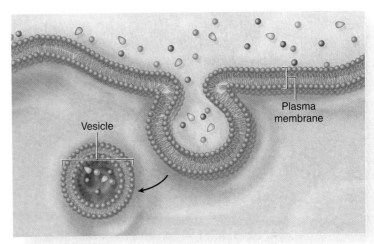

Plasma membrane

Vesicle

(b) Pinocytosis

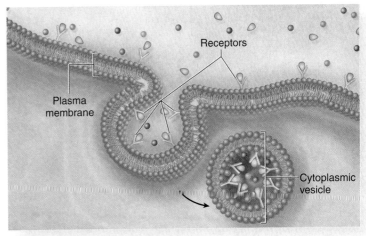

Receptors

Plasma membrane

Cytoplasmic vesicle

(c) Receptor-mediated endocytosis

Three Forms of Endocytosis
Figure 2.7

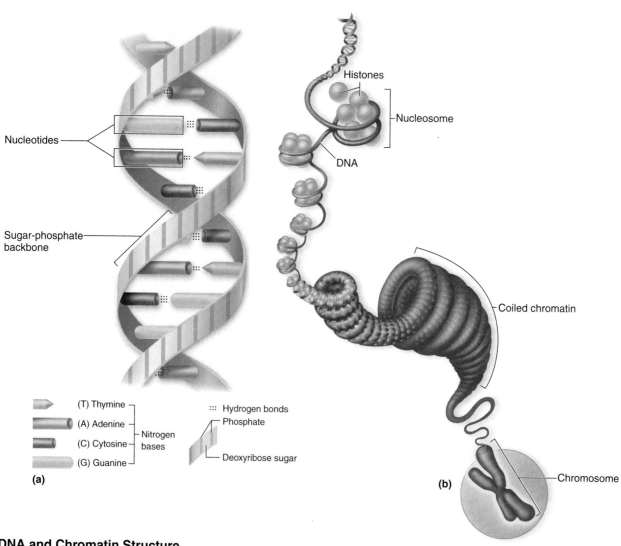

Nucleotides

Sugar-phosphate
backbone

Histones

Nucleosome

DNA

Coiled chromatin

Chromosome

(T) Thymine
(A) Adenine
(C) Cytosine — Nitrogen bases
(G) Guanine

::: Hydrogen bonds
Phosphate
Deoxyribose sugar

(a)

(b)

DNA and Chromatin Structure
Figure 2.18

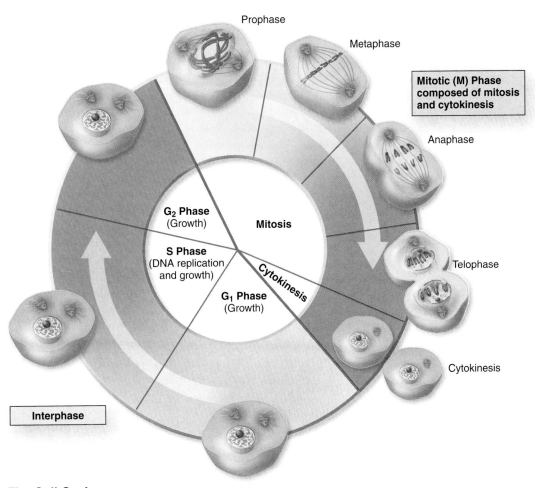

Prophase

Metaphase

Mitotic (M) Phase composed of mitosis and cytokinesis

Anaphase

G₂ Phase
(Growth)

Mitosis

S Phase
(DNA replication
and growth)

Cytokinesis

Telophase

G₁ Phase
(Growth)

Cytokinesis

Interphase

The Cell Cycle
Figure 2.19

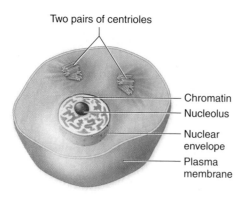

(a) Interphase

(b) Prophase

Interphase and Mitosis
Figure 2.20a,b

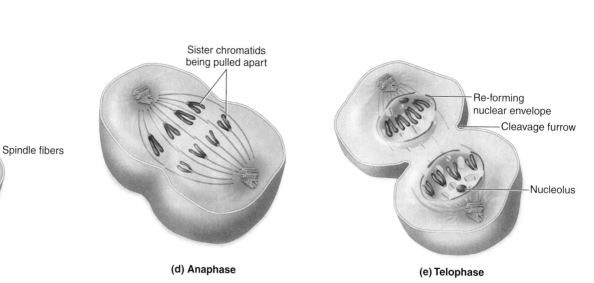

(c) Metaphase

(d) Anaphase

(e) Telophase

Interphase and Mitosis
Figure 2.20c,d,e

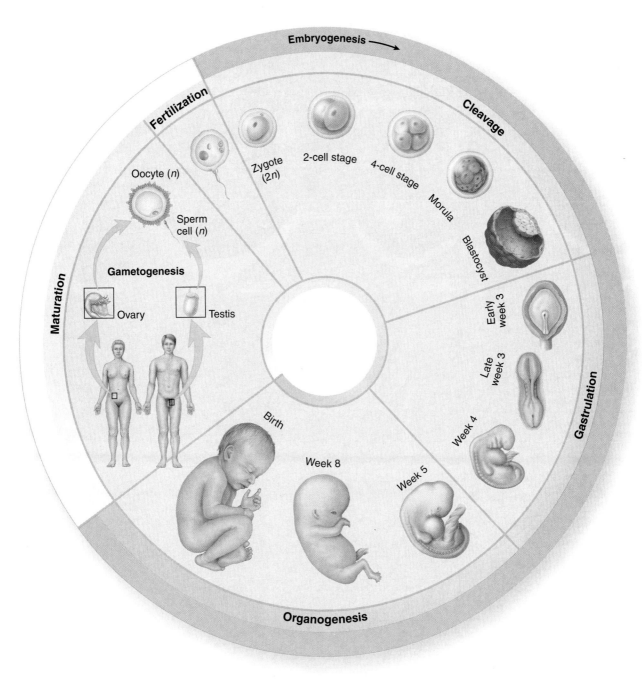

Developmental History of a Human
Figure 3.1

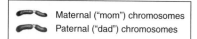

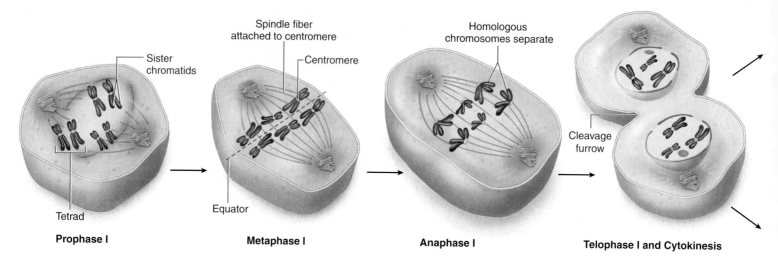

Meiosis I
Figure 3.2 left

Sister chromatids separate

Cells separate into four haploid daughter cells

Sister chromatids separate

Single-stranded chromosomes

Prophase II

Metaphase II

Anaphase II

Telophase II and Cytokinesis

Meiosis II
Figure 3.2 right

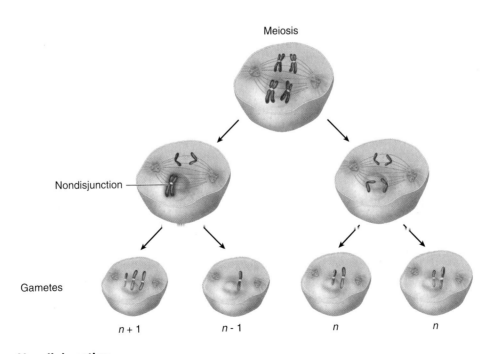

Meiosis

Nondisjunction

Gametes

$n + 1$ $n - 1$ n n

Nondisjunction
Clinical View p. 60

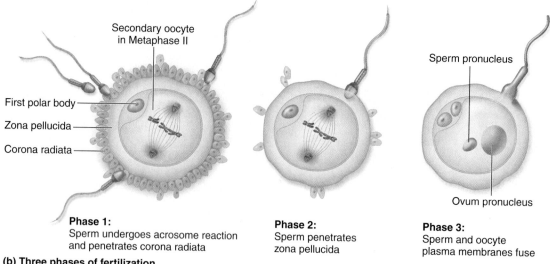

Secondary oocyte
in Metaphase II

First polar body

Zona pellucida

Corona radiata

Sperm pronucleus

Ovum pronucleus

Phase 1:
Sperm undergoes acrosome reaction
and penetrates corona radiata

Phase 2:
Sperm penetrates
zona pellucida

Phase 3:
Sperm and oocyte
plasma membranes fuse

(b) Three phases of fertilization

Three Phases of Fertilization
Figure 3.3b

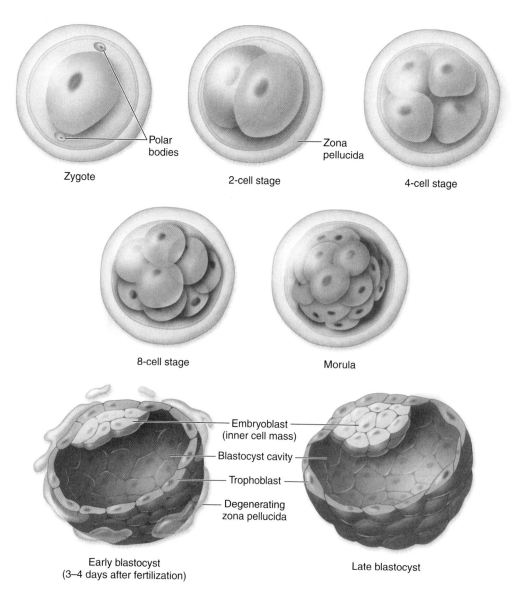

Polar bodies

Zona pellucida

Zygote

2-cell stage

4-cell stage

8-cell stage

Morula

Embryoblast
(inner cell mass)

Blastocyst cavity

Trophoblast

Degenerating
zona pellucida

Early blastocyst
(3–4 days after fertilization)

Late blastocyst

Cleavage in the Pre-embryo
Figure 3.4

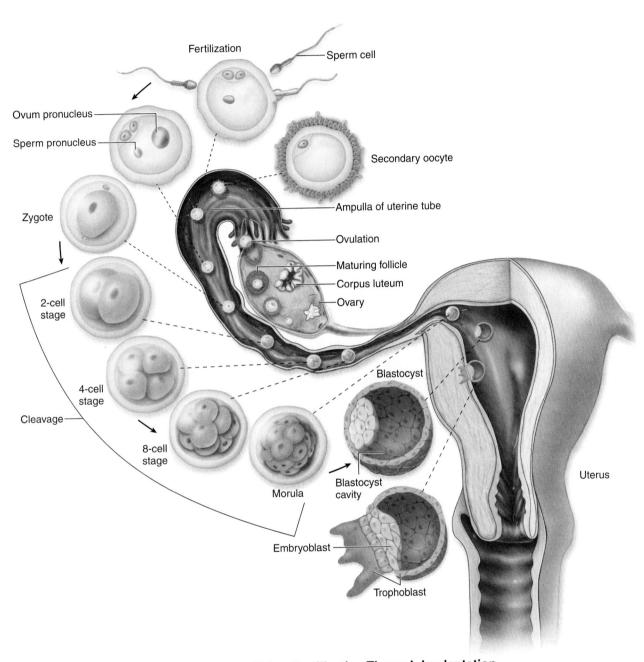

Fertilization

Sperm cell

Ovum pronucleus

Sperm pronucleus

Secondary oocyte

Zygote

Ampulla of uterine tube

Ovulation

Maturing follicle

Corpus luteum

Ovary

2-cell stage

4-cell stage

Cleavage

8-cell stage

Blastocyst

Morula

Blastocyst cavity

Uterus

Embryoblast

Trophoblast

Transit of Pre-embryo Through Uterine Tube: Fertilization Through Implantation
Figure 3.5

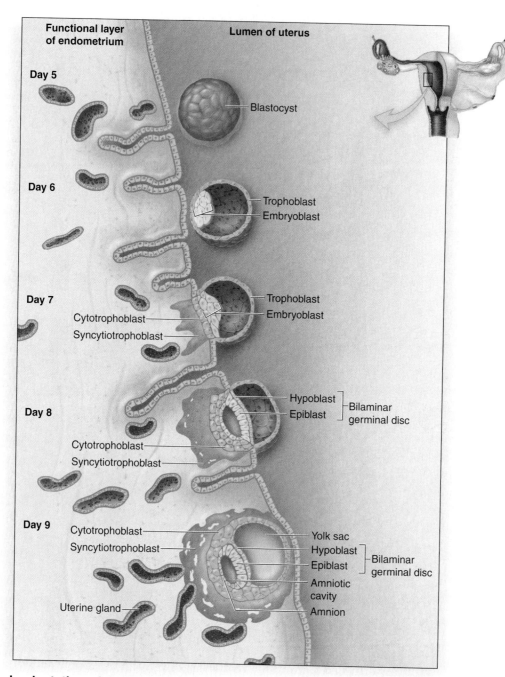

Implantation of the Blastocyst
Figure 3.6

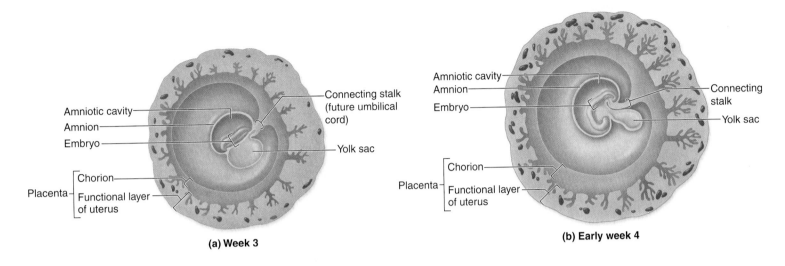

Amniotic cavity
Amnion
Embryo

Connecting stalk
(future umbilical
cord)

Yolk sac

Chorion
Placenta
Functional layer
of uterus

(a) Week 3

Amniotic cavity
Amnion
Embryo

Connecting
stalk

Yolk sac

Chorion
Placenta
Functional layer
of uterus

(b) Early week 4

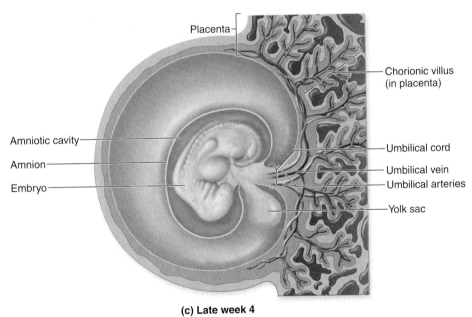

Placenta

Chorionic villus
(in placenta)

Amniotic cavity

Amnion

Embryo

Umbilical cord

Umbilical vein

Umbilical arteries

Yolk sac

(c) Late week 4

Formation of Extraembryonic Membranes
Figure 3.7

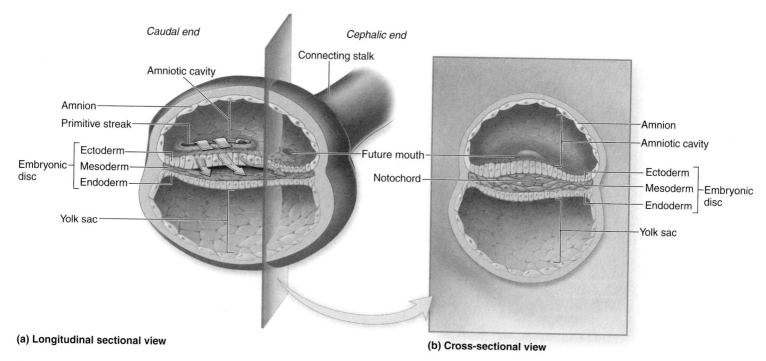

(a) Longitudinal sectional view

(b) Cross-sectional view

Gastrulation
Figure 3.8

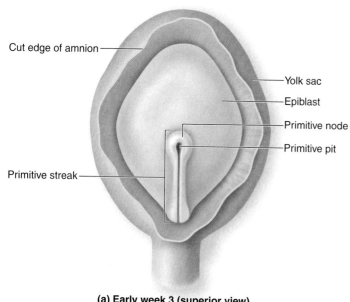

(a) Early week 3 (superior view)

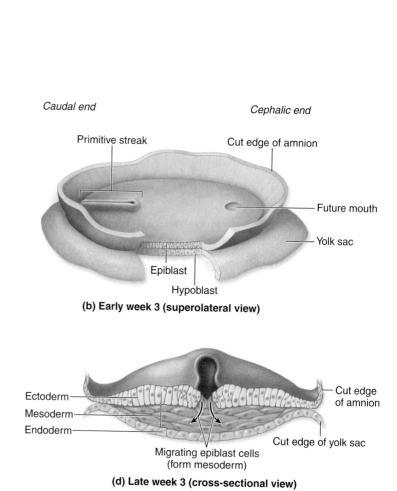

(b) Early week 3 (superolateral view)

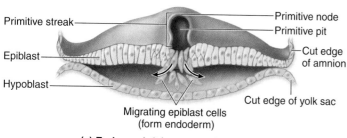

(c) Early week 3 (cross-sectional view)

(d) Late week 3 (cross-sectional view)

Primitive Streak
Figure 3.9

Cephalocaudal folding

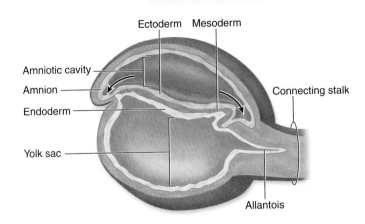

Ectoderm   Mesoderm
Amniotic cavity
Amnion
Endoderm
Yolk sac
Connecting stalk
Allantois

Week 3

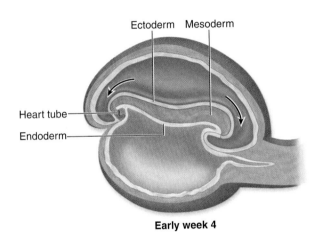

Ectoderm   Mesoderm
Heart tube
Endoderm

Early week 4

Transverse folding

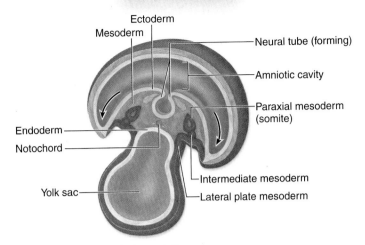

Ectoderm
Mesoderm
Neural tube (forming)
Amniotic cavity
Endoderm
Notochord
Paraxial mesoderm (somite)
Yolk sac
Intermediate mesoderm
Lateral plate mesoderm

Late week 3

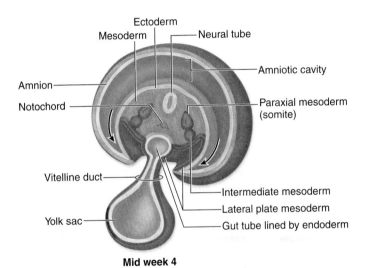

Ectoderm
Mesoderm
Neural tube
Amnion
Amniotic cavity
Notochord
Paraxial mesoderm (somite)
Vitelline duct
Yolk sac
Intermediate mesoderm
Lateral plate mesoderm
Gut tube lined by endoderm

Mid week 4

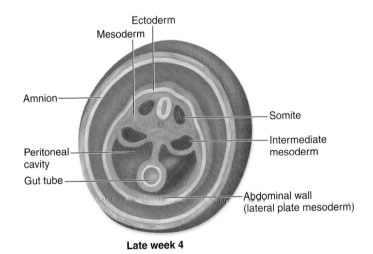

Ectoderm
Mesoderm
Neural tube
Amnion
Somite
Intermediate mesoderm
Peritoneal cavity
Gut tube
Abdominal wall (lateral plate mesoderm)

Late week 4

Ectoderm   Mesoderm
Endoderm
Midgut
Yolk sac

Late week 4

Folding of the Embryonic Disc
Figure 3.10

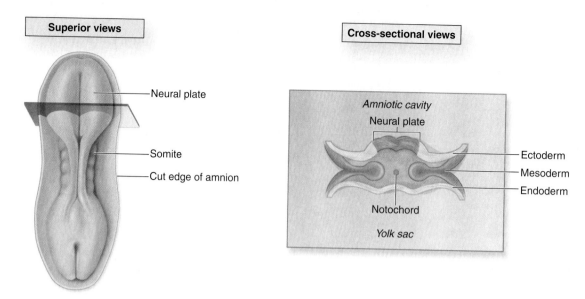

Neural plate

Somite

Cut edge of amnion

Amniotic cavity
Neural plate

Ectoderm

Mesoderm

Endoderm

Notochord

Yolk sac

(a) Mid week 3: neural plate forms

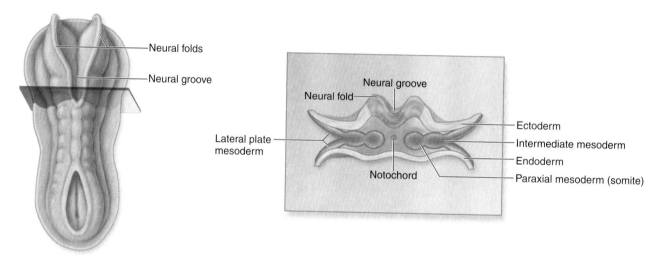

Neural folds

Neural groove

Neural groove

Neural fold

Lateral plate mesoderm

Ectoderm

Intermediate mesoderm

Endoderm

Notochord

Paraxial mesoderm (somite)

(b) Late week 3: neural folds and neural groove form

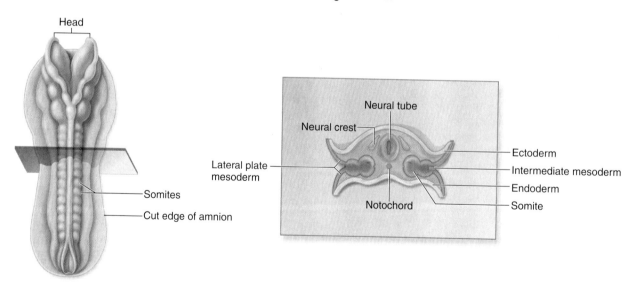

Head

Somites

Cut edge of amnion

Neural tube

Neural crest

Lateral plate mesoderm

Ectoderm

Intermediate mesoderm

Endoderm

Notochord

Somite

(c) Early week 4: neural folds fuse to form neural tube

Neurulation
Figure 3.11

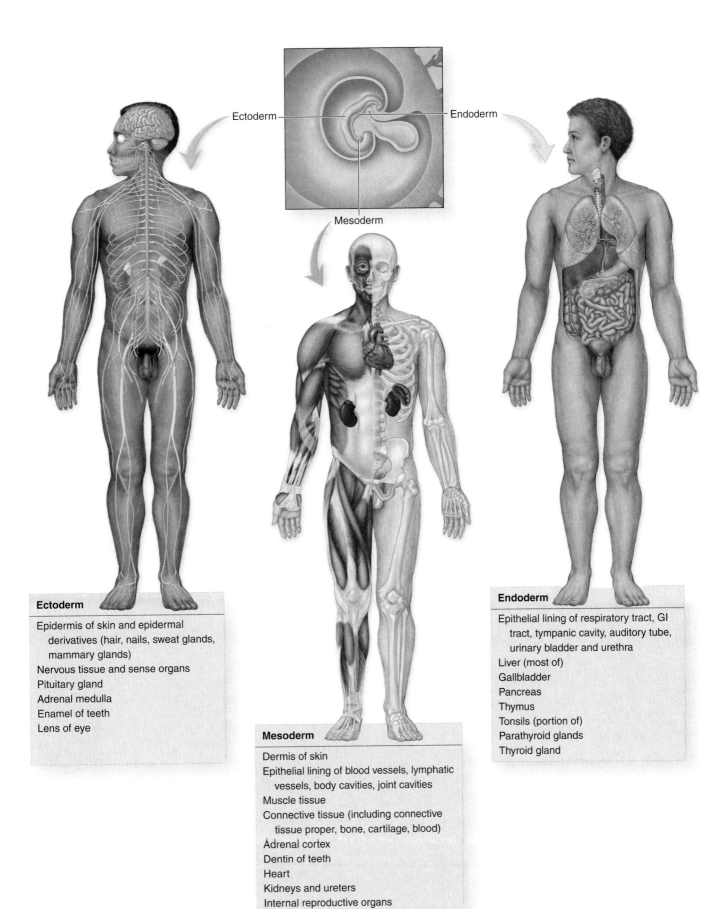

Ectoderm

Epidermis of skin and epidermal
   derivatives (hair, nails, sweat glands,
   mammary glands)
Nervous tissue and sense organs
Pituitary gland
Adrenal medulla
Enamel of teeth
Lens of eye

Mesoderm

Dermis of skin
Epithelial lining of blood vessels, lymphatic
   vessels, body cavities, joint cavities
Muscle tissue
Connective tissue (including connective
   tissue proper, bone, cartilage, blood)
Adrenal cortex
Dentin of teeth
Heart
Kidneys and ureters
Internal reproductive organs
Spleen

Endoderm

Epithelial lining of respiratory tract, GI
   tract, tympanic cavity, auditory tube,
   urinary bladder and urethra
Liver (most of)
Gallbladder
Pancreas
Thymus
Tonsils (portion of)
Parathyroid glands
Thyroid gland

Three Primary Germ Layers and Their Derivatives
Figure 3.12

31

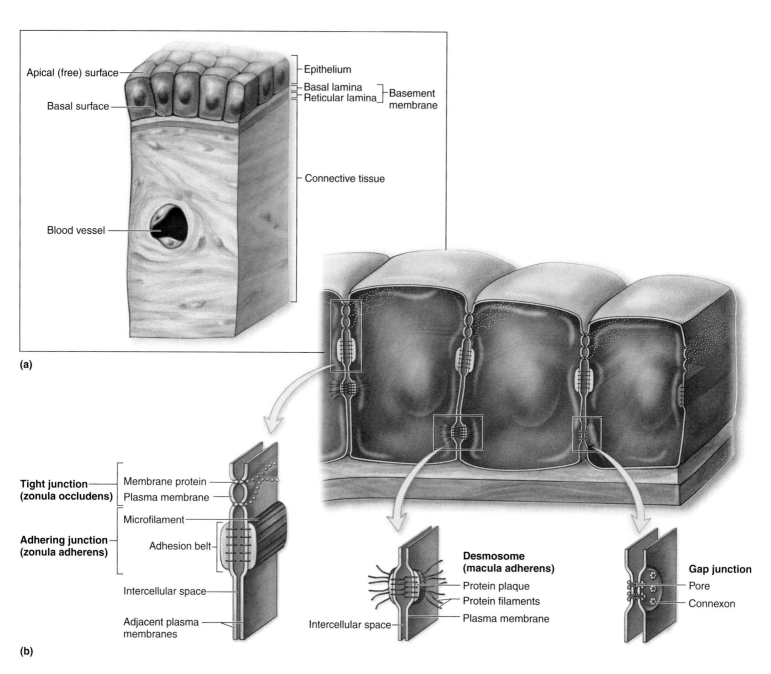

(a)

Apical (free) surface

Basal surface

Blood vessel

Epithelium

Basal lamina
Reticular lamina ⎤ Basement
                      membrane

Connective tissue

(b)

Tight junction (zonula occludens)
Membrane protein
Plasma membrane

Adhering junction (zonula adherens)
Microfilament
Adhesion belt

Intercellular space

Adjacent plasma membranes

Desmosome (macula adherens)
Protein plaque
Protein filaments
Plasma membrane
Intercellular space

Gap junction
Pore
Connexon

Polarity and Intercellular Junctions
Figure 4.1

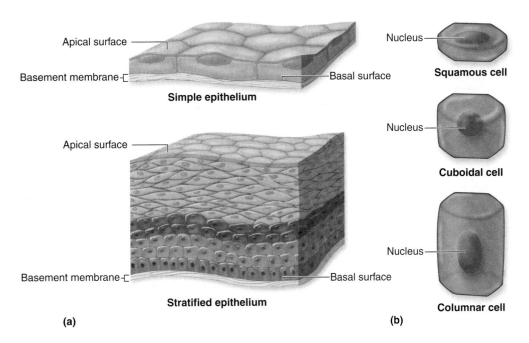

Apical surface

Basement membrane

Basal surface

Simple epithelium

Apical surface

Basement membrane

Basal surface

Stratified epithelium

Nucleus

Squamous cell

Nucleus

Cuboidal cell

Nucleus

Columnar cell

(a)

(b)

Classification of Epithelia
Figure 4.2

33

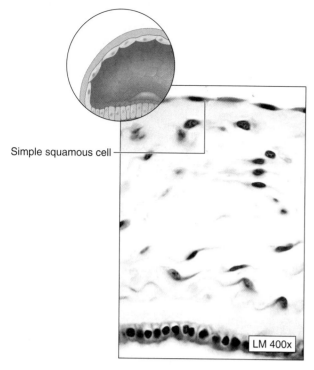

Simple squamous cell ⎯

LM 400x

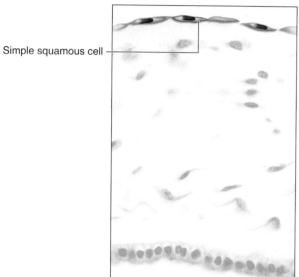

Simple squamous cell ⎯

(a) Simple squamous epithelium

Simple Squamous Epithelium
Table 4.3 a
Table 4.3a: © The McGraw-Hill Companies, Inc./Photo by Dr. Alvin Telser

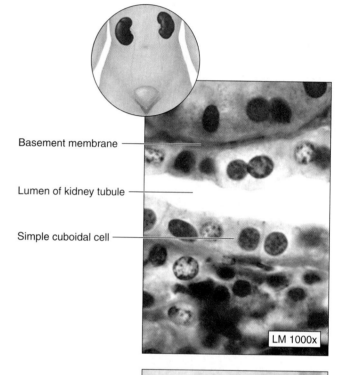

Basement membrane ⎯

Lumen of kidney tubule ⎯

Simple cuboidal cell ⎯

LM 1000x

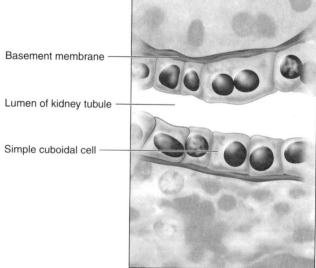

Basement membrane ⎯

Lumen of kidney tubule ⎯

Simple cuboidal cell ⎯

(a) Simple cuboidal epithelium

Simple Cuboidal Epithelium
Table 4.3 b
Table 4.3b: © The McGraw-Hill Companies, Inc./Photo by Dr. Alvin Telser

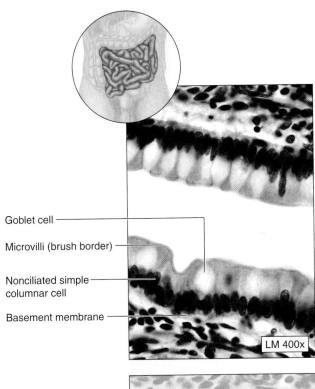

Goblet cell

Microvilli (brush border)

Nonciliated simple columnar cell

Basement membrane

LM 400x

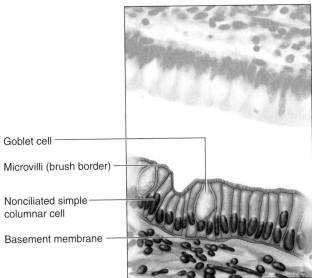

Goblet cell

Microvilli (brush border)

Nonciliated simple columnar cell

Basement membrane

Nonciliated Simple Columnar Epithelium
Table 4.3 c
Table 4.3c: © The McGraw-Hill Companies, Inc./Photo by Dr. Alvin Telser

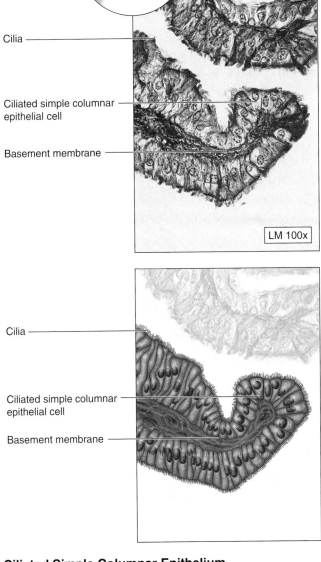

Cilia

Ciliated simple columnar epithelial cell

Basement membrane

LM 100x

Cilia

Ciliated simple columnar epithelial cell

Basement membrane

Ciliated Simple Columnar Epithelium
Table 4.3 d
Table 4.3d: © Ed Reschke

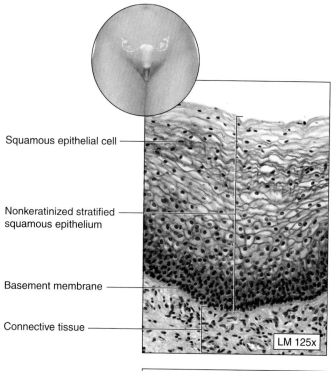

Squamous epithelial cell

Nonkeratinized stratified squamous epithelium

Basement membrane

Connective tissue

LM 125x

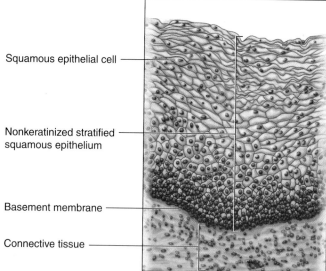

Squamous epithelial cell

Nonkeratinized stratified squamous epithelium

Basement membrane

Connective tissue

Nonkeratinized Stratified Squamous Epithelium
Table 4.4 a

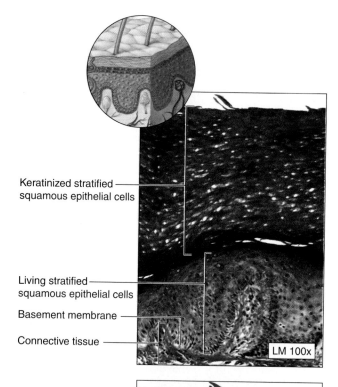

Keratinized stratified squamous epithelial cells

Living stratified squamous epithelial cells

Basement membrane

Connective tissue

LM 100x

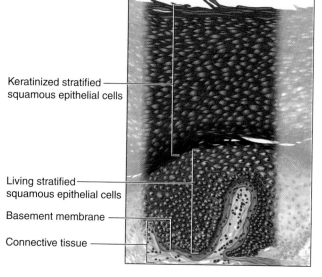

Keratinized stratified squamous epithelial cells

Living stratified squamous epithelial cells

Basement membrane

Connective tissue

Keratinized Stratified Squamous Epithelium
Table 4.4 b

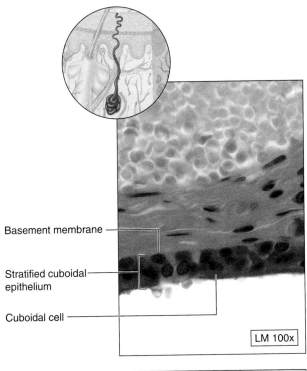

Basement membrane

Stratified cuboidal epithelium

Cuboidal cell

LM 100x

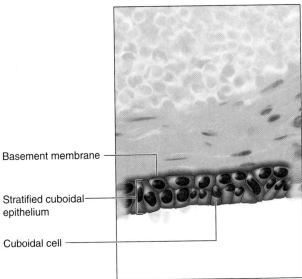

Basement membrane

Stratified cuboidal epithelium

Cuboidal cell

Stratified Cuboidal Epithelium
Table 4.4 c
Table 4.4c: © The McGraw-Hill Companies, Inc./Photo by Dr. Alvin Telser

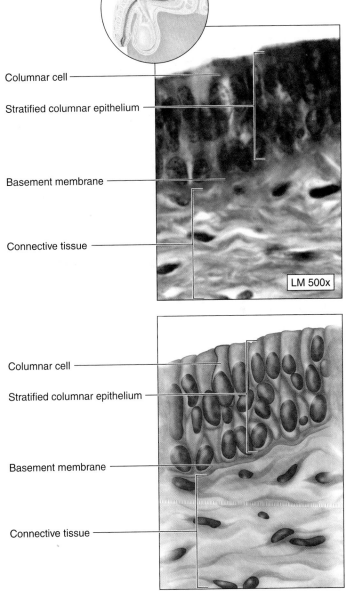

Columnar cell

Stratified columnar epithelium

Basement membrane

Connective tissue

LM 500x

Columnar cell

Stratified columnar epithelium

Basement membrane

Connective tissue

Stratified Columnar Epithelium
Table 4.4 d
Table 4.4d: © The McGraw-Hill Companies, Inc./Photo by Dr. Alvin Telser

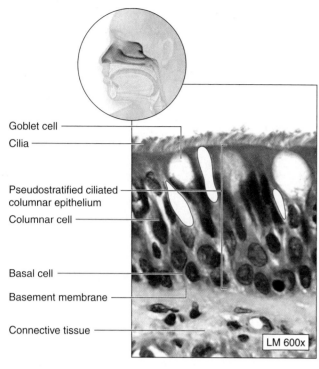

Goblet cell
Cilia

Pseudostratified ciliated
columnar epithelium
Columnar cell

Basal cell

Basement membrane

Connective tissue

LM 600x

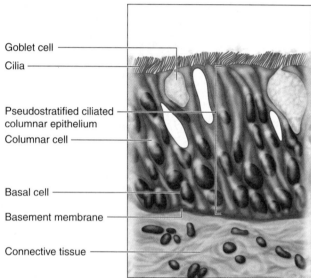

Goblet cell
Cilia

Pseudostratified ciliated
columnar epithelium
Columnar cell

Basal cell

Basement membrane

Connective tissue

Pseudostratified Columnar Epithelium
Table 4.5 a

Table 4.5a: © The McGraw-Hill Companies, Inc./Photo by Dr. Alvin Telser

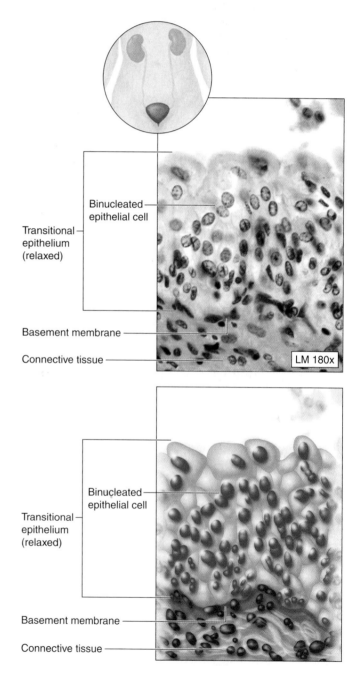

Binucleated
epithelial cell

Transitional
epithelium
(relaxed)

Basement membrane

Connective tissue

LM 180x

Binucleated
epithelial cell

Transitional
epithelium
(relaxed)

Basement membrane

Connective tissue

Transitional Epithelium
Table 4.5 b

Table 4.5b: © The McGraw-Hill Companies, Inc./Photo by Dr. Alvin Telser

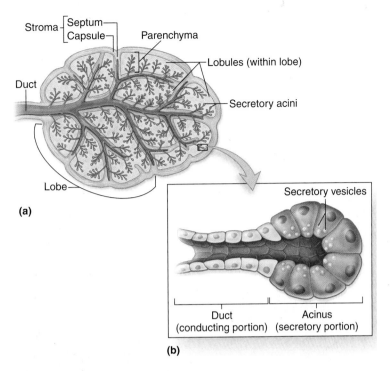

(a)

(b)

General Structure of Exocrine Glands
Figure 4.4

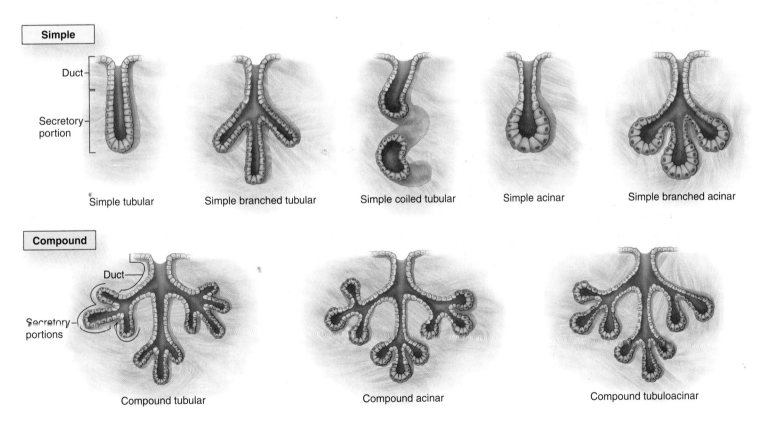

Simple

Duct

Secretory portion

Simple tubular　　Simple branched tubular　　Simple coiled tubular　　Simple acinar　　Simple branched acinar

Compound

Duct

Secretory portions

Compound tubular　　　　Compound acinar　　　　Compound tubuloacinar

Structural Classification of Multicellular Exocrine Glands
Figure 4.5

39

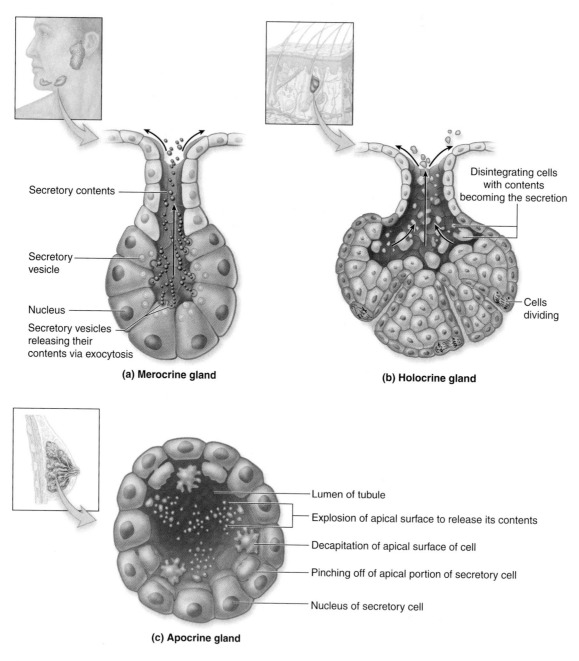

Secretory contents

Secretory vesicle

Nucleus

Secretory vesicles releasing their contents via exocytosis

(a) Merocrine gland

Disintegrating cells with contents becoming the secretion

Cells dividing

(b) Holocrine gland

Lumen of tubule

Explosion of apical surface to release its contents

Decapitation of apical surface of cell

Pinching off of apical portion of secretory cell

Nucleus of secretory cell

(c) Apocrine gland

Modes of Exocrine Secretion
Figure 4.6

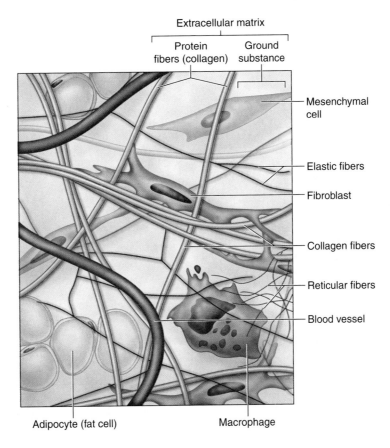

Extracellular matrix
Protein fibers (collagen)
Ground substance
Mesenchymal cell
Elastic fibers
Fibroblast
Collagen fibers
Reticular fibers
Blood vessel
Adipocyte (fat cell)
Macrophage

Connective Tissue Components and Organization
Figure 4.7

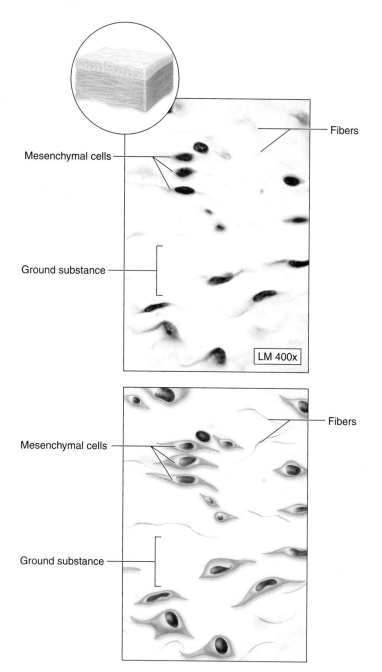

Mesenchyme
Table 4.6 a
Table 4.6a: © The McGraw-Hill Companies, Inc./Photo by Dr. Alvin Telser

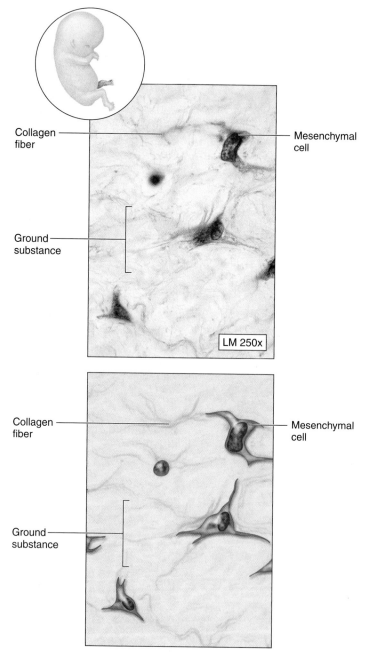

Mucous Connective Tissue
Table 4.6 b
Table 4.6b: © Ed Reschke

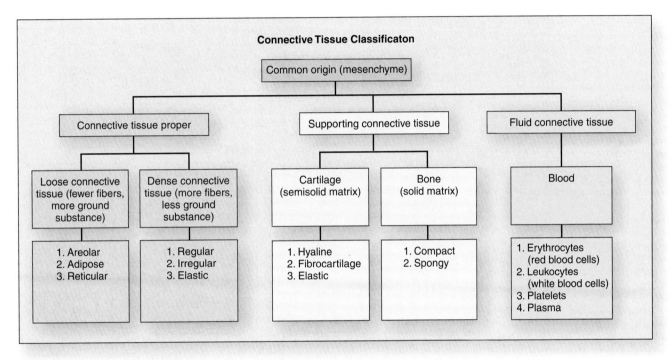

Connective Tissue Classification
Figure 4.8

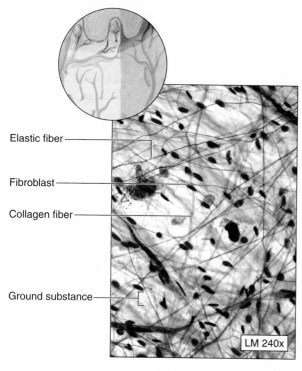

Elastic fiber

Fibroblast

Collagen fiber

Ground substance

LM 240x

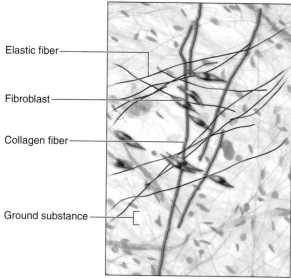

Elastic fiber

Fibroblast

Collagen fiber

Ground substance

Areolar Connective Tissue
Table 4.9 a
Table 4.9a: © The McGraw-Hill Companies, Inc./Photo by Dr. Alvin Telser

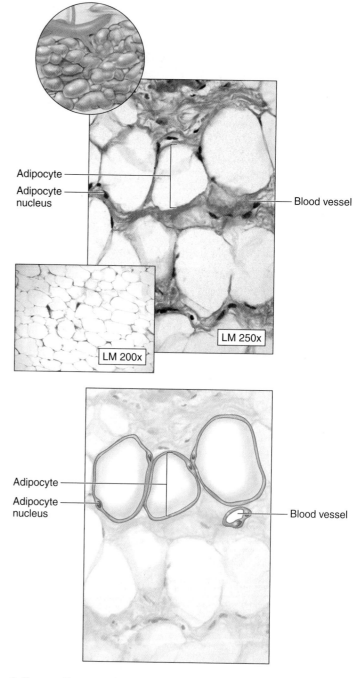

Adipocyte

Adipocyte nucleus

Blood vessel

LM 250x

LM 200x

Adipocyte

Adipocyte nucleus

Blood vessel

Adipose Connective Tissue
Table 4.9 b
Table 4.9b: © The McGraw-Hill Companies, Inc./Photo by Dr. Alvin Telser

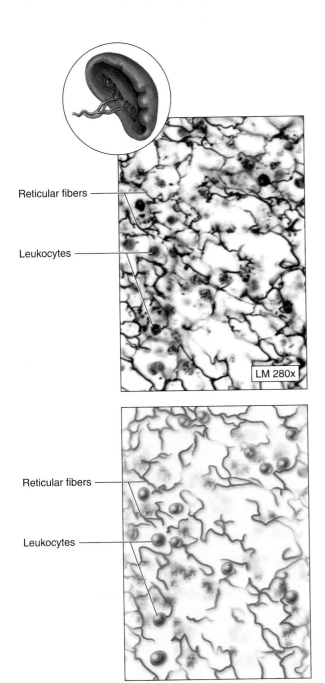

Reticular fibers

Leukocytes

LM 280x

Reticular fibers

Leukocytes

Reticular Connective Tissue
Table 4.9 c
Table 4.9c: © The McGraw-Hill Companies, Inc./Photo by Dr. Alvin Telser

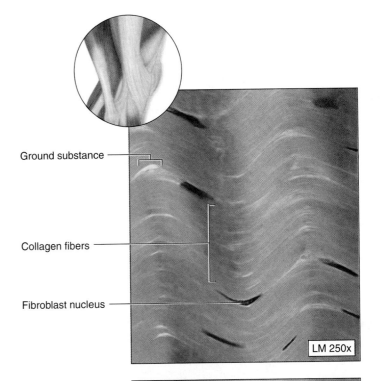

Ground substance

Collagen fibers

Fibroblast nucleus

LM 250x

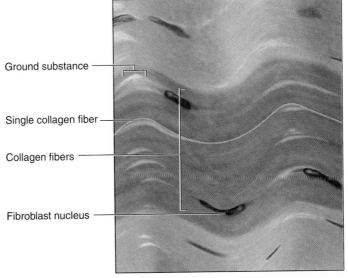

Ground substance

Single collagen fiber

Collagen fibers

Fibroblast nucleus

Dense Regular Connective Tissue
Table 4.10 a
Table 4.10a: © Ed Reschke

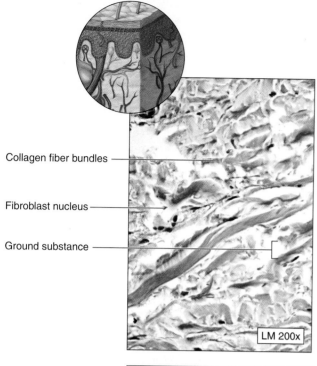

Collagen fiber bundles

Fibroblast nucleus

Ground substance

LM 200x

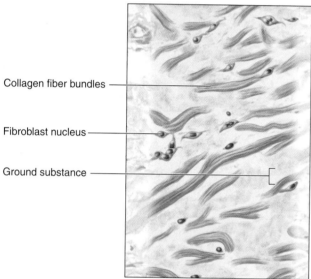

Collagen fiber bundles

Fibroblast nucleus

Ground substance

Dense Irregular Connective Tissue
Table 4.10 b

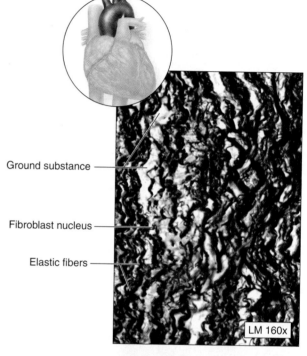

Ground substance

Fibroblast nucleus

Elastic fibers

LM 160x

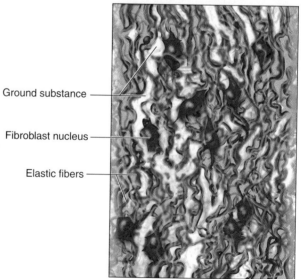

Ground substance

Fibroblast nucleus

Elastic fibers

Elastic Connective Tissue
Table 4.10 c

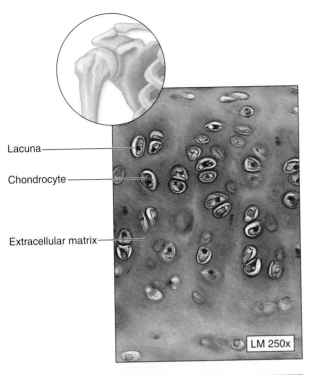

Lacuna

Chondrocyte

Extracellular matrix

LM 250x

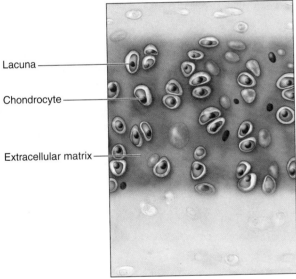

Lacuna

Chondrocyte

Extracellular matrix

Hyaline Cartilage
Table 4.11 a
Table 4.11a: © Ed Reschke/Peter Arnold

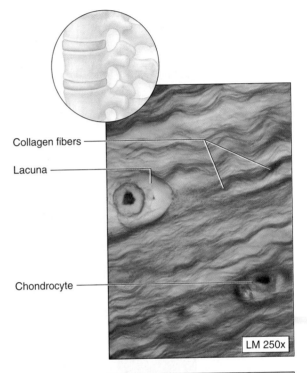

Collagen fibers

Lacuna

Chondrocyte

LM 250x

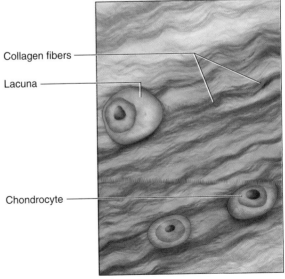

Collagen fibers

Lacuna

Chondrocyte

Fibrocartilage
Table 4.11 b
Table 4.11b: © Ed Reschke

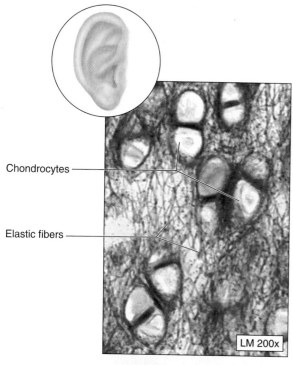

Chondrocytes

Elastic fibers

LM 200x

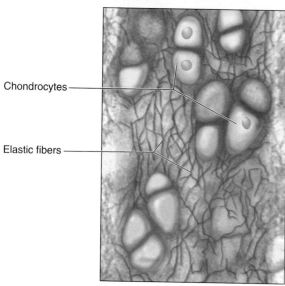

Chondrocytes

Elastic fibers

Elastic Cartilage
Table 4.11 c
Table 4.11c: © The McGraw-Hill Companies, Inc./Photo by Dr. Alvin Telser

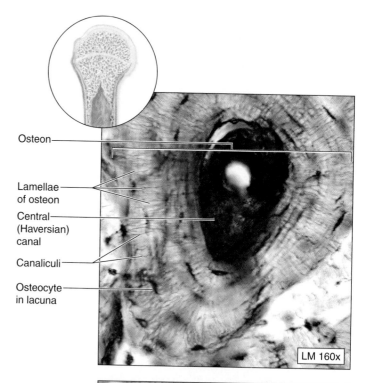

Osteon

Lamellae
of osteon

Central
(Haversian)
canal

Canaliculi

Osteocyte
in lacuna

LM 160x

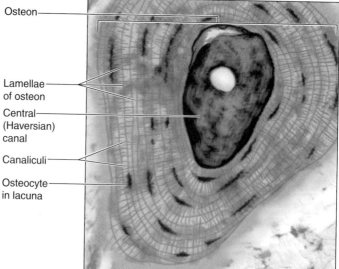

Osteon

Lamellae
of osteon

Central
(Haversian)
canal

Canaliculi

Osteocyte
in lacuna

Bone
Table 4.12
Table 4.12: © The McGraw-Hill Companies, Inc./Photo by Dr. Alvin Telser

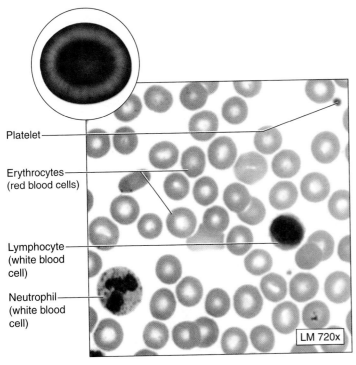

Platelet

Erythrocytes
(red blood cells)

Lymphocyte
(white blood
cell)

Neutrophil
(white blood
cell)

LM 720x

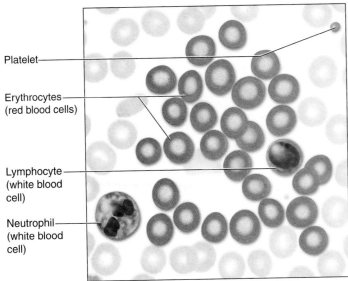

Platelet

Erythrocytes
(red blood cells)

Lymphocyte
(white blood
cell)

Neutrophil
(white blood
cell)

Blood
Table 4.13
Table 4.13: © The McGraw-Hill Companies, Inc./Photo by Dr. Alvin Telser

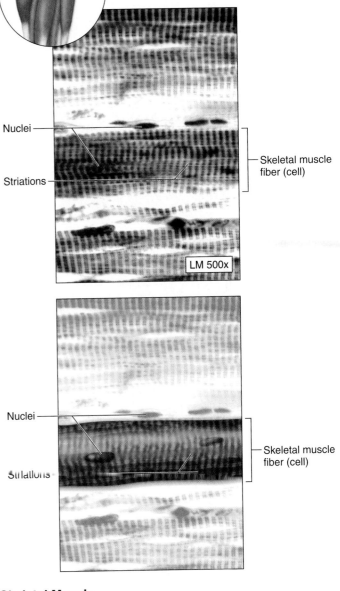

Nuclei

Striations

Skeletal muscle
fiber (cell)

LM 500x

Nuclei

Striations

Skeletal muscle
fiber (cell)

Skeletal Muscle
Table 4.14 a
Table 4.14a: © The McGraw-Hill Companies, Inc./Photo by Dr. Alvin Telser

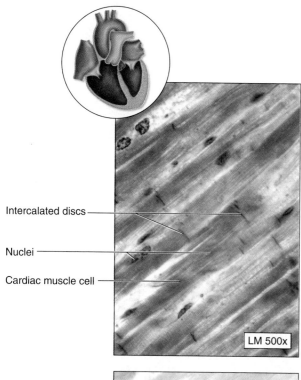

Intercalated discs

Nuclei

Cardiac muscle cell

LM 500x

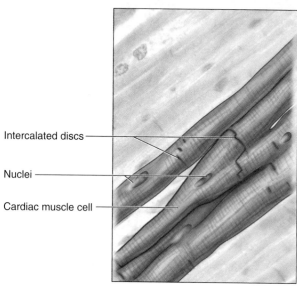

Intercalated discs

Nuclei

Cardiac muscle cell

Cardiac Muscle
Table 4.14 b
Table 4.14b: © The McGraw-Hill Companies, Inc./Photo by Dr. Alvin Telser

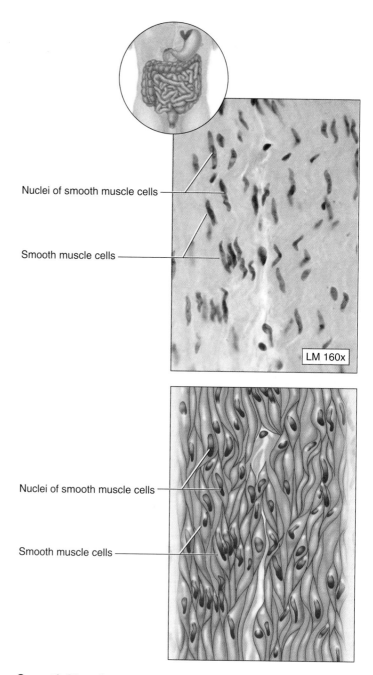

Nuclei of smooth muscle cells

Smooth muscle cells

LM 160x

Nuclei of smooth muscle cells

Smooth muscle cells

Smooth Muscle
Table 4.14 c
Table 4.14c: © The McGraw-Hill Companies, Inc./Photo by Dr. Alvin Telser

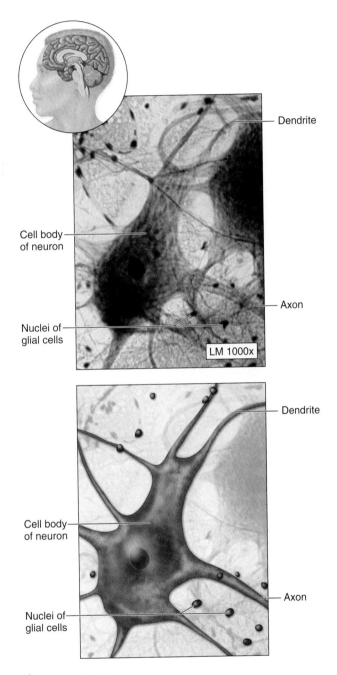

Dendrite

Cell body
of neuron

Axon

Nuclei of
glial cells

LM 1000x

Dendrite

Cell body
of neuron

Axon

Nuclei of
glial cells

Nervous Tissue
Table 4.15
Table 4.15: © Carolina Biological Supply/Phototake

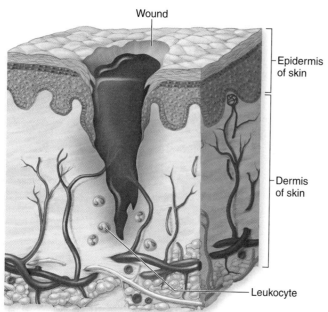

Wound

Epidermis of skin

Dermis of skin

Leukocyte

(1) Cut blood vessels bleed into the wound.

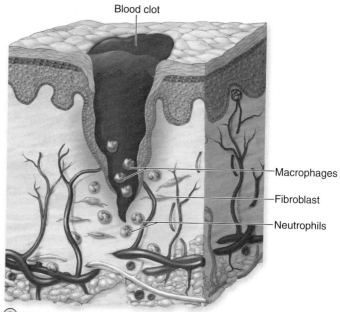

Blood clot

Macrophages

Fibroblast

Neutrophils

(2) Blood clot forms, and leukocytes clean wound.

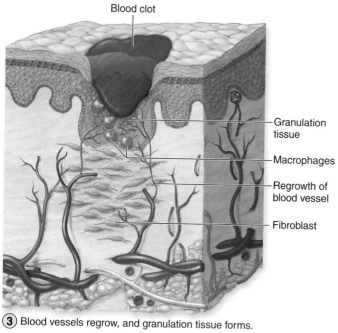

Blood clot

Granulation tissue

Macrophages

Regrowth of blood vessel

Fibroblast

(3) Blood vessels regrow, and granulation tissue forms.

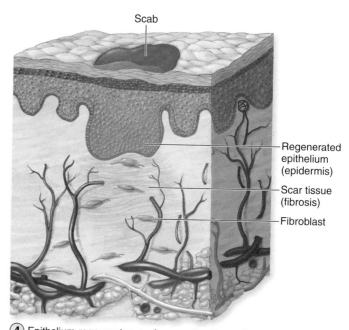

Scab

Regenerated epithelium (epidermis)

Scar tissue (fibrosis)

Fibroblast

(4) Epithelium regenerates, and connective tissue fibrosis occurs.

Stages in Wound Healing
Figure 4.9

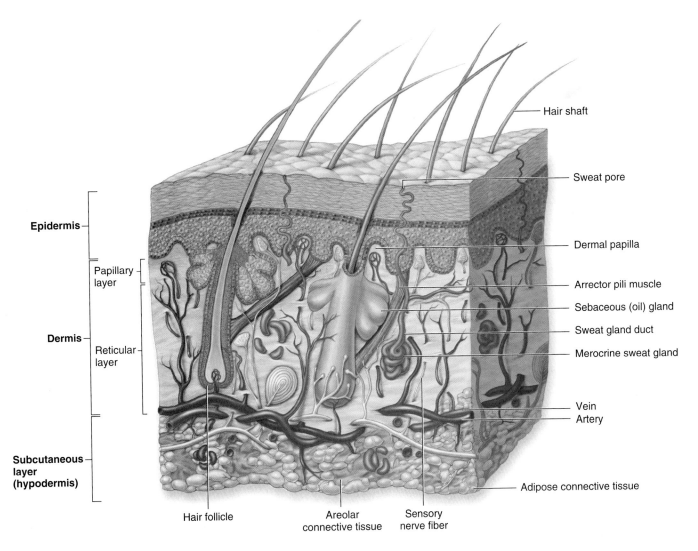

Epidermis

Dermis
- Papillary layer
- Reticular layer

Subcutaneous layer (hypodermis)

Hair shaft

Sweat pore

Dermal papilla

Arrector pili muscle

Sebaceous (oil) gland

Sweat gland duct

Merocrine sweat gland

Vein

Artery

Adipose connective tissue

Hair follicle

Areolar connective tissue

Sensory nerve fiber

Layers of the Integument
Figure 5.1

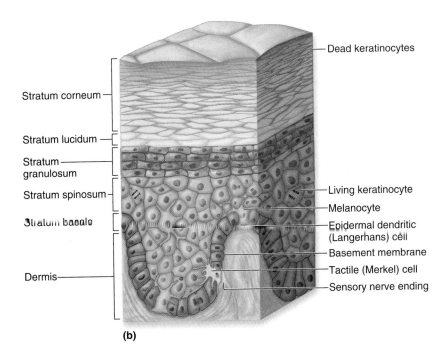

Stratum corneum

Stratum lucidum

Stratum granulosum

Stratum spinosum

Stratum basale

Dermis

Dead keratinocytes

Living keratinocyte

Melanocyte

Epidermal dendritic (Langerhans) cell

Basement membrane

Tactile (Merkel) cell

Sensory nerve ending

(b)

Epidermal Strata
Figure 5.2b

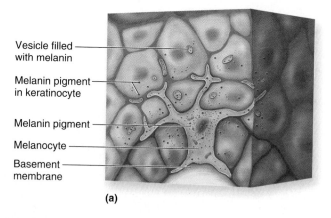

Vesicle filled with melanin

Melanin pigment in keratinocyte

Melanin pigment

Melanocyte

Basement membrane

(a)

Production of Melanin by Melanocytes
Figure 5.4a

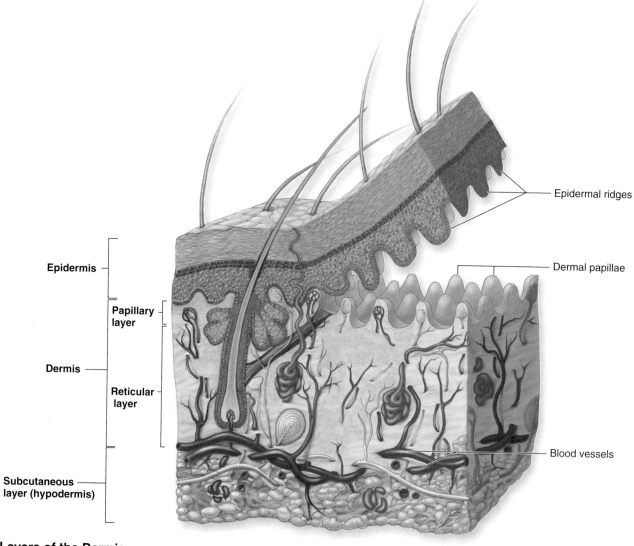

Epidermal ridges

Dermal papillae

Epidermis

Papillary layer

Dermis

Reticular layer

Blood vessels

Subcutaneous layer (hypodermis)

Layers of the Dermis
Figure 5.6

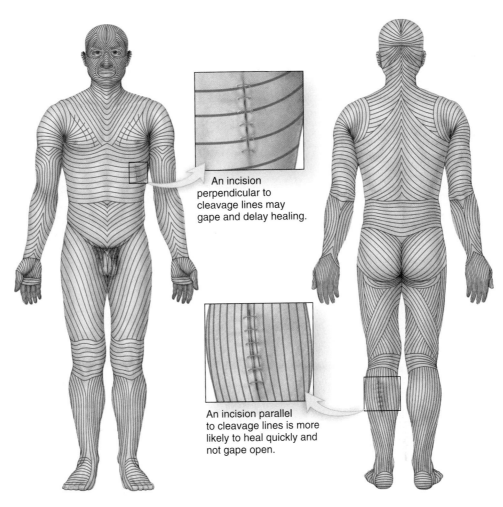

An incision
perpendicular to
cleavage lines may
gape and delay healing.

An incision parallel
to cleavage lines is more
likely to heal quickly and
not gape open.

Lines of Cleavage
Figure 5.7

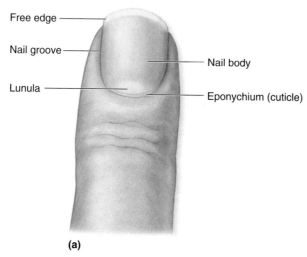

Free edge ——————
Nail groove ——————
Lunula ——————
                                        —————— Nail body
                                        —————— Eponychium (cuticle)

(a)

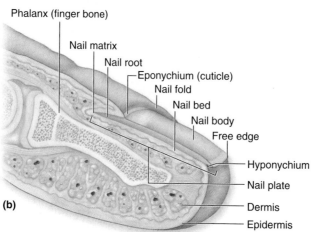

Phalanx (finger bone)
              Nail matrix
                     Nail root
                          ┌— Eponychium (cuticle)
                             Nail fold
                               Nail bed
                                 Nail body
                                   Free edge
                                        —————— Hyponychium
                                        —————— Nail plate
(b)
                                        —————— Dermis
                                        —————— Epidermis

Structure of a Fingernail
Figure 5.8

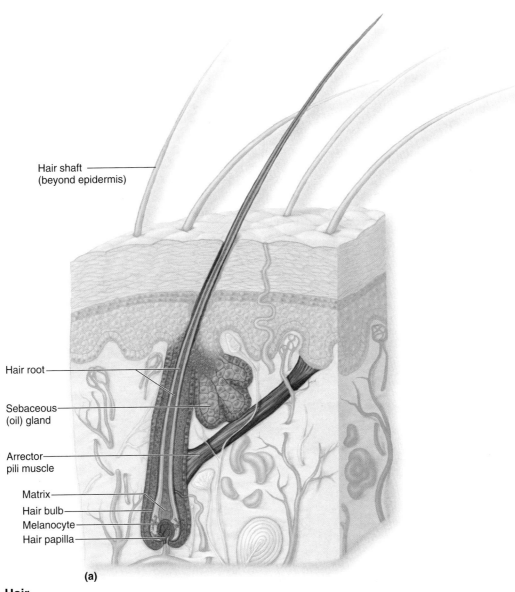

Hair shaft
(beyond epidermis)

Hair root

Sebaceous
(oil) gland

Arrector
pili muscle

Matrix

Hair bulb

Melanocyte

Hair papilla

(a)

Hair
Figure 5.9a

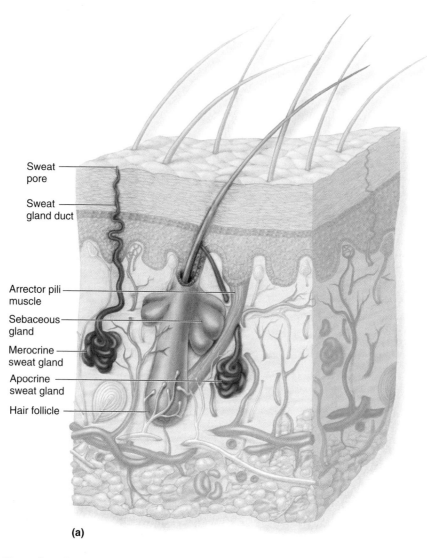

Sweat pore

Sweat gland duct

Arrector pili muscle

Sebaceous gland

Merocrine sweat gland

Apocrine sweat gland

Hair follicle

(a)

Exocrine Glands of the Skin
Figure 5.10a

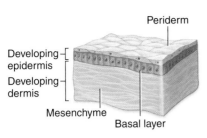
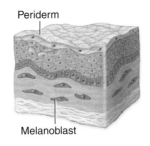
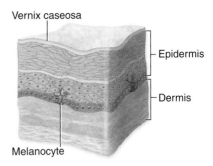

7 – 8 weeks   **11 – 12 weeks**   **Birth**

Integument Development
Figure 5.11

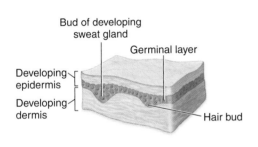
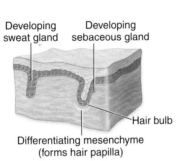
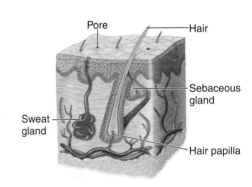

12 weeks   **15 weeks**   **Birth**

Hair and Gland Development
Figure 5.12

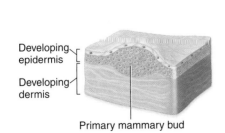
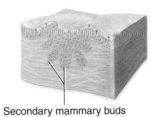
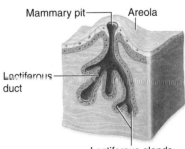

6 weeks   **16 weeks**   **28 weeks**

Mammary Gland Development
Figure 5.13

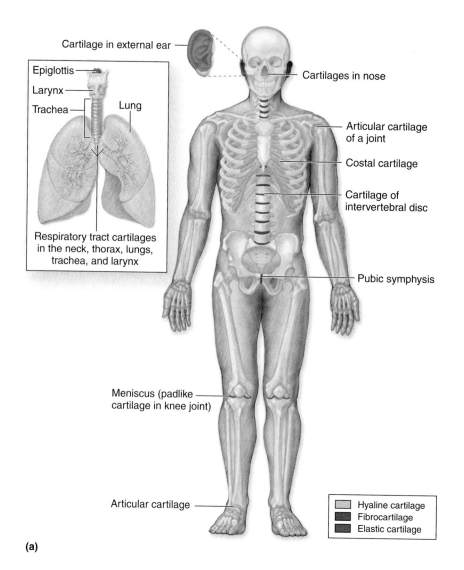

Cartilage in external ear

Epiglottis

Larynx

Trachea

Lung

Respiratory tract cartilages
in the neck, thorax, lungs,
trachea, and larynx

Cartilages in nose

Articular cartilage
of a joint

Costal cartilage

Cartilage of
intervertebral disc

Pubic symphysis

Meniscus (padlike
cartilage in knee joint)

Articular cartilage

Hyaline cartilage
Fibrocartilage
Elastic cartilage

(a)

Distribution of Cartilage in an Adult
Figure 6.1a

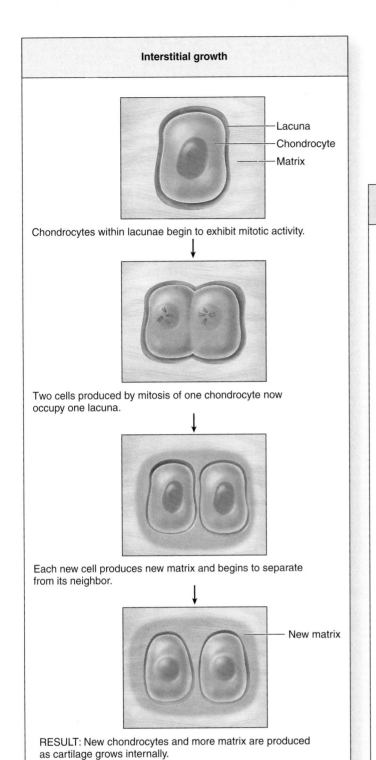

Interstitial growth

Lacuna
Chondrocyte
Matrix

Chondrocytes within lacunae begin to exhibit mitotic activity.

Two cells produced by mitosis of one chondrocyte now occupy one lacuna.

Each new cell produces new matrix and begins to separate from its neighbor.

New matrix

RESULT: New chondrocytes and more matrix are produced as cartilage grows internally.

(a)

Interstitial Growth
Figure 6.2a

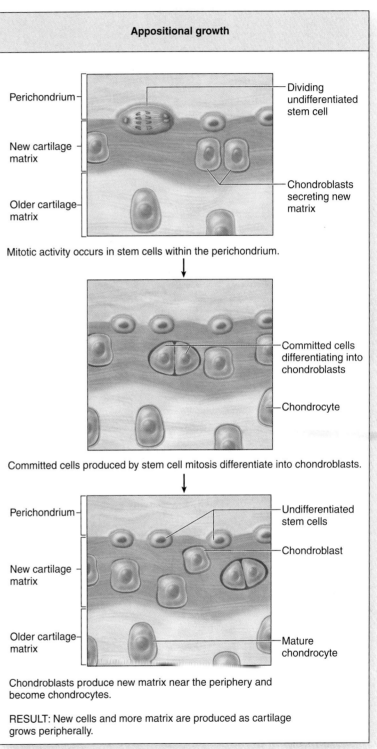

Appositional growth

Perichondrium

New cartilage matrix

Older cartilage matrix

Dividing undifferentiated stem cell

Chondroblasts secreting new matrix

Mitotic activity occurs in stem cells within the perichondrium.

Committed cells differentiating into chondroblasts

Chondrocyte

Committed cells produced by stem cell mitosis differentiate into chondroblasts.

Perichondrium

New cartilage matrix

Older cartilage matrix

Undifferentiated stem cells

Chondroblast

Mature chondrocyte

Chondroblasts produce new matrix near the periphery and become chondrocytes.

RESULT: New cells and more matrix are produced as cartilage grows peripherally.

(b)

Appositional Growth
Figure 6.2b

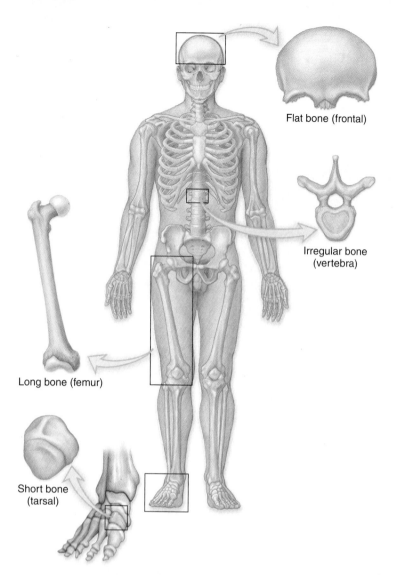

Classification of Bone by Shape
Figure 6.3

Flat bone (frontal)

Irregular bone (vertebra)

Long bone (femur)

Short bone (tarsal)

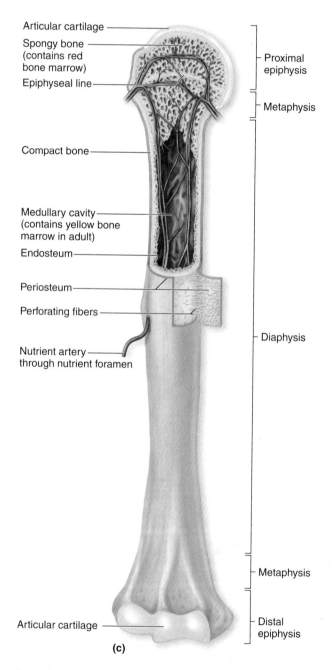

Articular cartilage

Spongy bone (contains red bone marrow)

Epiphyseal line

Compact bone

Medullary cavity (contains yellow bone marrow in adult)

Endosteum

Periosteum

Perforating fibers

Nutrient artery through nutrient foramen

Articular cartilage

Proximal epiphysis

Metaphysis

Diaphysis

Metaphysis

Distal epiphysis

(c)

Gross Anatomy of a Long Bone
Figure 6.4c

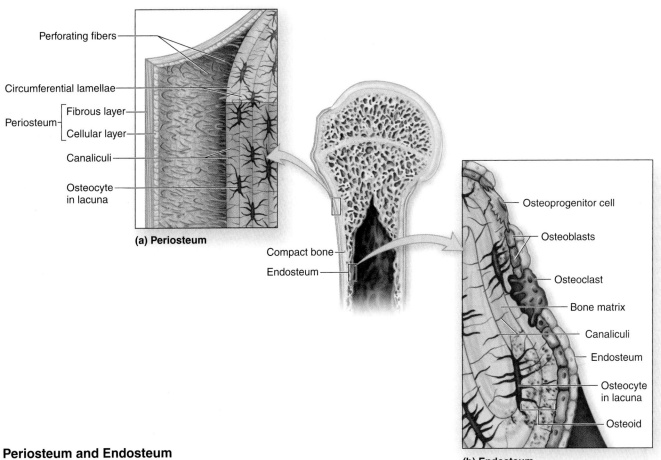

Perforating fibers

Circumferential lamellae

Periosteum
- Fibrous layer
- Cellular layer

Canaliculi

Osteocyte in lacuna

(a) Periosteum

Compact bone
Endosteum

Osteoprogenitor cell

Osteoblasts

Osteoclast

Bone matrix

Canaliculi

Endosteum

Osteocyte in lacuna

Osteoid

(b) Endosteum

Periosteum and Endosteum
Figure 6.5

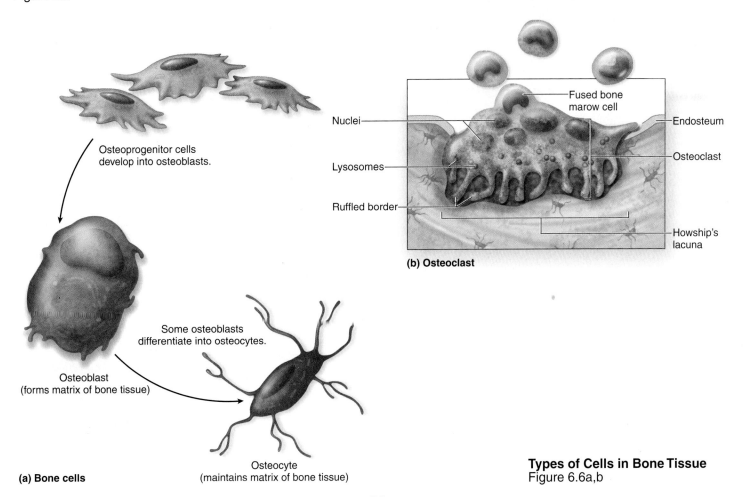

Osteoprogenitor cells develop into osteoblasts.

Some osteoblasts differentiate into osteocytes.

Osteoblast
(forms matrix of bone tissue)

Osteocyte
(maintains matrix of bone tissue)

(a) Bone cells

Nuclei

Lysosomes

Ruffled border

Fused bone marrow cell

Endosteum

Osteoclast

Howship's lacuna

(b) Osteoclast

Types of Cells in Bone Tissue
Figure 6.6a,b

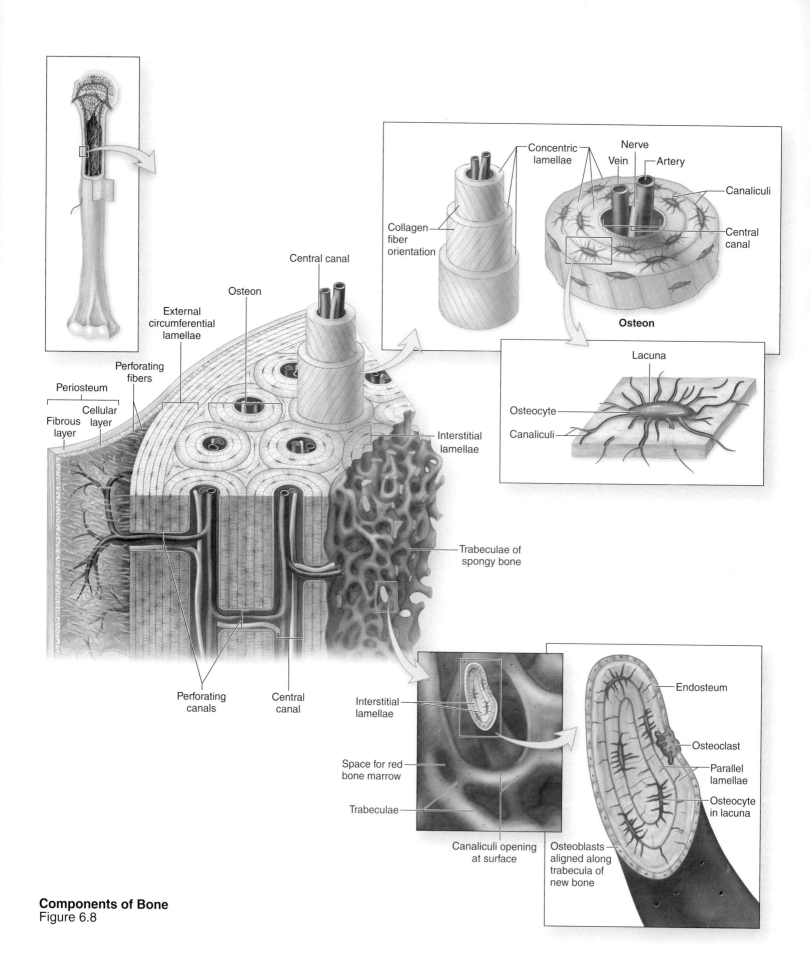

Nerve

Concentric lamellae

Vein Artery

Canaliculi

Collagen fiber orientation

Central canal

Osteon

Lacuna

Osteocyte

Canaliculi

Central canal

Osteon

External circumferential lamellae

Perforating fibers

Periosteum

Cellular layer

Fibrous layer

Interstitial lamellae

Trabeculae of spongy bone

Perforating canals

Central canal

Interstitial lamellae

Space for red bone marrow

Trabeculae

Canaliculi opening at surface

Endosteum

Osteoclast

Parallel lamellae

Osteocyte in lacuna

Osteoblasts aligned along trabecula of new bone

Components of Bone
Figure 6.8

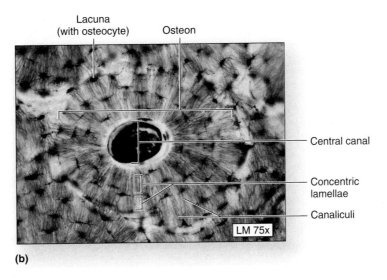

Lacuna
(with osteocyte)   Osteon

Central canal

Concentric
lamellae

Canaliculi

LM 75x

(b)

Microscopic Anatomy of Bone
Figure 6.9b
Figure 6.9b: © Carolina Biological Supply Company/Phototake

(1) Ossification centers form within thickened regions of mesenchyme.

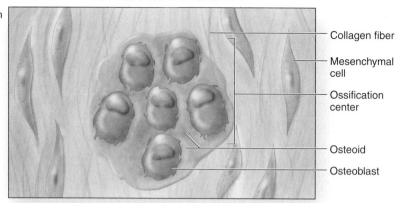

Collagen fiber

Mesenchymal cell

Ossification center

Osteoid

Osteoblast

(2) Bone matrix (osteoid) undergoes calcification.

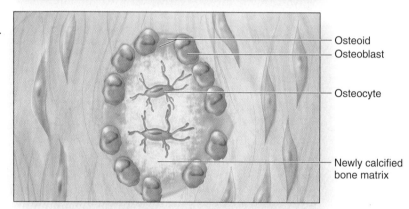

Osteoid
Osteoblast

Osteocyte

Newly calcified bone matrix

(3) Woven bone and surrounding periosteum form.

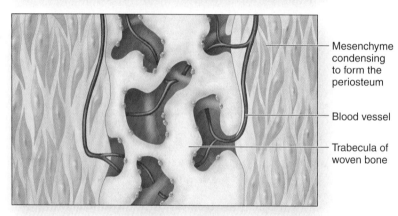

Mesenchyme condensing to form the periosteum

Blood vessel

Trabecula of woven bone

(4) Lamellar bone replaces woven bone, as compact and spongy bone form.

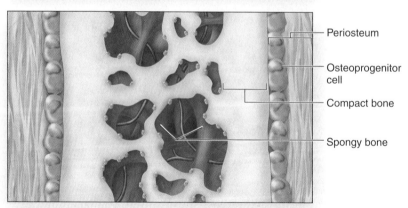

Periosteum

Osteoprogenitor cell

Compact bone

Spongy bone

Intramembranous Ossification
Figure 6.10

Endochondral Ossification

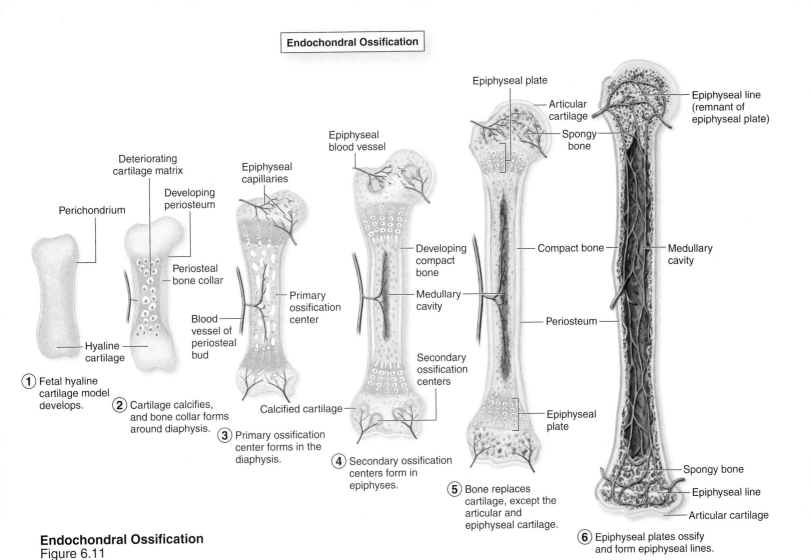

① Fetal hyaline cartilage model develops.

② Cartilage calcifies, and bone collar forms around diaphysis.

③ Primary ossification center forms in the diaphysis.

④ Secondary ossification centers form in epiphyses.

⑤ Bone replaces cartilage, except the articular and epiphyseal cartilage.

⑥ Epiphyseal plates ossify and form epiphyseal lines.

Endochondral Ossification
Figure 6.11

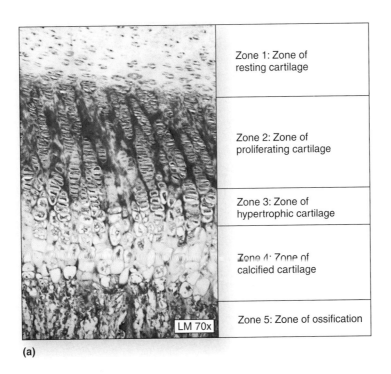

Zone 1: Zone of resting cartilage

Zone 2: Zone of proliferating cartilage

Zone 3: Zone of hypertrophic cartilage

Zone 4: Zone of calcified cartilage

Zone 5: Zone of ossification

LM 70x

(a)

Epiphyseal Plate
Figure 6.12a

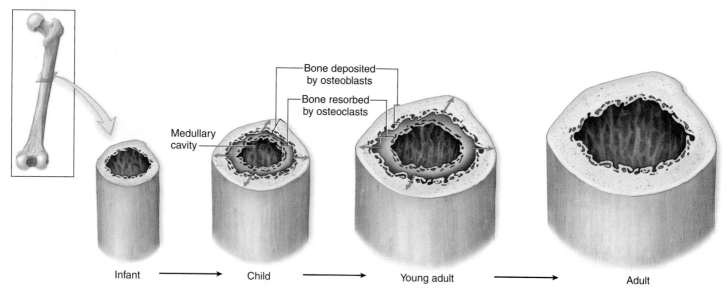

Bone deposited by osteoblasts

Bone resorbed by osteoclasts

Medullary cavity

Infant → Child → Young adult → Adult

Appositional Bone Growth
Figure 6.13

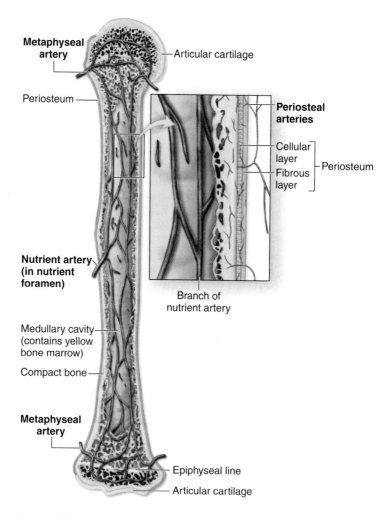

Metaphyseal artery

Articular cartilage

Periosteum

Periosteal arteries

Cellular layer

Fibrous layer

Periosteum

Nutrient artery (in nutrient foramen)

Branch of nutrient artery

Medullary cavity (contains yellow bone marrow)

Compact bone

Metaphyseal artery

Epiphyseal line

Articular cartilage

Arterial Supply to a Mature Bone
Figure 6.14

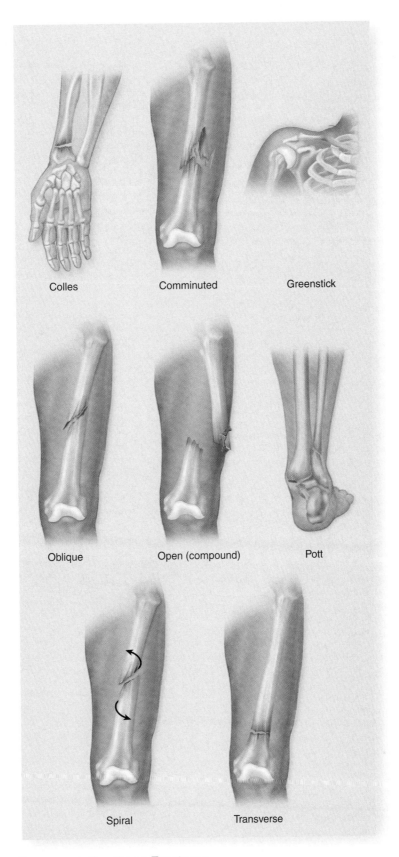

Colles

Comminuted

Greenstick

Oblique

Open (compound)

Pott

Spiral

Transverse

Representative Bone Fractures
Figure 6.15

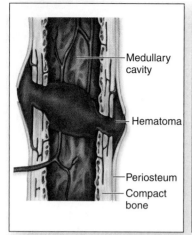
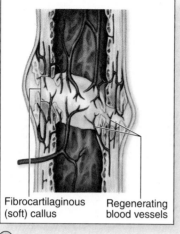
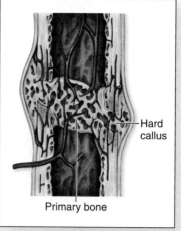
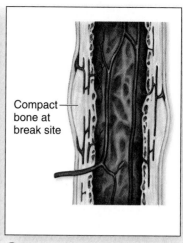

Medullary cavity

Hematoma

Periosteum

Compact bone

Fibrocartilaginous (soft) callus

Regenerating blood vessels

Hard callus

Primary bone

Compact bone at break site

① A fracture hematoma forms.

② A fibrocartilaginous (soft) callus forms.

③ A hard (bony) callus forms.

④ The bone is remodeled.

Fracture Repair
Figure 6.16

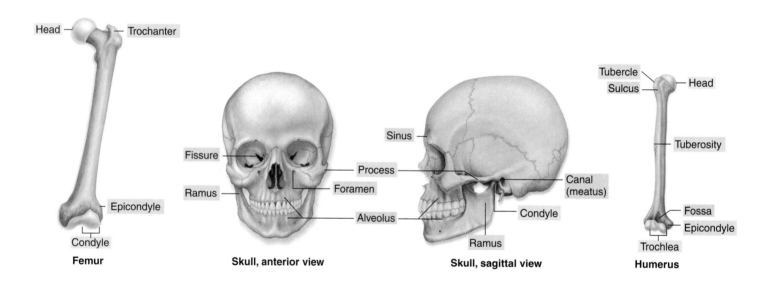

Head — Trochanter

Epicondyle

Condyle

Femur

Fissure

Ramus

Process

Foramen

Alveolus

Skull, anterior view

Sinus

Canal (meatus)

Condyle

Ramus

Skull, sagittal view

Tubercle

Sulcus

Head

Tuberosity

Fossa

Epicondyle

Trochlea

Humerus

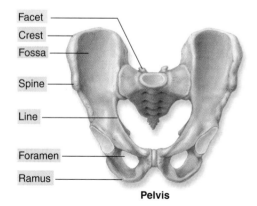

Facet

Crest

Fossa

Spine

Line

Foramen

Ramus

Pelvis

Bone Markings
Figure 6.17

70

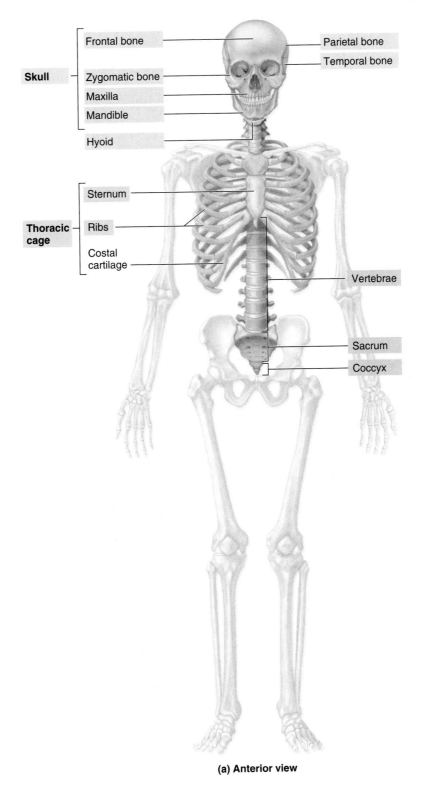

Frontal bone

Parietal bone

Temporal bone

Skull

Zygomatic bone

Maxilla

Mandible

Hyoid

Sternum

Thoracic cage

Ribs

Costal cartilage

Vertebrae

Sacrum

Coccyx

(a) Anterior view

Axial Skeleton—Anterior View
Figure 7.1a

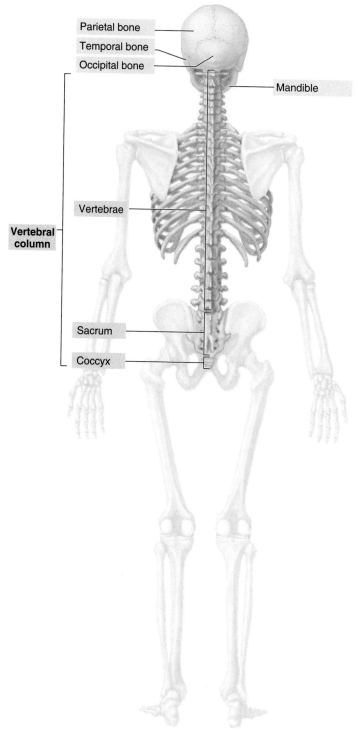

Parietal bone

Temporal bone

Occipital bone

Mandible

Vertebrae

Vertebral column

Sacrum

Coccyx

(b) Posterior view

Axial Skeleton—Posterior View
Figure 7.1b

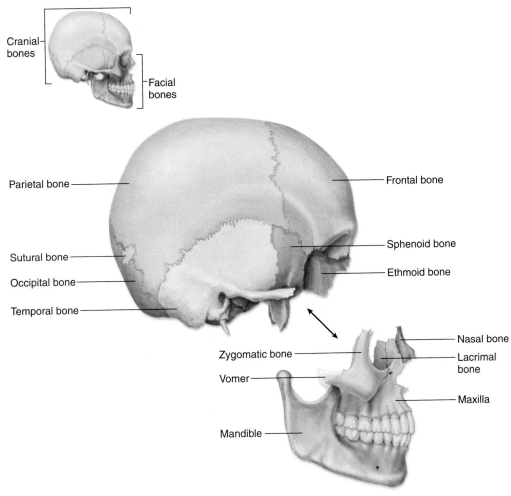

Cranial bones

Facial bones

Parietal bone

Frontal bone

Sphenoid bone

Sutural bone

Ethmoid bone

Occipital bone

Temporal bone

Nasal bone

Zygomatic bone

Lacrimal bone

Vomer

Maxilla

Mandible

Cranial and Facial Divisions of the Skull
Figure 7.2

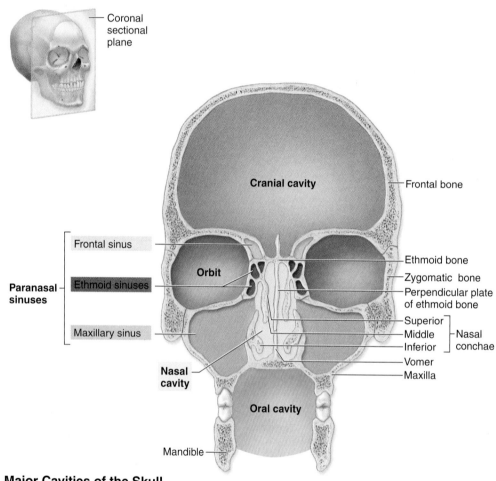

Coronal
sectional
plane

Cranial cavity

Frontal bone

**Paranasal
sinuses**

Frontal sinus

Orbit

Ethmoid bone

Zygomatic bone

Perpendicular plate
of ethmoid bone

Ethmoid sinuses

Maxillary sinus

Superior

Middle
Nasal
conchae
Inferior

**Nasal
cavity**

Vomer

Maxilla

Oral cavity

Mandible

Major Cavities of the Skull
Figure 7.3

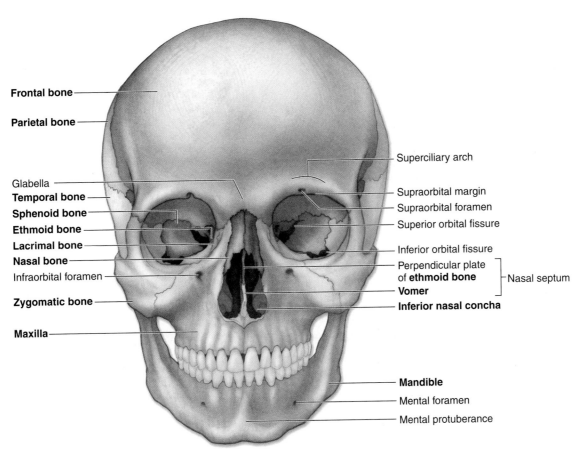

Frontal bone

Parietal bone

Superciliary arch

Glabella
Temporal bone
Sphenoid bone
Ethmoid bone
Lacrimal bone
Nasal bone
Infraorbital foramen

Zygomatic bone

Maxilla

Supraorbital margin
Supraorbital foramen
Superior orbital fissure

Inferior orbital fissure
Perpendicular plate
of ethmoid bone
Vomer
Inferior nasal concha

Nasal septum

Mandible
Mental foramen
Mental protuberance

Anterior View of Skull
Figure 7.4

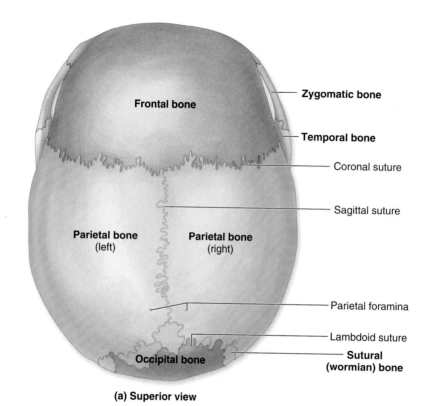

Frontal bone

Zygomatic bone

Temporal bone

Coronal suture

Sagittal suture

Parietal bone
(left)

Parietal bone
(right)

Parietal foramina

Lambdoid suture

**Sutural
(wormian) bone**

Occipital bone

(a) Superior view

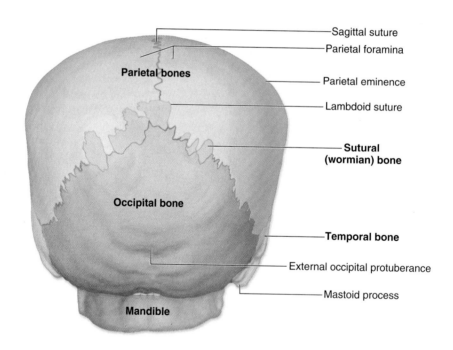

Sagittal suture

Parietal foramina

Parietal bones

Parietal eminence

Lambdoid suture

**Sutural
(wormian) bone**

Occipital bone

Temporal bone

External occipital protuberance

Mastoid process

Mandible

(b) Posterior View

Superior and Posterior Views of Skull
Figure 7.5

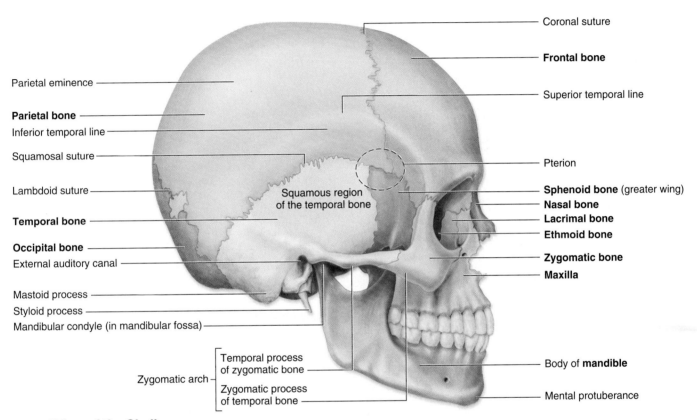

Coronal suture

Frontal bone

Superior temporal line

Pterion

Sphenoid bone (greater wing)
Nasal bone
Lacrimal bone
Ethmoid bone

Zygomatic bone

Maxilla

Body of **mandible**

Mental protuberance

Parietal eminence

Parietal bone
Inferior temporal line

Squamosal suture

Lambdoid suture

Temporal bone

Occipital bone
External auditory canal

Mastoid process
Styloid process
Mandibular condyle (in mandibular fossa)

Squamous region
of the temporal bone

Zygomatic arch

Temporal process
of zygomatic bone

Zygomatic process
of temporal bone

Lateral View of the Skull
Figure 7.6

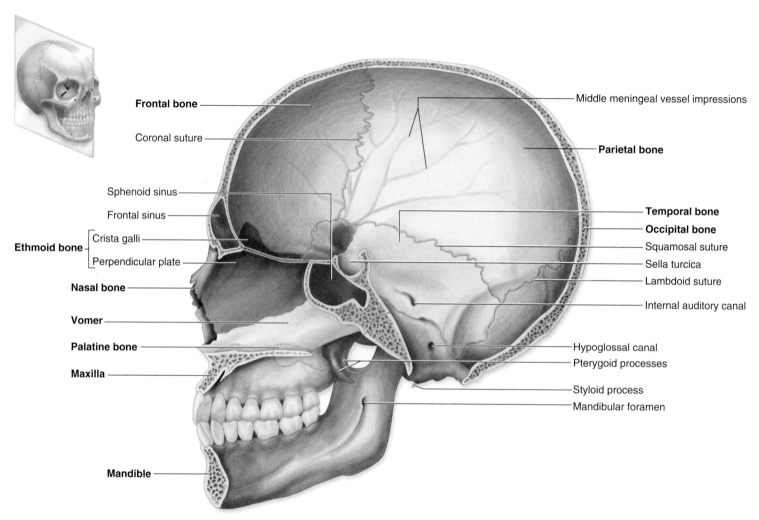

Frontal bone

Coronal suture

Sphenoid sinus

Frontal sinus

Ethmoid bone {
Crista galli
Perpendicular plate

Nasal bone

Vomer

Palatine bone

Maxilla

Mandible

Middle meningeal vessel impressions

Parietal bone

Temporal bone

Occipital bone

Squamosal suture

Sella turcica

Lambdoid suture

Internal auditory canal

Hypoglossal canal

Pterygoid processes

Styloid process

Mandibular foramen

Sagittal Section of the Skull
Figure 7.7

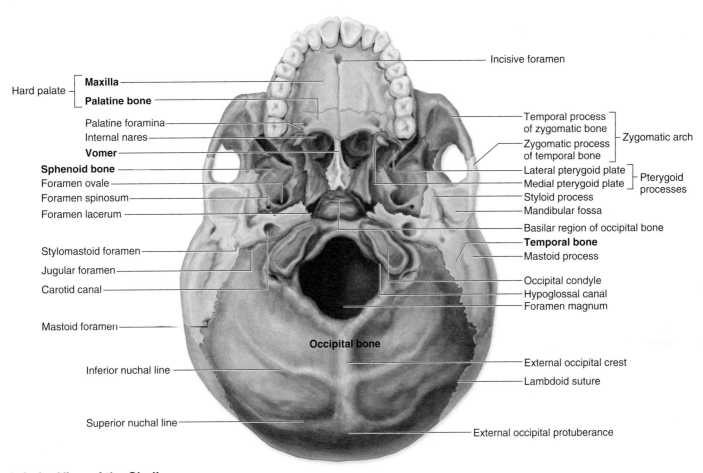

Incisive foramen

Hard palate — Maxilla

Palatine bone

Palatine foramina

Internal nares

Vomer

Sphenoid bone

Foramen ovale

Foramen spinosum

Foramen lacerum

Stylomastoid foramen

Jugular foramen

Carotid canal

Mastoid foramen

Inferior nuchal line

Superior nuchal line

Temporal process
of zygomatic bone — Zygomatic arch

Zygomatic process
of temporal bone

Lateral pterygoid plate — Pterygoid
processes
Medial pterygoid plate

Styloid process

Mandibular fossa

Basilar region of occipital bone

Temporal bone

Mastoid process

Occipital condyle

Hypoglossal canal

Foramen magnum

Occipital bone

External occipital crest

Lambdoid suture

External occipital protuberance

Inferior View of the Skull
Figure 7.8

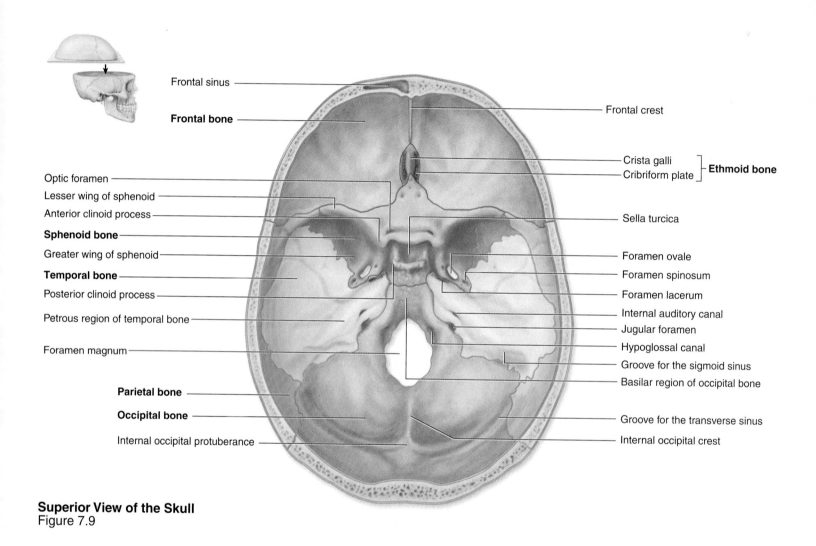

Frontal sinus

Frontal bone

Optic foramen
Lesser wing of sphenoid
Anterior clinoid process
Sphenoid bone
Greater wing of sphenoid
Temporal bone
Posterior clinoid process
Petrous region of temporal bone
Foramen magnum

Parietal bone
Occipital bone
Internal occipital protuberance

Frontal crest

Crista galli
Cribriform plate
Ethmoid bone

Sella turcica

Foramen ovale
Foramen spinosum
Foramen lacerum
Internal auditory canal
Jugular foramen
Hypoglossal canal
Groove for the sigmoid sinus
Basilar region of occipital bone

Groove for the transverse sinus
Internal occipital crest

Superior View of the Skull
Figure 7.9

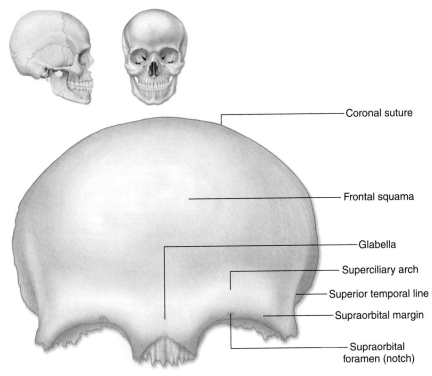

Coronal suture

Frontal squama

Glabella

Superciliary arch

Superior temporal line

Supraorbital margin

Supraorbital foramen (notch)

Frontal bone, anterior view

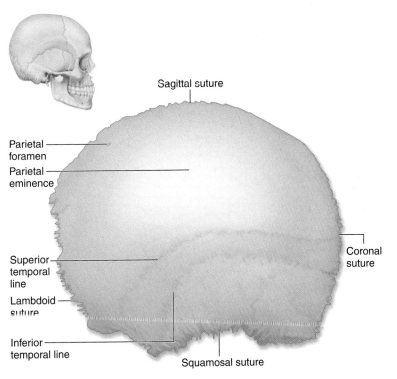

Sagittal suture

Parietal foramen

Parietal eminence

Coronal suture

Superior temporal line

Lambdoid suture

Inferior temporal line

Squamosal suture

Parietal bone, lateral view

Frontal Bone and Parietal Bone
Figure 7.10, 7.11

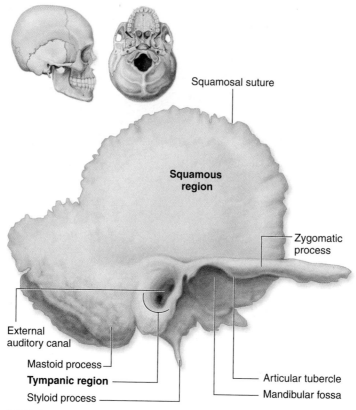

Squamosal suture

Squamous region

Zygomatic process

External auditory canal

Mastoid process

Tympanic region

Styloid process

Articular tubercle

Mandibular fossa

(a) Right temporal bone, external (lateral) view

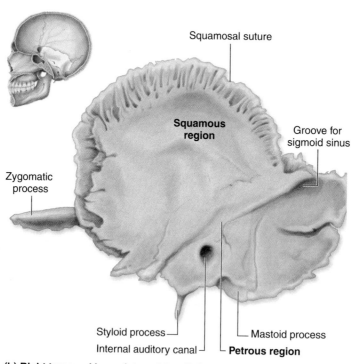

Squamosal suture

Squamous region

Groove for sigmoid sinus

Zygomatic process

Styloid process

Internal auditory canal

Mastoid process

Petrous region

(b) Right temporal bone, internal (medial) view

Temporal Bone
Figure 7.12

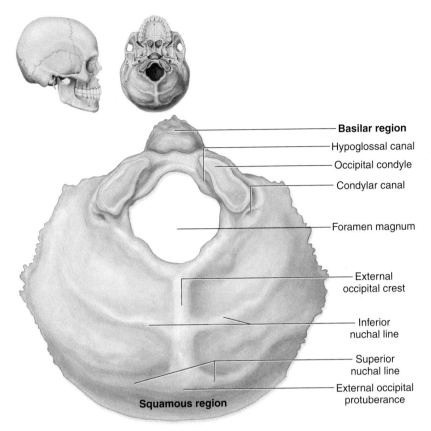

Basilar region
Hypoglossal canal
Occipital condyle
Condylar canal

Foramen magnum

External
occipital crest

Inferior
nuchal line

Superior
nuchal line

External occipital
protuberance

Squamous region

(a) Occipital bone, external (inferior) view

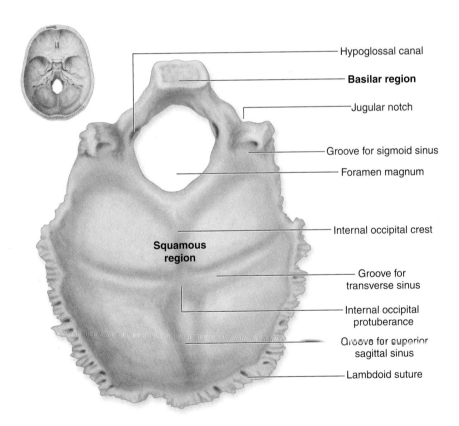

Hypoglossal canal

Basilar region

Jugular notch

Groove for sigmoid sinus

Foramen magnum

Internal occipital crest

**Squamous
region**

Groove for
transverse sinus

Internal occipital
protuberance

Groove for superior
sagittal sinus

Lambdoid suture

(b) Occipital bone, internal (superior) view

Occipital Bone
Figure 7.13

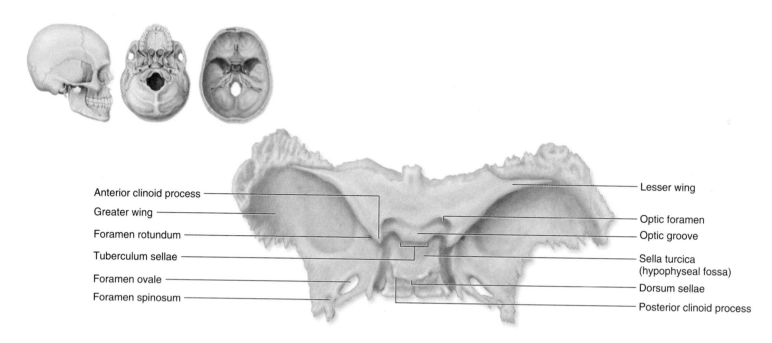

Anterior clinoid process — Lesser wing

Greater wing — Optic foramen

Foramen rotundum — Optic groove

Tuberculum sellae — Sella turcica (hypophyseal fossa)

Foramen ovale — Dorsum sellae

Foramen spinosum — Posterior clinoid process

(a) Sphenoid bone, superior view

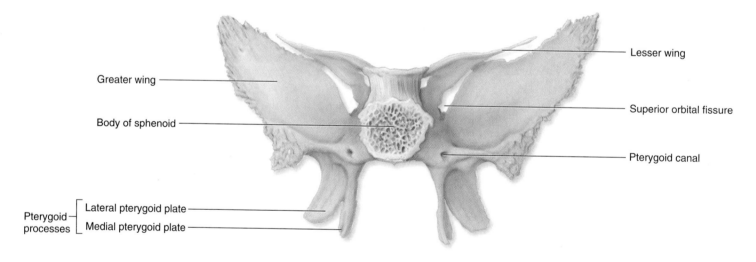

Greater wing — Lesser wing

Body of sphenoid — Superior orbital fissure

Pterygoid canal

Pterygoid processes { Lateral pterygoid plate
Medial pterygoid plate

(b) Sphenoid bone, posterior view

Sphenoid Bone
Figure 7.14

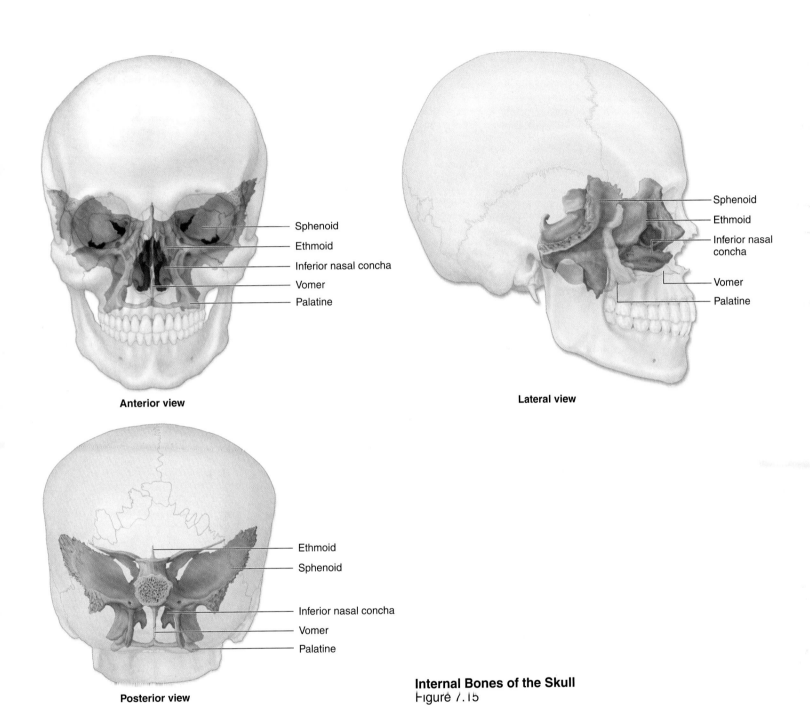

Anterior view

- Sphenoid
- Ethmoid
- Inferior nasal concha
- Vomer
- Palatine

Lateral view

- Sphenoid
- Ethmoid
- Inferior nasal concha
- Vomer
- Palatine

Posterior view

- Ethmoid
- Sphenoid
- Inferior nasal concha
- Vomer
- Palatine

Internal Bones of the Skull
Figure 7.15

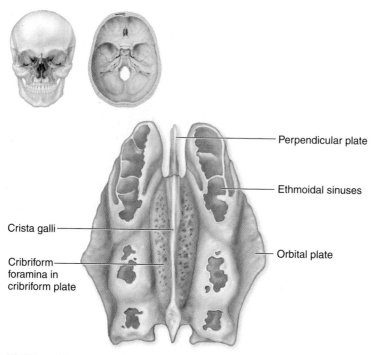

Crista galli

Cribriform
foramina in
cribriform plate

Perpendicular plate

Ethmoidal sinuses

Orbital plate

(a) Ethmoid bone, superior view

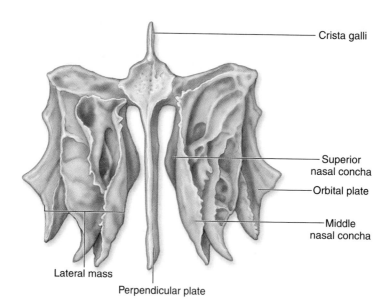

Crista galli

Superior
nasal concha

Orbital plate

Middle
nasal concha

Lateral mass

Perpendicular plate

(b) Ethmoid bone, anterior view

Ethmoid Bone
Figure 7.16a, b

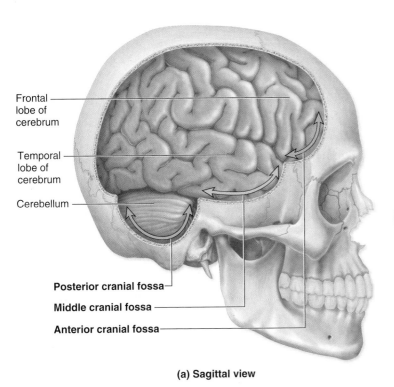

Frontal lobe of cerebrum

Temporal lobe of cerebrum

Cerebellum

Posterior cranial fossa

Middle cranial fossa

Anterior cranial fossa

(a) Sagittal view

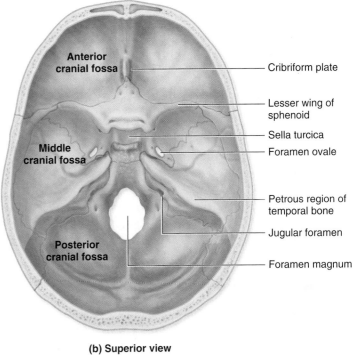

Anterior cranial fossa

Cribriform plate

Lesser wing of sphenoid

Sella turcica

Foramen ovale

Middle cranial fossa

Petrous region of temporal bone

Jugular foramen

Posterior cranial fossa

Foramen magnum

(b) Superior view

Cranial Fossae
Figure 7.17

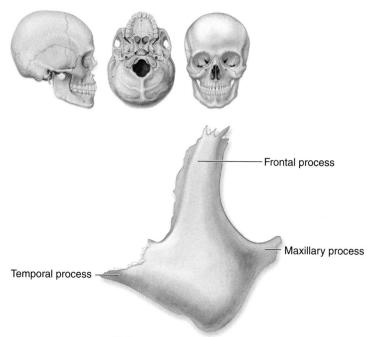

Frontal process

Maxillary process

Temporal process

Right zygomatic bone, lateral view

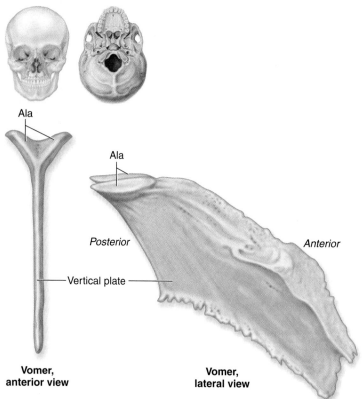

Ala

Ala

Posterior

Anterior

Vertical plate

**Vomer,
anterior view**

**Vomer,
lateral view**

Zygomatic Bone and Vomer
Figure 7.18, 7.19

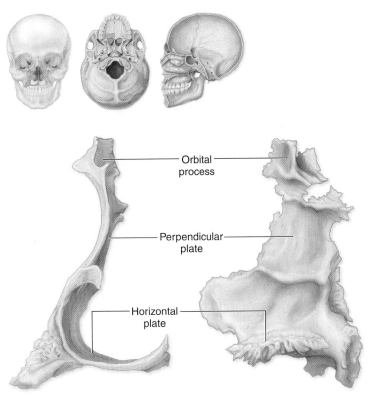

Orbital
process

Perpendicular
plate

Horizontal
plate

Right palatine bone, anterior view   **Right palatine bone, medial view**

Palatine Bone
Figure 7.20

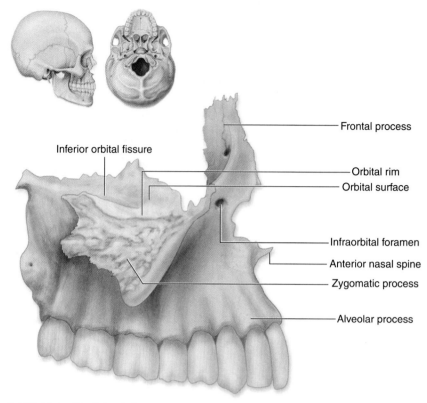

Inferior orbital fissure

Frontal process

Orbital rim

Orbital surface

Infraorbital foramen

Anterior nasal spine

Zygomatic process

Alveolar process

(a) Right maxilla, lateral view

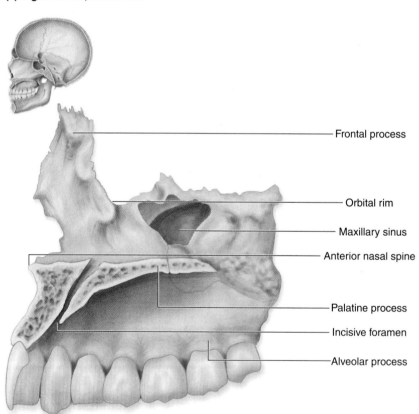

Frontal process

Orbital rim

Maxillary sinus

Anterior nasal spine

Palatine process

Incisive foramen

Alveolar process

(b) Right maxilla, medial view

Maxilla
Figure 7.21

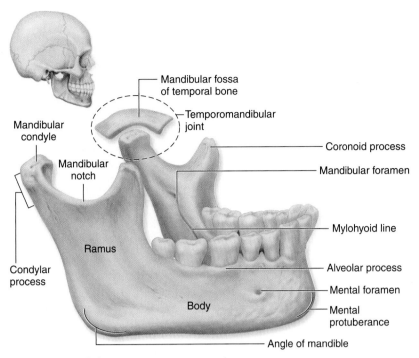

Mandibular fossa
of temporal bone

Temporomandibular
joint

Mandibular
condyle

Coronoid process

Mandibular foramen

Mandibular
notch

Mylohyoid line

Ramus

Condylar
process

Alveolar process

Mental foramen

Body

Mental
protuberance

Angle of mandible

Mandible, lateral view

Mandible
Figure 7.22

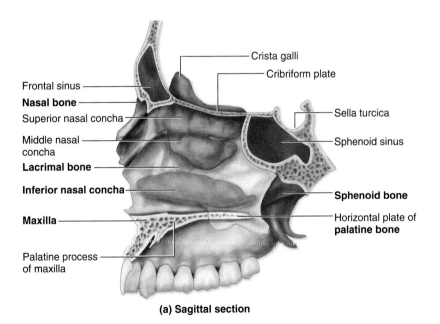

Crista galli

Cribriform plate

Frontal sinus

Nasal bone

Superior nasal concha

Sella turcica

Middle nasal
concha

Sphenoid sinus

Lacrimal bone

Inferior nasal concha

Sphenoid bone

Maxilla

Horizontal plate of
palatine bone

Palatine process
of maxilla

(a) Sagittal section

Nasal Cavity—Sagittal Section
Figure 7.23a

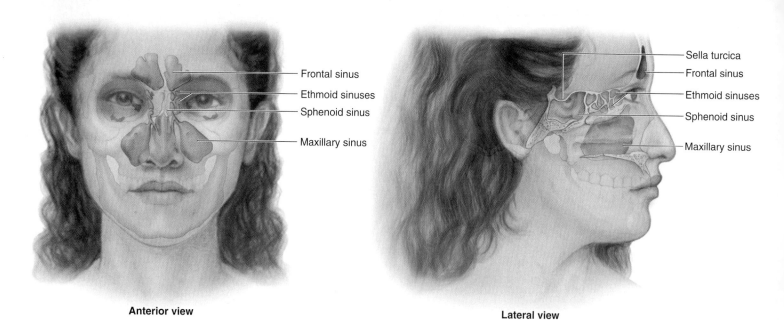

Frontal sinus
Ethmoid sinuses
Sphenoid sinus

Maxillary sinus

Anterior view

Sella turcica
Frontal sinus
Ethmoid sinuses
Sphenoid sinus

Maxillary sinus

Lateral view

Paranasal Sinuses
Figure 7.24

Roof of orbit

Lesser wing of
sphenoid bone ⌐   ⌐ Frontal bone

Zygomatic process of frontal bone

Greater wing of sphenoid bone   **Lateral wall**

Orbital surface of zygomatic bone

Optic foramen

Superior orbital fissure

Medial wall ⌐ Frontal process of maxilla
Lacrimal bone
Lateral mass of ethmoid bone

Inferior orbital fissure

Perpendicular plate
of palatine bone   ⌐ Orbital surface   Zygomatic
of maxilla   bone

Floor of orbit

Left Orbit
Figure 7.25

92

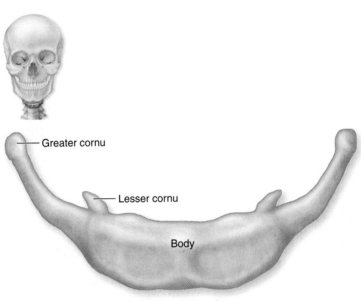

Hyoid, anterior view

Hyoid Bone
Figure 7.26

Table 7.5	Sex Differences in the Skull	
View	Female Skull	Male Skull
Anterior View		
Lateral View		

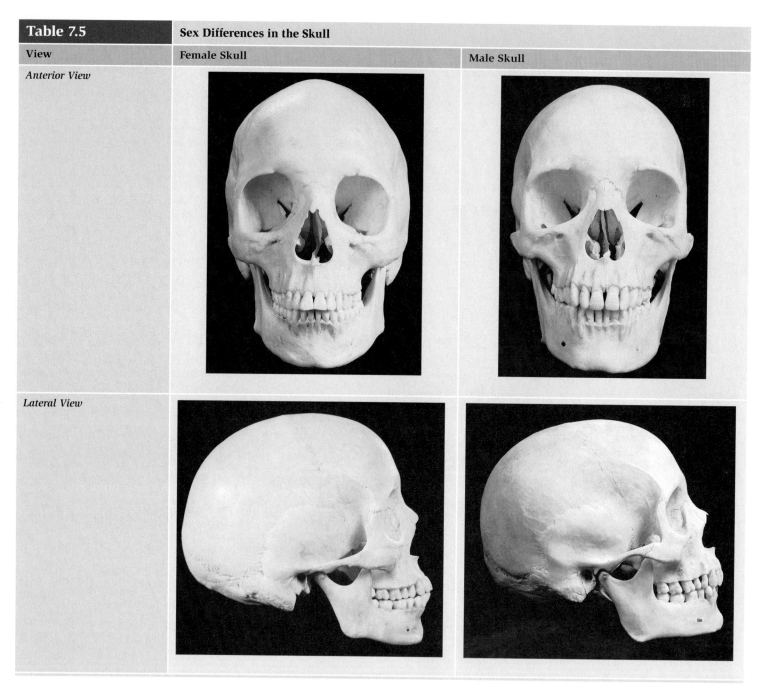

Sex Differences in the Skull
Table 7.5
Table 7.5: © Ralph T. Hutchings/imagingbody.com

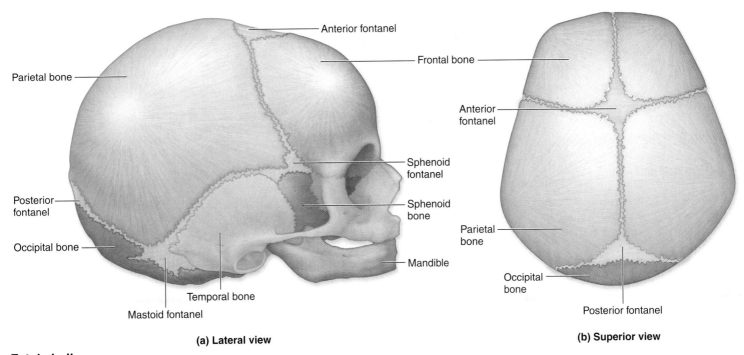

Parietal bone

Posterior fontanel

Occipital bone

Anterior fontanel

Frontal bone

Sphenoid fontanel

Sphenoid bone

Mandible

Mastoid fontanel

Temporal bone

(a) Lateral view

Anterior fontanel

Parietal bone

Occipital bone

Posterior fontanel

(b) Superior view

Fetal skull
Figure 7.27

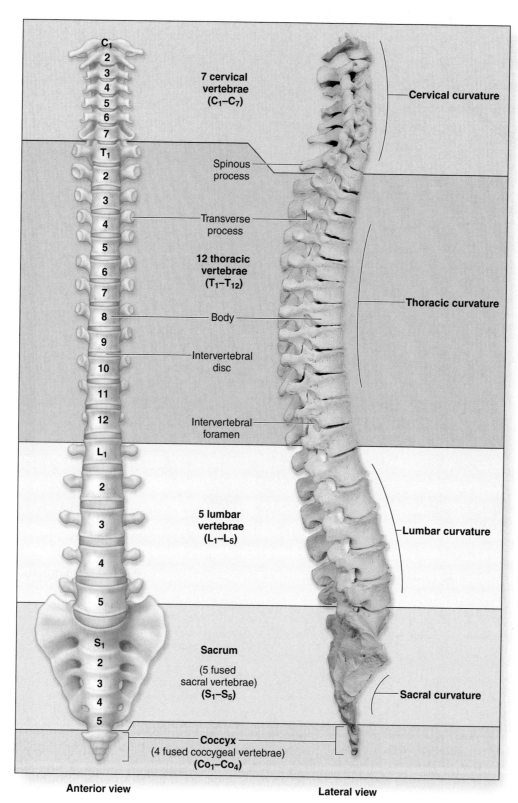

C₁
2
3
4
5
6
7

7 cervical vertebrae (C₁–C₇)

T₁
2
3
4
5
6
7
8
9
10
11
12

Spinous process

Transverse process

12 thoracic vertebrae (T₁–T₁₂)

Body

Intervertebral disc

Intervertebral foramen

L₁
2
3
4
5

5 lumbar vertebrae (L₁–L₅)

S₁
2
3
4
5

Sacrum

(5 fused sacral vertebrae) (S₁–S₅)

Coccyx

(4 fused coccygeal vertebrae) (Co₁–Co₄)

Cervical curvature

Thoracic curvature

Lumbar curvature

Sacral curvature

Anterior view

Lateral view

Vertebral Column
Figure 7.28

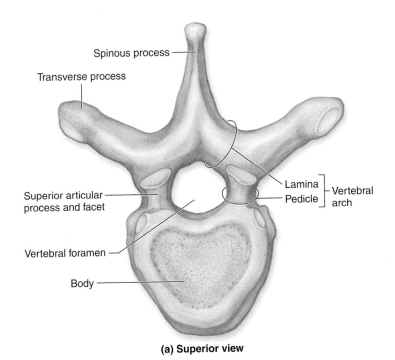

Spinous process

Transverse process

Superior articular
process and facet

Vertebral foramen

Body

Lamina
Pedicle — Vertebral arch

(a) Superior view

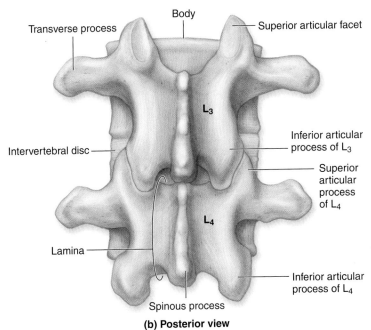

Body

Transverse process

Superior articular facet

Intervertebral disc

L₃

L₄

Lamina

Spinous process

Inferior articular
process of L₃

Superior
articular
process
of L₄

Inferior articular
process of L₄

(b) Posterior view

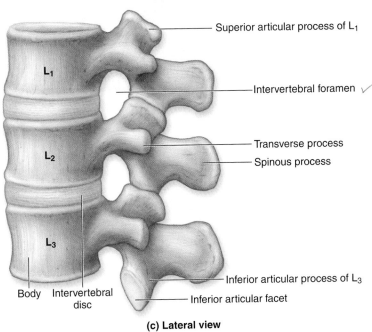

Superior articular process of L₁

L₁

L₂

L₃

Intervertebral foramen ✓

Transverse process

Spinous process

Inferior articular process of L₃

Inferior articular facet

Body   Intervertebral
disc

(c) Lateral view

Vertebral Anatomy
Figure 7.29

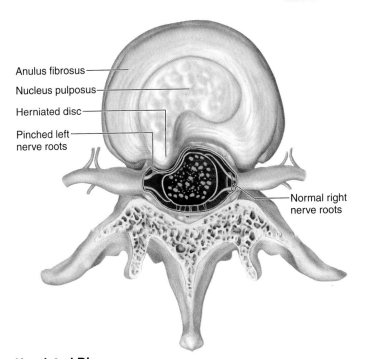

Anulus fibrosus

Nucleus pulposus

Herniated disc

Pinched left
nerve roots

Normal right
nerve roots

Herniated Disc
Clinical View p. 207

Table 7.6	Characteristic Features of Cervical, Thoracic, and Lumbar Vertebrae	
View	**(a) Cervical Vertebra**	**(b) Thoracic Vertebra**
Superior View		
Lateral View		

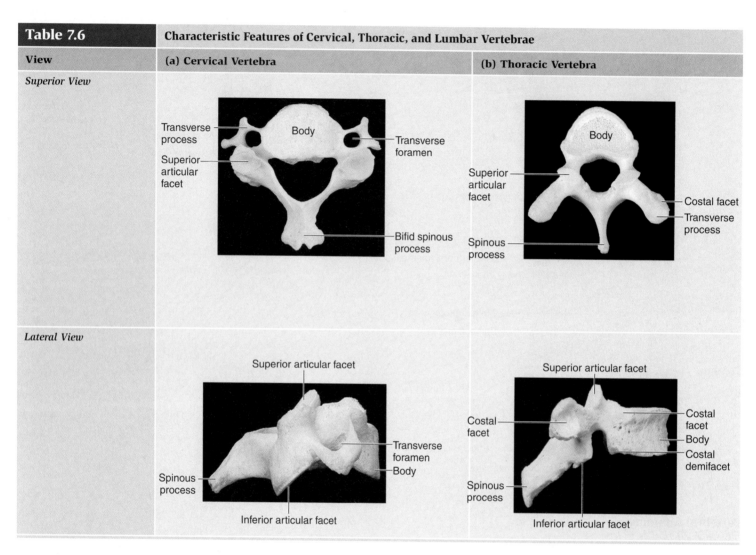

**Characteristic Features of Cervical, Thoracic
and Lumbar Vertebrae**
Table 7.6

Table 7.6	
View	**(c) Lumbar Vertebra**
Superior View	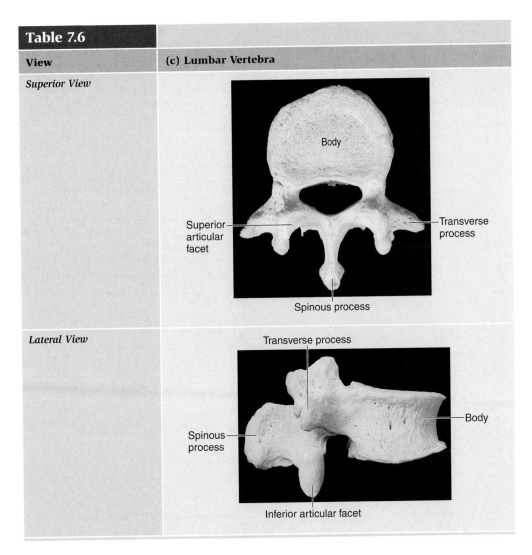
Lateral View	

Characteristic Features of Cervical, Thoracic and Lumbar Vertebrae
Table 7.6 (continued)

Table 7.6: © The McGraw-Hill Companies, Inc./Photo by Christine Eckel

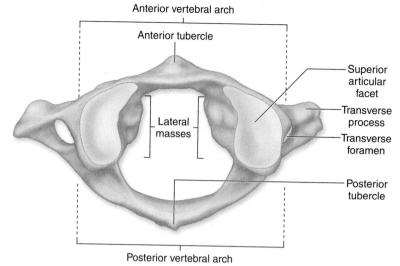

Anterior vertebral arch

Anterior tubercle

Superior articular facet

Transverse process

Transverse foramen

Lateral masses

Posterior tubercle

Posterior vertebral arch

(a) Atlas (C₁), superior view

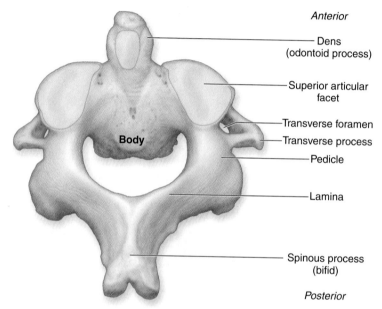

Anterior

Dens (odontoid process)

Superior articular facet

Transverse foramen

Transverse process

Pedicle

Body

Lamina

Spinous process (bifid)

Posterior

(b) Axis (C₂), posterosuperior view

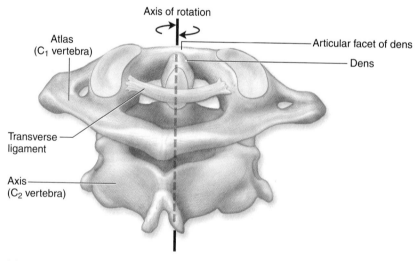

Axis of rotation

Atlas (C₁ vertebra)

Articular facet of dens

Dens

Transverse ligament

Axis (C₂ vertebra)

(c) Axis and atlas, posterosuperior view

Cervical Vertebrae C₁ and C₂
Figure 7.30

100

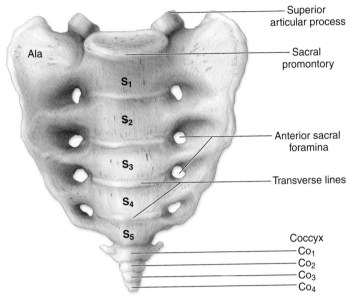

Superior
articular process

Ala

Sacral
promontory

S₁

S₂

Anterior sacral
foramina

S₃

Transverse lines

S₄

S₅

Coccyx
Co₁
Co₂
Co₃
Co₄

(a) Sacrum and coccyx, anterior view

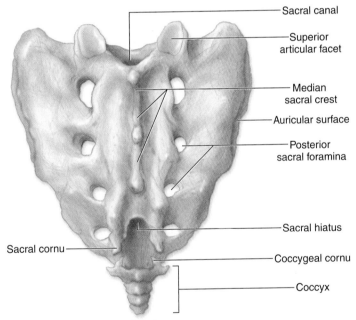

Sacral canal

Superior
articular facet

Median
sacral crest

Auricular surface

Posterior
sacral foramina

Sacral hiatus

Sacral cornu

Coccygeal cornu

Coccyx

(b) Sacrum and coccyx, posterior view

Sacrum and Coccyx
Figure 7.31

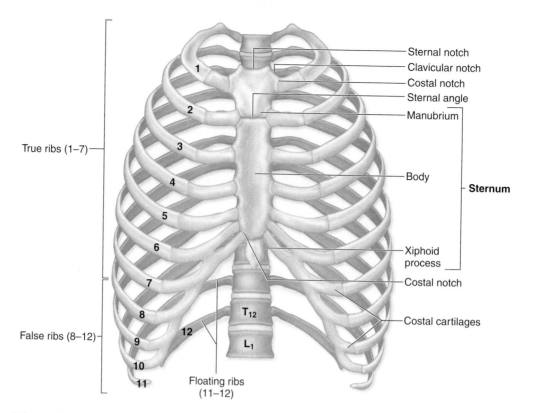

True ribs (1–7)

1
2
3
4
5
6
7
8
9
10
11
12

T₁₂
L₁

Sternal notch
Clavicular notch
Costal notch
Sternal angle
Manubrium

Body

Sternum

Xiphoid process

Costal notch

Costal cartilages

False ribs (8–12)

Floating ribs (11–12)

Thoracic Cage
Figure 7.32

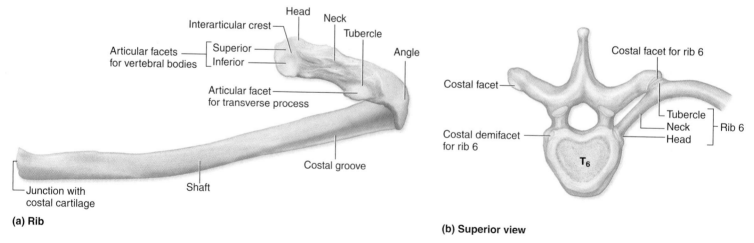

Head   Neck

Interarticular crest   Tubercle

Articular facets
for vertebral bodies   Superior   Angle
                        Inferior

Articular facet
for transverse process

Costal groove

Junction with
costal cartilage   Shaft

(a) Rib

Costal facet for rib 6

Costal facet

Tubercle
Neck   Rib 6
Head

Costal demifacet
for rib 6

T₆

(b) Superior view

Rib Anatomy and Articulation with Thoracic Vertebrae
Figure 7.33a, b

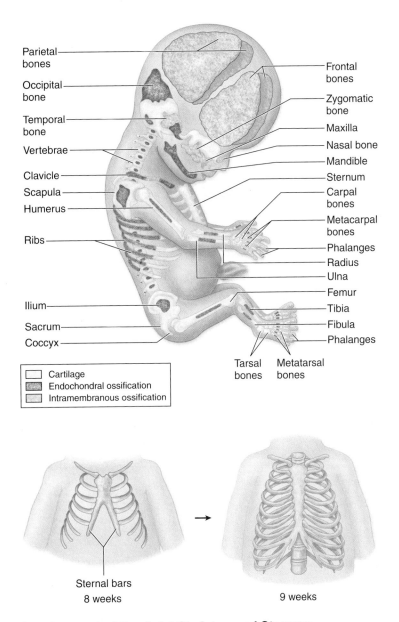

Parietal bones	Frontal bones
Occipital bone	Zygomatic bone
Temporal bone	Maxilla
Vertebrae	Nasal bone
	Mandible
Clavicle	Sternum
Scapula	Carpal bones
Humerus	Metacarpal bones
Ribs	Phalanges
	Radius
	Ulna
	Femur
Ilium	Tibia
Sacrum	Fibula
Coccyx	Phalanges

Tarsal bones   Metatarsal bones

☐ Cartilage
▨ Endochondral ossification
▨ Intramembranous ossification

Sternal bars
8 weeks

9 weeks

Development of the Axial Skeleton and Sternum Development
Figure 7.34, 7.35

103

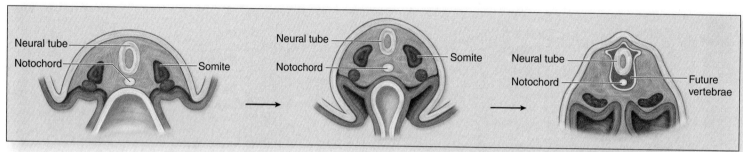

(a) Week 4: Sclerotome portions of somites surround the neural tube and form the future vertebrae and ribs.

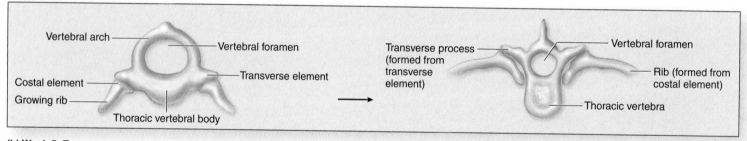

(b) Week 5: Transverse elements of thoracic vertebrae form transverse processes, while costal elements of thoracic vertebrae form ribs.

Rib and Vertebrae Development
Figure 7.36

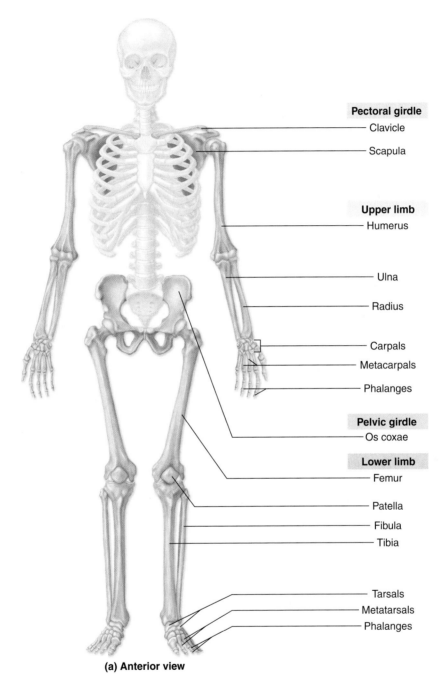

Pectoral girdle
Clavicle
Scapula

Upper limb
Humerus

Ulna

Radius

Carpals
Metacarpals
Phalanges

Pelvic girdle
Os coxae

Lower limb
Femur

Patella
Fibula
Tibia

Tarsals
Metatarsals
Phalanges

(a) Anterior view

Appendicular Skeleton—Anterior View
Figure 8.1a

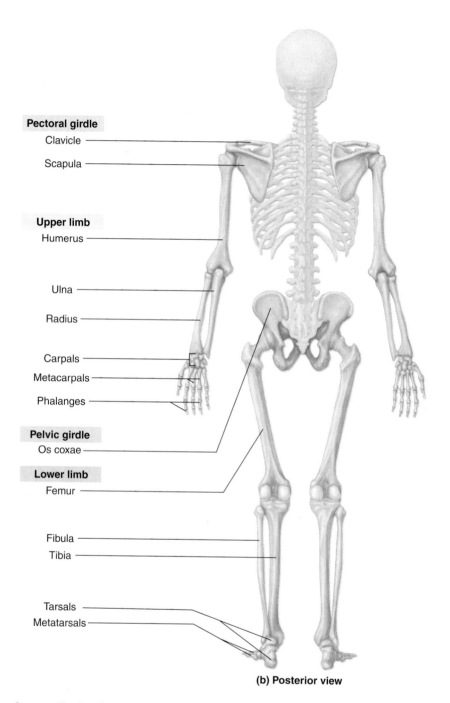

Pectoral girdle
Clavicle
Scapula

Upper limb
Humerus

Ulna

Radius

Carpals
Metacarpals
Phalanges

Pelvic girdle
Os coxae

Lower limb
Femur

Fibula
Tibia

Tarsals
Metatarsals

(b) Posterior view

Appendicular Skeleton—Posterior View
Figure 8.1b

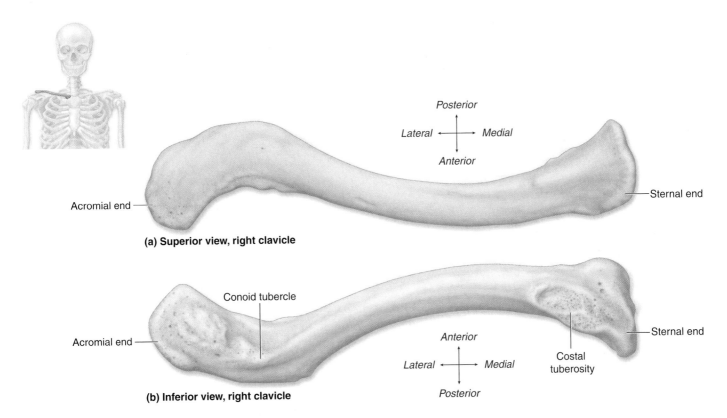

Posterior

Lateral ← → Medial

Anterior

Acromial end — Sternal end

(a) Superior view, right clavicle

Conoid tubercle

Acromial end — Sternal end

Anterior

Lateral ← → Medial

Posterior

Costal tuberosity

(b) Inferior view, right clavicle

Clavicle
Figure 8.2a,b

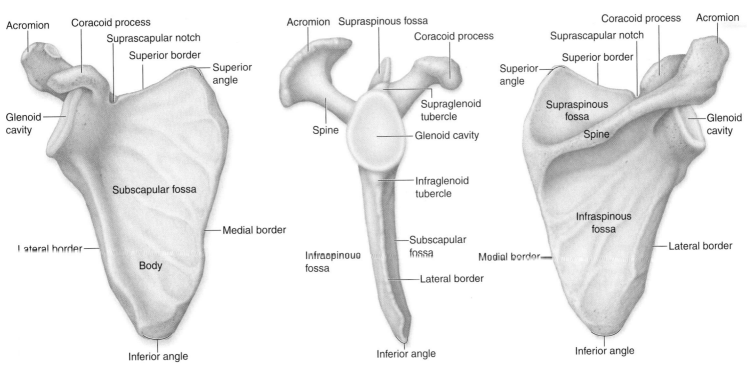

Acromion   Coracoid process
Suprascapular notch
Superior border
Superior angle

Glenoid cavity

Subscapular fossa

Lateral border

Medial border

Body

Inferior angle

(a) Right scapula, anterior view

Acromion   Supraspinous fossa
Coracoid process

Supraglenoid tubercle

Spine

Glenoid cavity

Infraglenoid tubercle

Subscapular fossa

Infraspinous fossa

Lateral border

Inferior angle

(b) Right scapula, lateral view

Coracoid process   Acromion
Suprascapular notch
Superior border

Superior angle

Supraspinous fossa

Glenoid cavity

Spine

Infraspinous fossa

Medial border

Lateral border

Inferior angle

(c) Right scapula, posterior view

Scapula
Figure 8.3

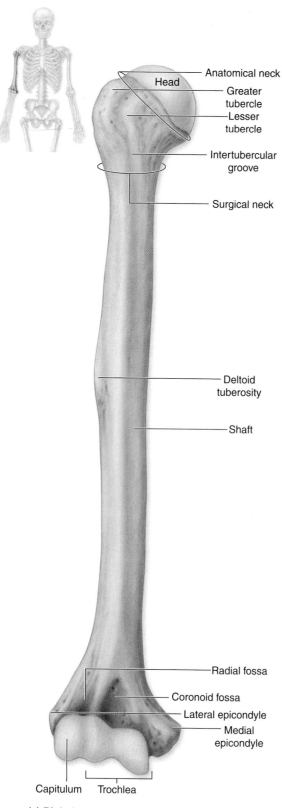

Anatomical neck

Head

Greater tubercle

Lesser tubercle

Intertubercular groove

Surgical neck

Deltoid tuberosity

Shaft

Radial fossa

Coronoid fossa

Lateral epicondyle

Medial epicondyle

Capitulum

Trochlea

(a) Right humerus, anterior view

Humerus—Anterior View
Figure 8.4a

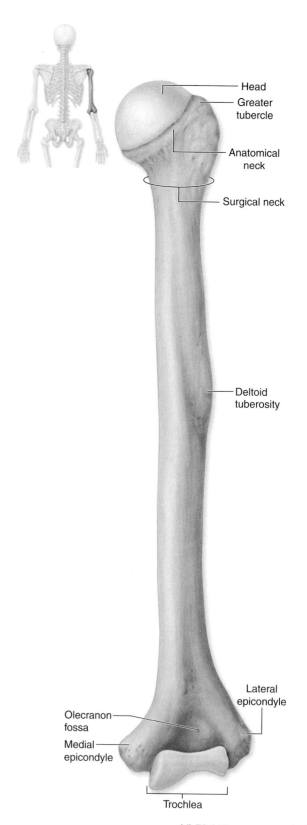

Head

Greater tubercle

Anatomical neck

Surgical neck

Deltoid tuberosity

Lateral epicondyle

Olecranon fossa

Medial epicondyle

Trochlea

(d) Right humerus, posterior view

Humerus—Posterior View
Figure 8.4d

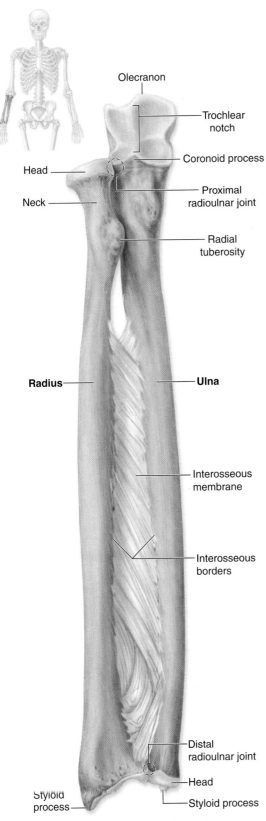

Olecranon

Trochlear notch

Coronoid process

Head

Proximal radioulnar joint

Neck

Radial tuberosity

Radius

Ulna

Interosseous membrane

Interosseous borders

Distal radioulnar joint

Head

Styloid process

Styloid process

(a) Right radius and ulna, anterior view

Radius and Ulna—Anterior View
Figure 8.5a

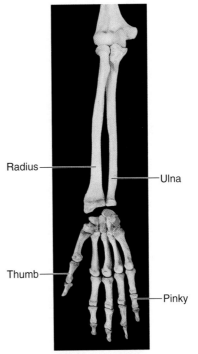

Radius

Ulna

Thumb

Pinky

(d) Supination of right forearm

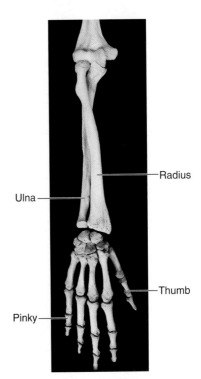

Ulna

Radius

Pinky

Thumb

(e) Pronation of right forearm

Radius and Ulna—Supination and Pronation
Figure 8.5d,e

Figure 8.5d,e: © The McGraw-Hill Companies, Inc./Photo by Christine Eckel.

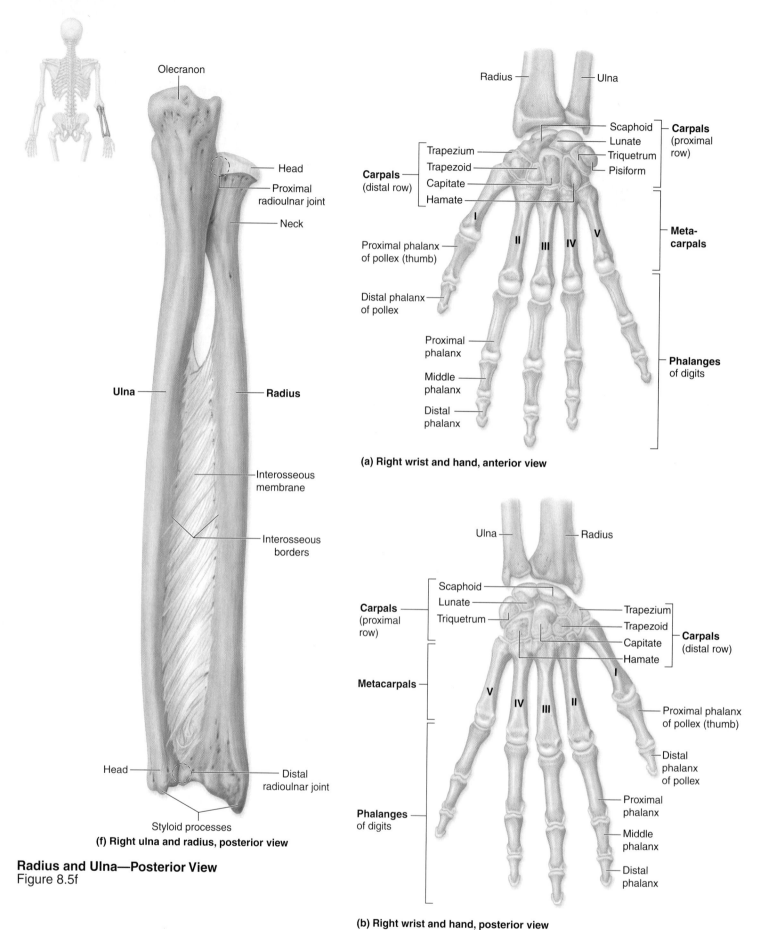

Olecranon

Head

Proximal
radioulnar joint

Neck

Ulna

Radius

Interosseous
membrane

Interosseous
borders

Head

Distal
radioulnar joint

Styloid processes

(f) Right ulna and radius, posterior view

Radius and Ulna—Posterior View
Figure 8.5f

Radius

Ulna

Scaphoid
Lunate

Carpals
(proximal
row)

Triquetrum
Pisiform

Trapezium
Trapezoid
Capitate
Hamate

Carpals
(distal row)

I

Proximal phalanx
of pollex (thumb)

II   III   IV   V

**Meta-
carpals**

Distal phalanx
of pollex

Proximal
phalanx

Middle
phalanx

Distal
phalanx

Phalanges
of digits

(a) Right wrist and hand, anterior view

Ulna

Radius

Carpals
(proximal
row)

Scaphoid
Lunate
Triquetrum

Trapezium
Trapezoid
Capitate
Hamate

Carpals
(distal row)

Metacarpals

V   IV   III   II

I

Proximal phalanx
of pollex (thumb)

Distal
phalanx
of pollex

Proximal
phalanx

Phalanges
of digits

Middle
phalanx

Distal
phalanx

(b) Right wrist and hand, posterior view

Right Wrist and Hand
Figure 8.6

110

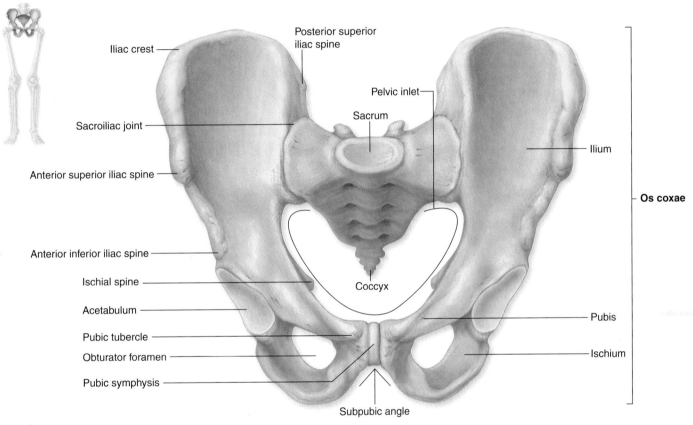

Iliac crest

Posterior superior
iliac spine

Pelvic inlet

Sacrum

Sacroiliac joint

Ilium

Anterior superior iliac spine

Os coxae

Anterior inferior iliac spine

Coccyx

Ischial spine

Acetabulum

Pubis

Pubic tubercle

Obturator foramen

Ischium

Pubic symphysis

Subpubic angle

Pelvis
Figure 8.7

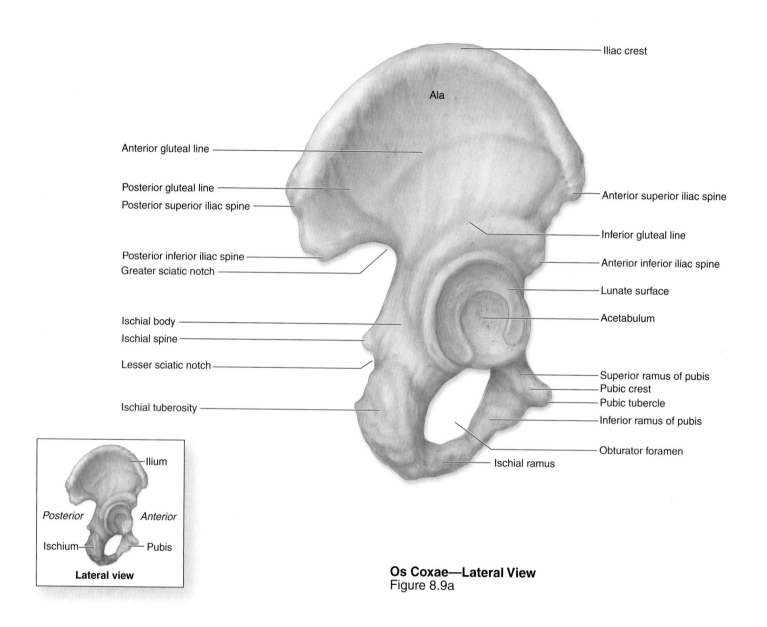

Iliac crest

Ala

Anterior gluteal line

Posterior gluteal line

Posterior superior iliac spine

Anterior superior iliac spine

Inferior gluteal line

Posterior inferior iliac spine

Anterior inferior iliac spine

Greater sciatic notch

Lunate surface

Ischial body

Acetabulum

Ischial spine

Lesser sciatic notch

Superior ramus of pubis

Pubic crest

Pubic tubercle

Ischial tuberosity

Inferior ramus of pubis

Obturator foramen

Ischial ramus

Ilium

Posterior

Anterior

Ischium

Pubis

Lateral view

Os Coxae—Lateral View
Figure 8.9a

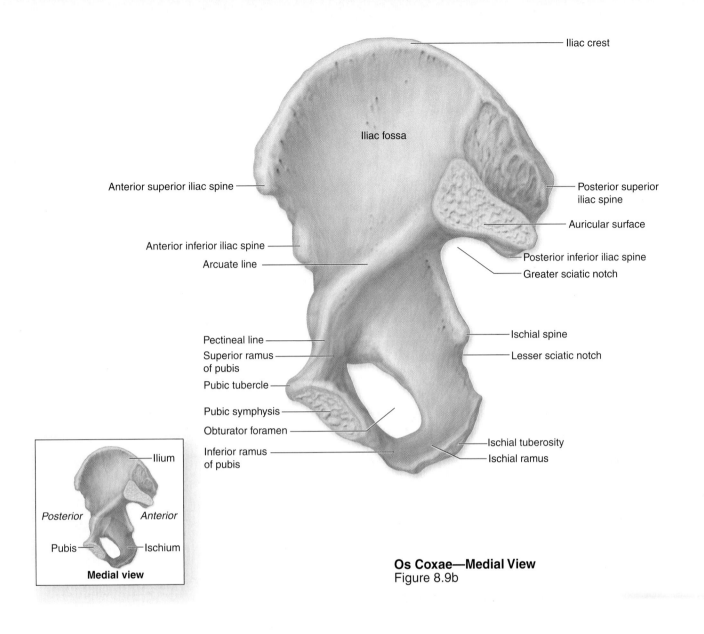

Iliac crest

Iliac fossa

Anterior superior iliac spine

Posterior superior iliac spine

Auricular surface

Anterior inferior iliac spine

Posterior inferior iliac spine

Arcuate line

Greater sciatic notch

Pectineal line

Ischial spine

Superior ramus of pubis

Lesser sciatic notch

Pubic tubercle

Pubic symphysis

Obturator foramen

Ischial tuberosity

Inferior ramus of pubis

Ischial ramus

Ilium

Posterior

Anterior

Pubis

Ischium

Medial view

Os Coxae—Medial View
Figure 8.9b

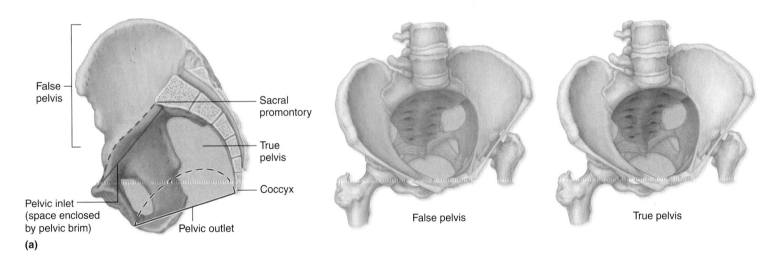

False pelvis

Sacral promontory

True pelvis

Pelvic inlet (space enclosed by pelvic brim)

Coccyx

Pelvic outlet

(a)

False pelvis

True pelvis

Features of the Pelvis—True and False Pelvis
Figure 8.10a

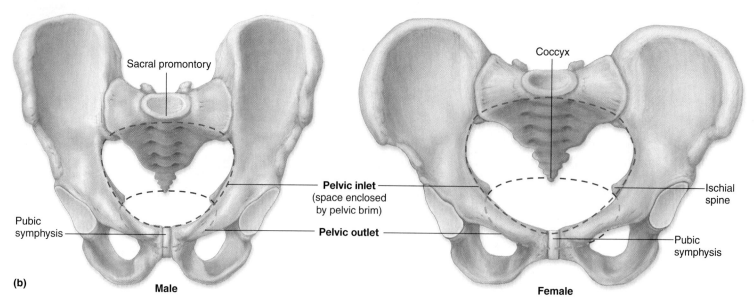

Features of the Pelvis—Male vs. Female Pelvis
Figure 8.10b

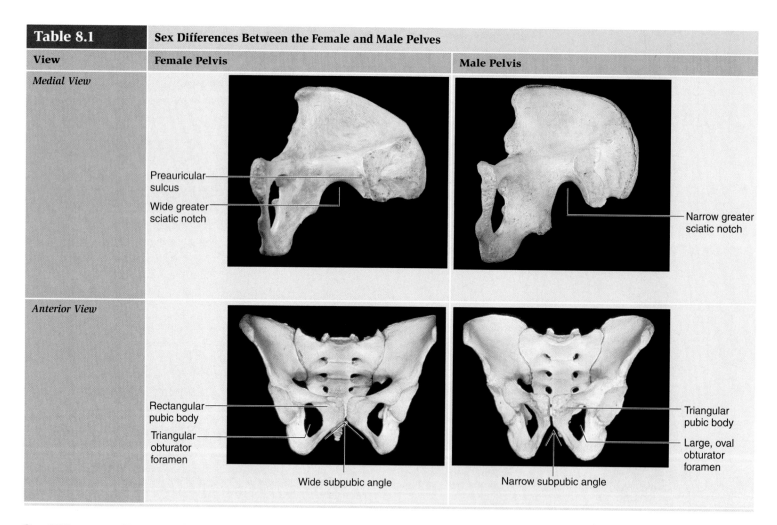

Table 8.1	Sex Differences Between the Female and Male Pelves	
View	Female Pelvis	Male Pelvis
Medial View	Preauricular sulcus / Wide greater sciatic notch	Narrow greater sciatic notch
Anterior View	Rectangular pubic body / Triangular obturator foramen / Wide subpubic angle	Triangular pubic body / Large, oval obturator foramen / Narrow subpubic angle

Sex Differences Between the Female
and Male Pelvis
Table 8.1

Table 8.1: (upper left, upper right): © David Hunt/Smithsonian Institution; (lower left, lower right): © L. Bassett/Visuals Unlimited

Femur—Anterior View
Figure 8.11a

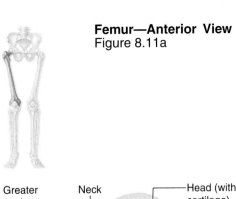

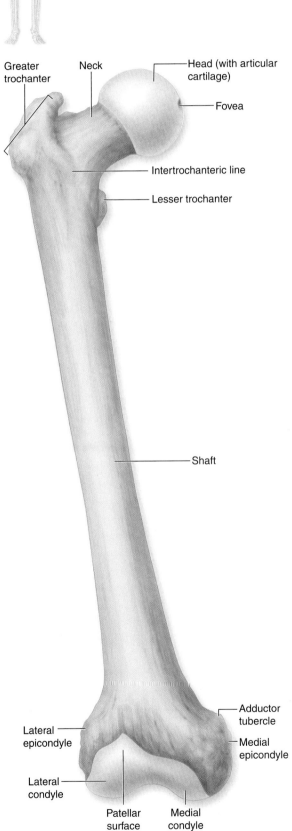

Greater trochanter

Neck

Head (with articular cartilage)

Fovea

Intertrochanteric line

Lesser trochanter

Shaft

Lateral epicondyle

Adductor tubercle

Medial epicondyle

Lateral condyle

Patellar surface

Medial condyle

(a) Right femur, anterior view

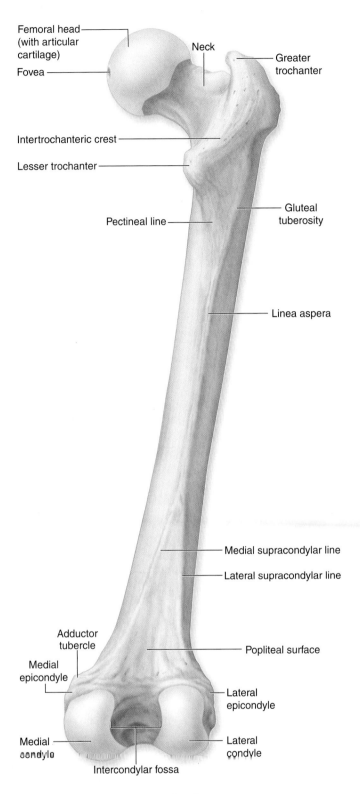

Femoral head (with articular cartilage)

Neck

Fovea

Greater trochanter

Intertrochanteric crest

Lesser trochanter

Pectineal line

Gluteal tuberosity

Linea aspera

Medial supracondylar line

Lateral supracondylar line

Adductor tubercle

Medial epicondyle

Popliteal surface

Lateral epicondyle

Medial condyle

Lateral condyle

Intercondylar fossa

(d) Right femur, posterior view

Femur—Posterior View
Figure 8.11d

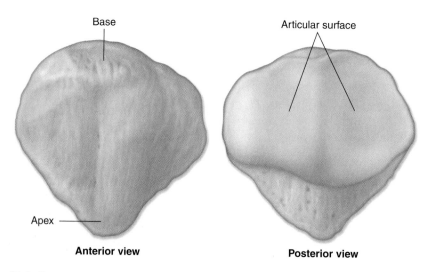

Base

Articular surface

Apex

Anterior view

Posterior view

Patella
Figure 8.12

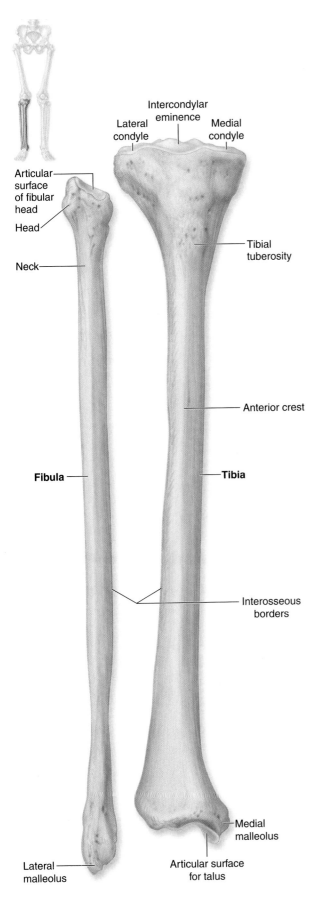

Intercondylar eminence

Lateral condyle

Medial condyle

Articular surface of fibular head

Head

Neck

Tibial tuberosity

Anterior crest

Fibula

Tibia

Interosseous borders

Medial malleolus

Lateral malleolus

Articular surface for talus

(a) Right tibia and fibula, anterior view

Tibia and Fibula—Anterior View
Figure 8.13a

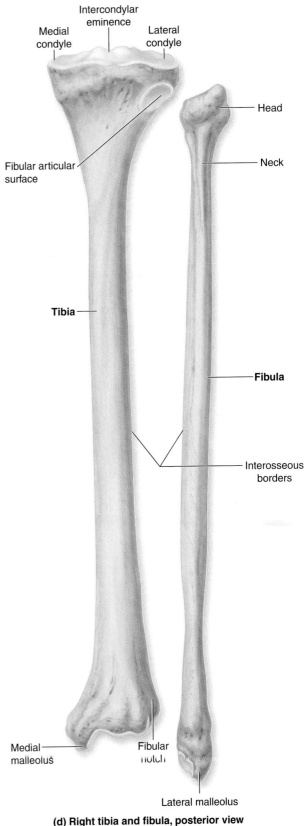

Intercondylar eminence

Medial condyle

Lateral condyle

Head

Neck

Fibular articular surface

Tibia

Fibula

Interosseous borders

Medial malleolus

Fibular notch

Lateral malleolus

(d) Right tibia and fibula, posterior view

Tibia and Fibula—Posterior View
Figure 8.13d

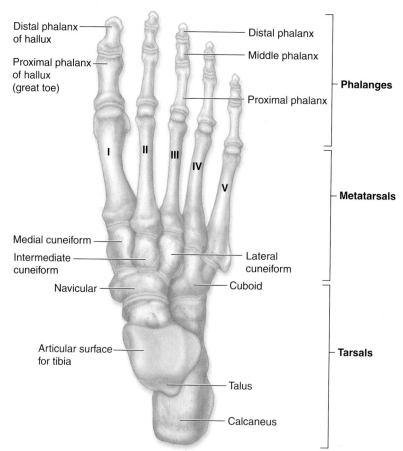

Distal phalanx of hallux

Proximal phalanx of hallux (great toe)

Distal phalanx

Middle phalanx

Proximal phalanx

Phalanges

I II III IV V

Metatarsals

Medial cuneiform

Intermediate cuneiform

Navicular

Lateral cuneiform

Cuboid

Articular surface for tibia

Tarsals

Talus

Calcaneus

(a) Right foot, superior view

Bones of the Tarsals, Metatarsals, and Phalanges
Figure 8.14

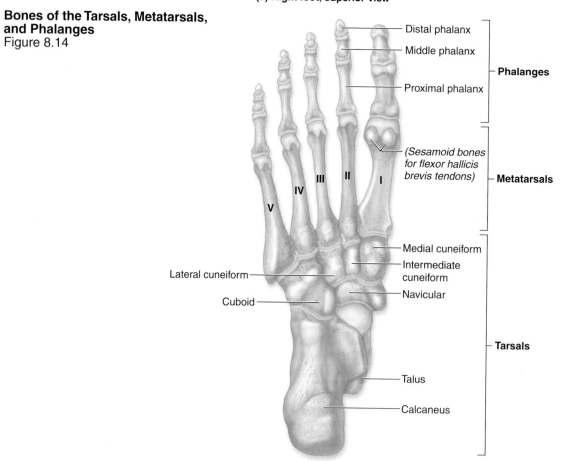

Distal phalanx

Middle phalanx

Proximal phalanx

Phalanges

V IV III II I

(Sesamoid bones for flexor hallicis brevis tendons)

Metatarsals

Lateral cuneiform

Cuboid

Medial cuneiform

Intermediate cuneiform

Navicular

Tarsals

Talus

Calcaneus

(b) Right foot, inferior view

118

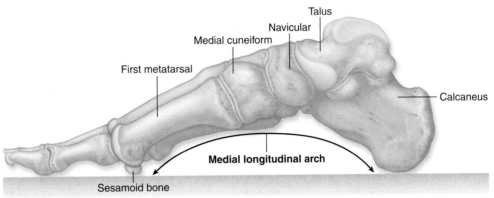

Talus

Navicular

Medial cuneiform

First metatarsal

Calcaneus

Medial longitudinal arch

Sesamoid bone

(a) Right foot, medial view

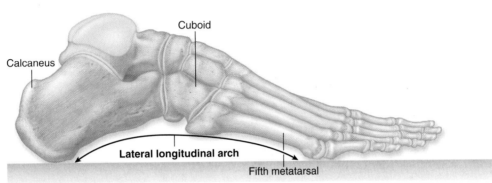

Cuboid

Calcaneus

Lateral longitudinal arch

Fifth metatarsal

(b) Right foot, lateral view

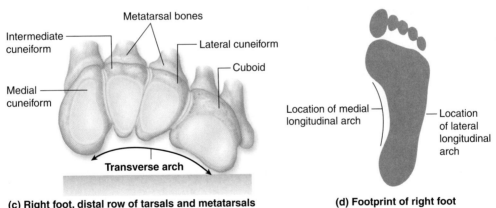

Metatarsal bones

Intermediate cuneiform

Lateral cuneiform

Cuboid

Medial cuneiform

Transverse arch

(c) Right foot, distal row of tarsals and metatarsals

Location of medial longitudinal arch

Location of lateral longitudinal arch

(d) Footprint of right foot

Arches of the Foot
Figure 8.15

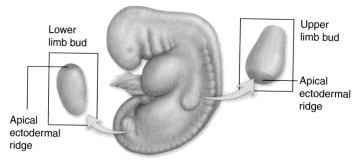

(a) Week 4: Upper and lower limb buds form.

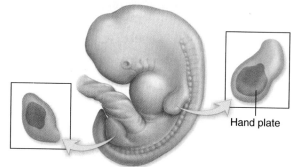

(b) Week 5: Hand plate forms.

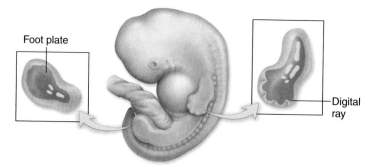

Development of the Appendicular Skeleton
Figure 8.16

(c) Week 6: Digital rays appear in hand plate. Foot plate forms.

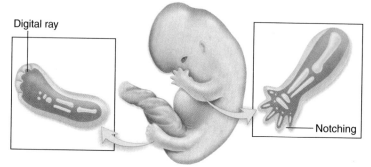

(d) Week 7: Notching develops between digital rays of hand plate. Digital rays appear in foot plate.

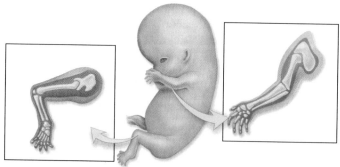

(e) Week 8: Separate fingers and toes formed.

120

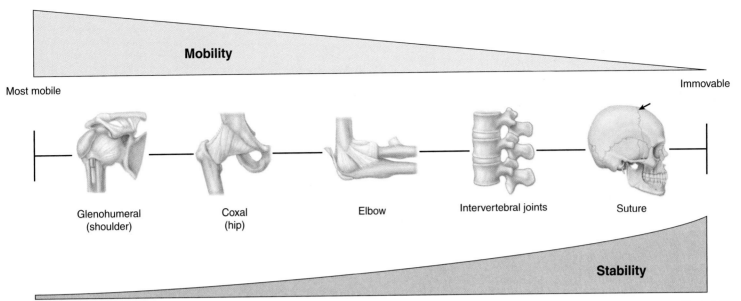

Relationship Between Mobility and Stability in Joints
Figure 9.1

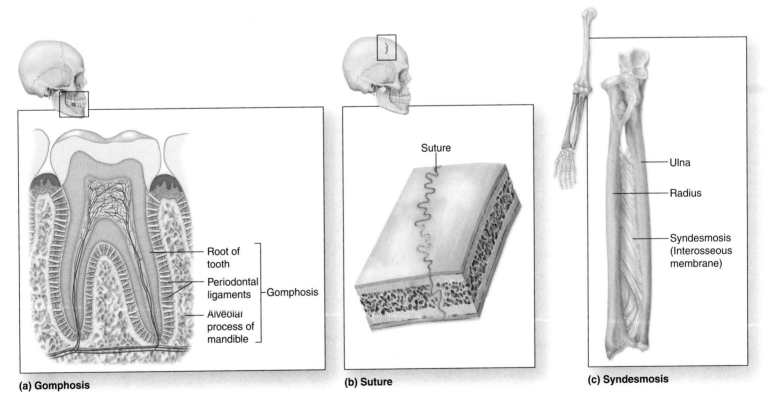

(a) Gomphosis

- Root of tooth
- Periodontal ligaments ⎤
- Alveolar process of mandible ⎦ Gomphosis

(b) Suture

Suture

(c) Syndesmosis

- Ulna
- Radius
- Syndesmosis (Interosseous membrane)

Fibrous Joints
Figure 9.2

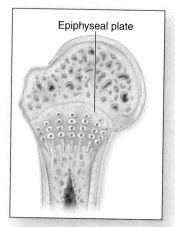

Epiphyseal plate

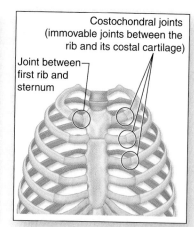

Costochondral joints
(immovable joints between the
rib and its costal cartilage)

Joint between
first rib and
sternum

(a) Synchondroses (contain hyaline cartilage)

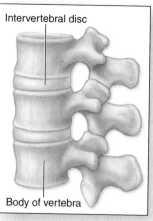

Intervertebral disc

Body of vertebra

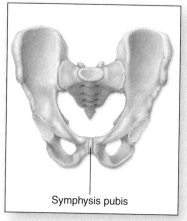

Symphysis pubis

(b) Symphyses (contain fibrocartilage)

Cartilaginous Joints
Figure 9.3

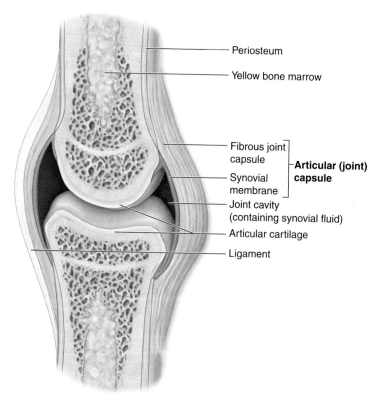

Periosteum

Yellow bone marrow

Fibrous joint
capsule

Synovial
membrane

Joint cavity
(containing synovial fluid)

Articular cartilage

Ligament

**Articular (joint)
capsule**

Typical synovial joint

Synovial Joint
Figure 9.4

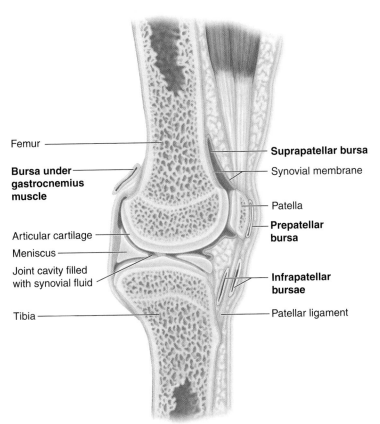

Femur

Bursa under gastrocnemius muscle

Articular cartilage

Meniscus

Joint cavity filled with synovial fluid

Tibia

Suprapatellar bursa

Synovial membrane

Patella

Prepatellar bursa

Infrapatellar bursae

Patellar ligament

(a) Bursae of the knee joint, sagittal section

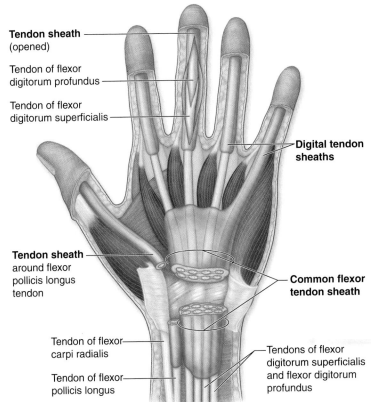

Tendon sheath (opened)

Tendon of flexor digitorum profundus

Tendon of flexor digitorum superficialis

Digital tendon sheaths

Tendon sheath around flexor pollicis longus tendon

Common flexor tendon sheath

Tendon of flexor carpi radialis

Tendon of flexor pollicis longus

Tendons of flexor digitorum superficialis and flexor digitorum profundus

(b) Tendon sheaths of wrist and hand, anterior view

Bursae and Tendon Sheaths
Figure 9.5

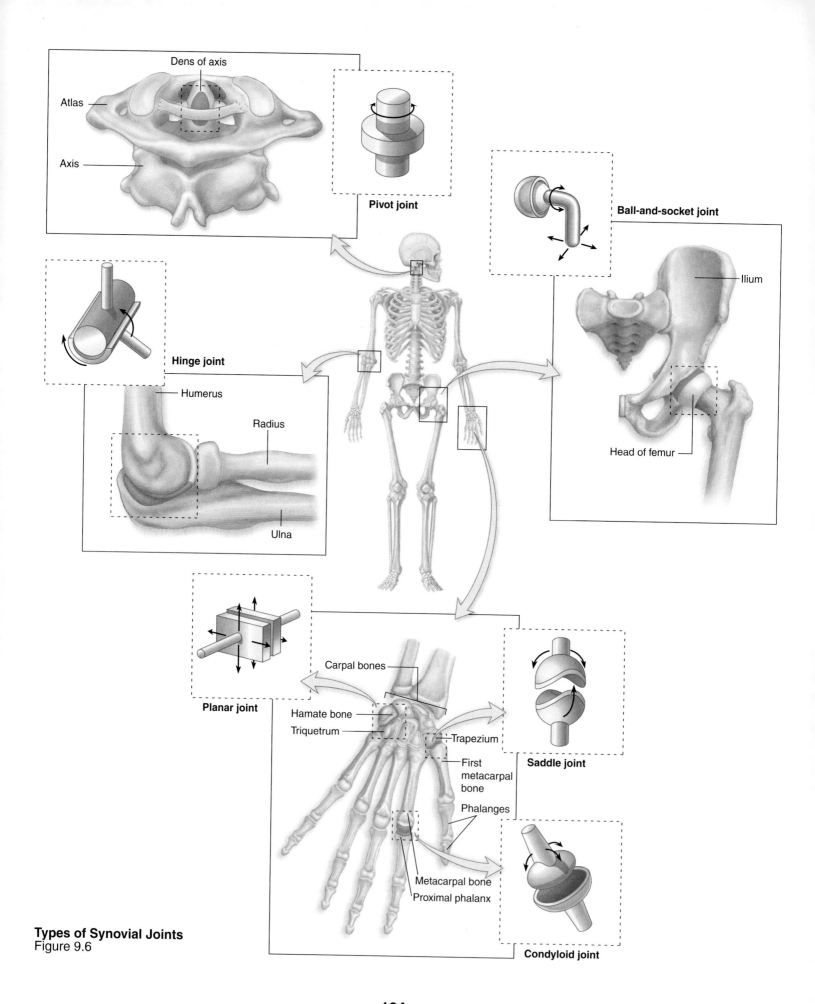

Dens of axis

Atlas

Axis

Pivot joint

Ball-and-socket joint

Ilium

Head of femur

Hinge joint

Humerus

Radius

Ulna

Planar joint

Carpal bones

Hamate bone

Triquetrum

Trapezium

First
metacarpal
bone

Phalanges

Saddle joint

Metacarpal bone

Proximal phalanx

Condyloid joint

Types of Synovial Joints
Figure 9.6

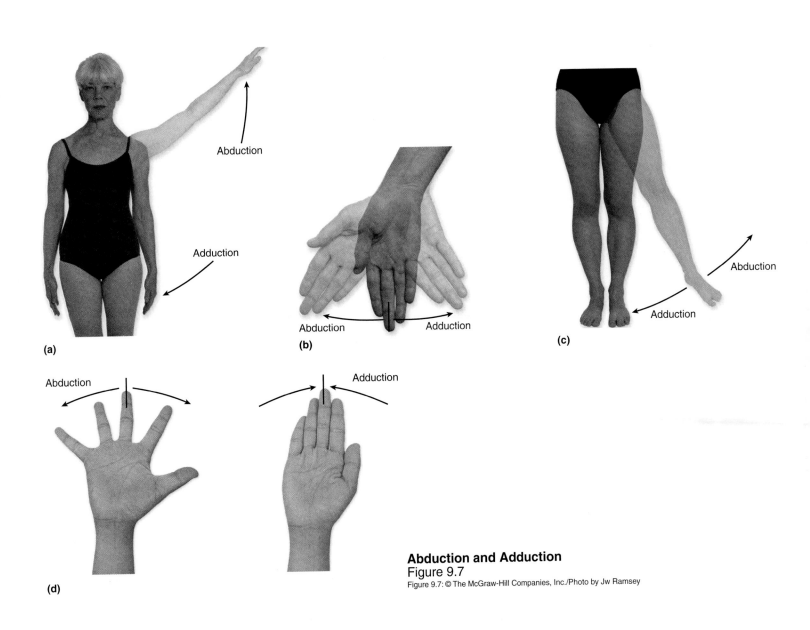

(a)

Abduction

Adduction

(b)

Abduction

Adduction

(c)

Abduction

Adduction

(d)

Abduction

Adduction

Abduction and Adduction
Figure 9.7
Figure 9.7: © The McGraw-Hill Companies, Inc./Photo by Jw Ramsey

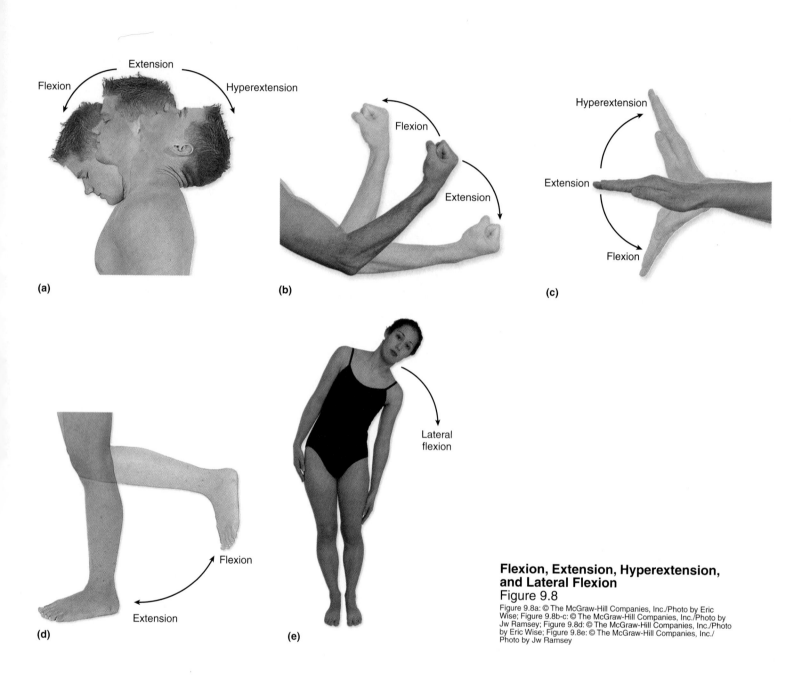

(a)

(b)

(c)

(d)

(e)

Flexion, Extension, Hyperextension, and Lateral Flexion
Figure 9.8

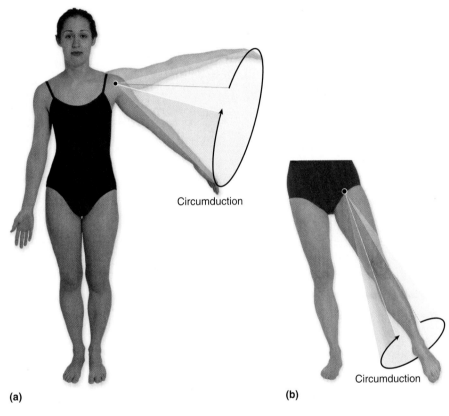

(a)

(b)

Circumduction

Circumduction

Circumduction
Figure 9.9
Figure 9.9: © The McGraw-Hill Companies, Inc./Photo by Jw Ramsey

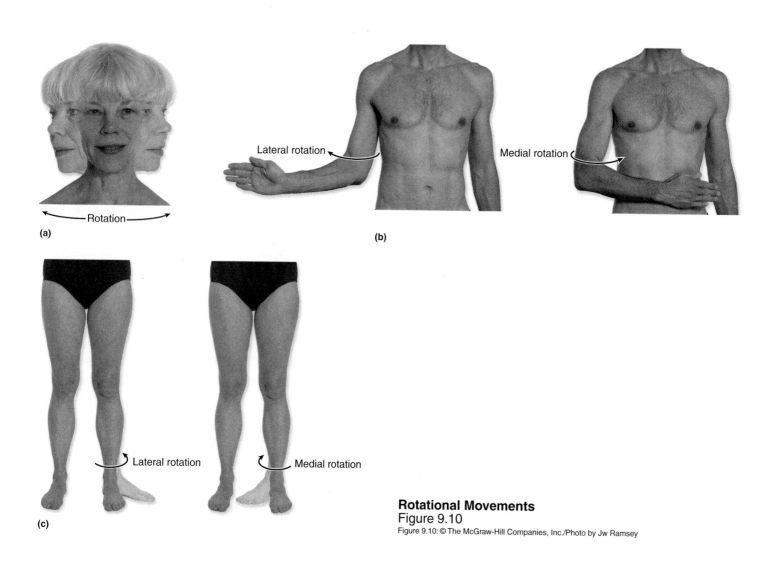

(a) — Rotation

(b) Lateral rotation — Medial rotation

(c) Lateral rotation — Medial rotation

Rotational Movements
Figure 9.10
Figure 9.10: © The McGraw-Hill Companies, Inc./Photo by Jw Ramsey

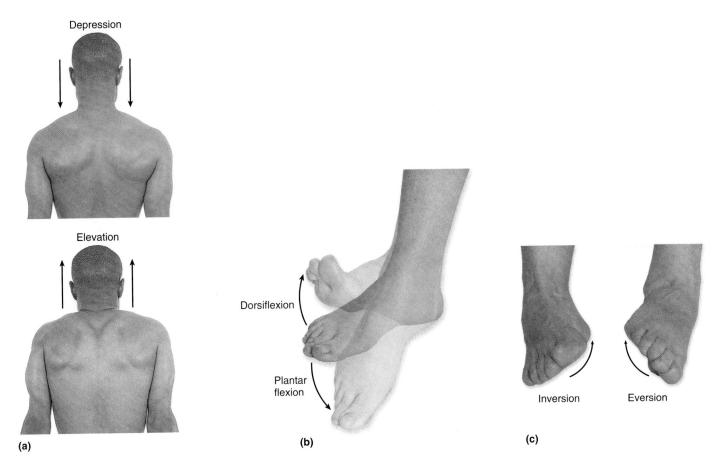

(a)

(b)

(c)

Depression, Elevation, Dorsiflexion, Plantar Flexion, Inversion and Eversion
Figure 9.11a,b,c

Figure 9.11a (both): © The McGraw-Hill Companies, Inc./Photo by Eric Wise; Figure 9.11b-c: © The McGraw-Hill Companies, Inc./Photo by Jw Ramsey

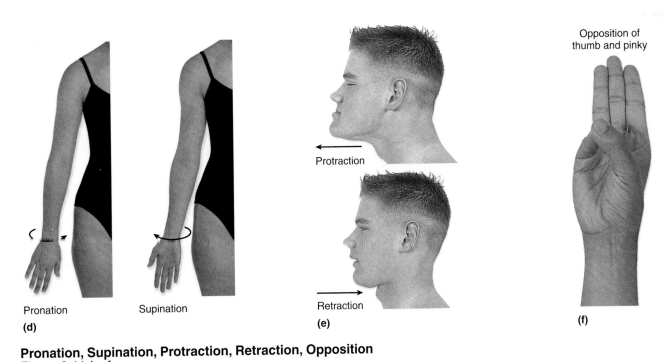

(d)

(e)

(f)

Pronation, Supination, Protraction, Retraction, Opposition
Figure 9.11d,e,f

9.11d: © The McGraw-Hill Companies, Inc./Photo by Jw Ramsey; Figure 9.11e (both), Figure 9.11f: © The McGraw-Hill Companies, Inc./Photo by Eric Wise

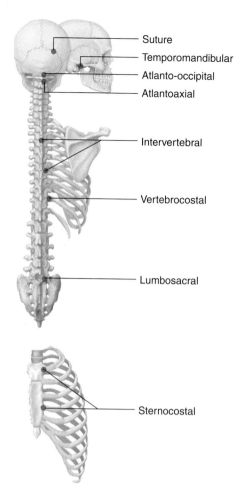

	Suture
	Temporomandibular
	Atlanto-occipital
	Atlantoaxial
	Intervertebral
	Vertebrocostal
	Lumbosacral
	Sternocostal

Axial Skeleton Joints
Table 9.3

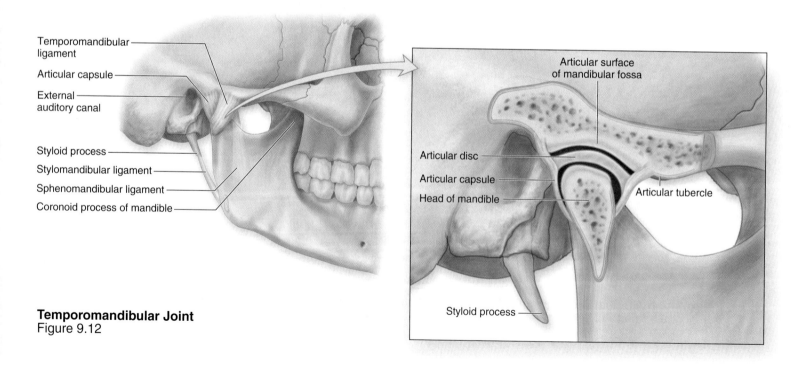

Temporomandibular ligament

Articular capsule

External auditory canal

Styloid process

Stylomandibular ligament

Sphenomandibular ligament

Coronoid process of mandible

Articular surface of mandibular fossa

Articular disc

Articular capsule

Head of mandible

Articular tubercle

Styloid process

Temporomandibular Joint
Figure 9.12

130

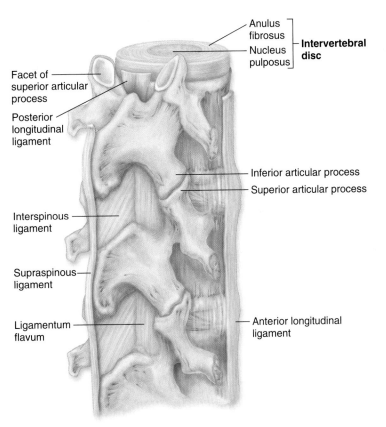

Anulus
fibrosus
Nucleus
pulposus ⎦ **Intervertebral disc**

Facet of
superior articular
process

Posterior
longitudinal
ligament

Inferior articular process

Superior articular process

Interspinous
ligament

Supraspinous
ligament

Ligamentum
flavum

Anterior longitudinal
ligament

Intervertebral Articulations
Figure 9.13

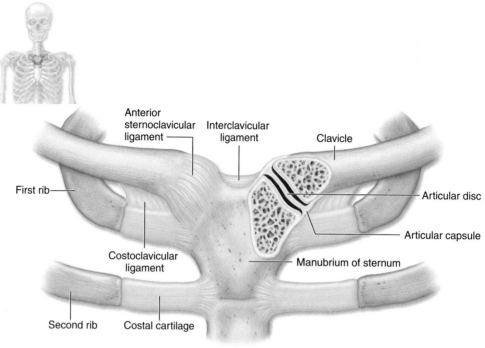

Anterior
sternoclavicular
ligament

Interclavicular
ligament

Clavicle

First rib

Articular disc

Articular capsule

Costoclavicular
ligament

Manubrium of sternum

Second rib   Costal cartilage

Anterior view

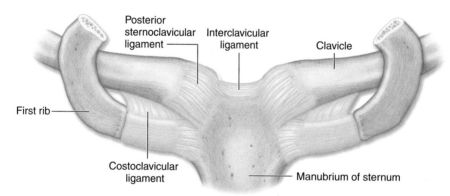

Posterior
sternoclavicular
ligament

Interclavicular
ligament

Clavicle

First rib

Costoclavicular
ligament

Manubrium of sternum

Posterior view

Sternoclavicular Joint
Figure 9.14

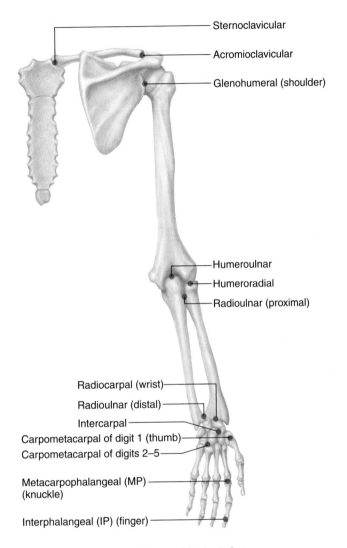

Sternoclavicular

Acromioclavicular

Glenohumeral (shoulder)

Humeroulnar

Humeroradial

Radioulnar (proximal)

Radiocarpal (wrist)

Radioulnar (distal)

Intercarpal

Carpometacarpal of digit 1 (thumb)

Carpometacarpal of digits 2–5

Metacarpophalangeal (MP) (knuckle)

Interphalangeal (IP) (finger)

Pectoral Girdle and Upper Limb Joints
Table 9.4

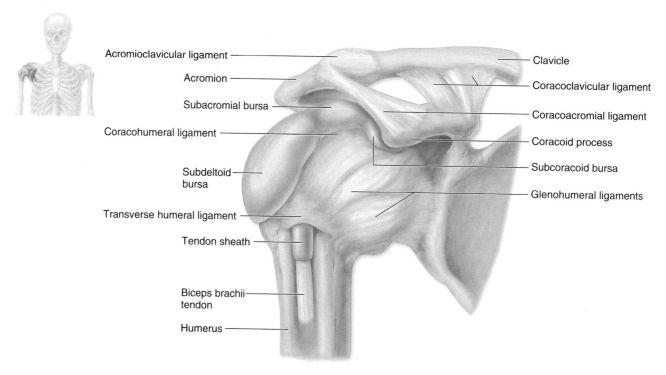

Glenohumeral Joint—Anterior View
Figure 9.15a

Acromioclavicular ligament
Acromion
Subacromial bursa
Coracohumeral ligament
Subdeltoid bursa
Transverse humeral ligament
Tendon sheath
Biceps brachii tendon
Humerus
Clavicle
Coracoclavicular ligament
Coracoacromial ligament
Coracoid process
Subcoracoid bursa
Glenohumeral ligaments

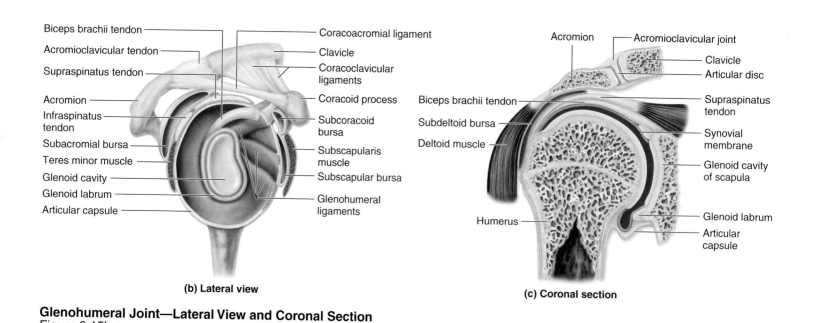

Biceps brachii tendon
Acromioclavicular tendon
Supraspinatus tendon
Acromion
Infraspinatus tendon
Subacromial bursa
Teres minor muscle
Glenoid cavity
Glenoid labrum
Articular capsule

Coracoacromial ligament
Clavicle
Coracoclavicular ligaments
Coracoid process
Subcoracoid bursa
Subscapularis muscle
Subscapular bursa
Glenohumeral ligaments

(b) Lateral view

Acromion
Biceps brachii tendon
Subdeltoid bursa
Deltoid muscle
Humerus

Acromioclavicular joint
Clavicle
Articular disc
Supraspinatus tendon
Synovial membrane
Glenoid cavity of scapula
Glenoid labrum
Articular capsule

(c) Coronal section

Glenohumeral Joint—Lateral View and Coronal Section
Figure 9.15b,c

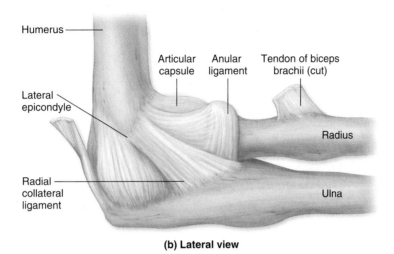

(b) Lateral view

Humerus

Articular capsule

Anular ligament

Tendon of biceps brachii (cut)

Lateral epicondyle

Radius

Radial collateral ligament

Ulna

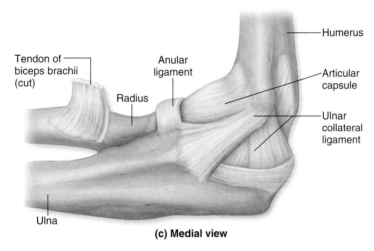

(c) Medial view

Tendon of biceps brachii (cut)

Anular ligament

Humerus

Radius

Articular capsule

Ulnar collateral ligament

Ulna

Elbow Joint—Lateral and Medial Views
Figure 9.16b, c

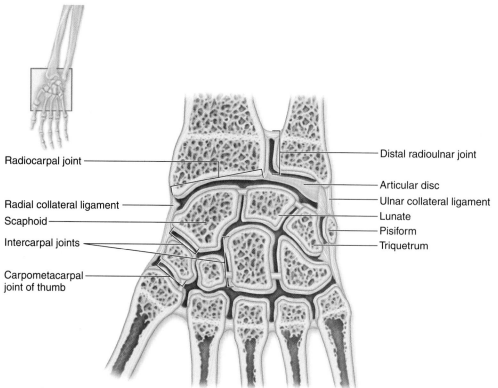

Radiocarpal joint

Radial collateral ligament

Scaphoid

Intercarpal joints

Carpometacarpal
joint of thumb

Distal radioulnar joint

Articular disc

Ulnar collateral ligament

Lunate

Pisiform

Triquetrum

Right radiocarpal joint, coronal section

Radiocarpal (Wrist) Articulation
Figure 9.17

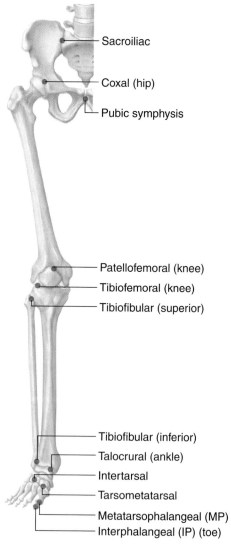

Sacroiliac

Coxal (hip)

Pubic symphysis

Patellofemoral (knee)

Tibiofemoral (knee)

Tibiofibular (superior)

Tibiofibular (inferior)

Talocrural (ankle)

Intertarsal

Tarsometatarsal

Metatarsophalangeal (MP)

Interphalangeal (IP) (toe)

Pelvic Girdle and Lower Limb Joints
Table 9.5

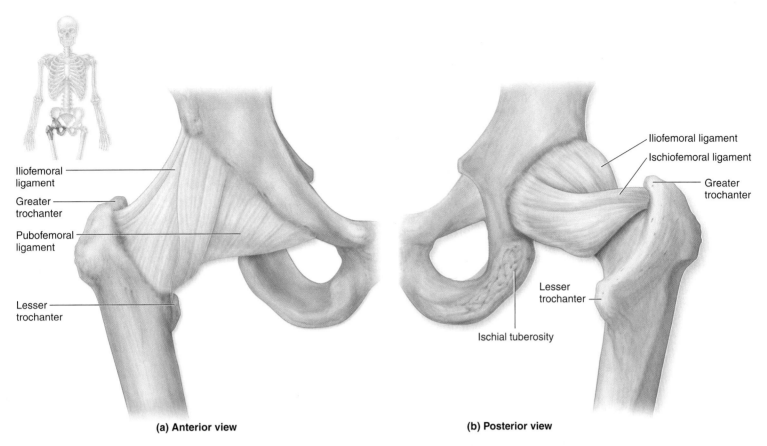

Iliofemoral
ligament

Greater
trochanter

Pubofemoral
ligament

Lesser
trochanter

(a) Anterior view

Iliofemoral ligament

Ischiofemoral ligament

Greater
trochanter

Lesser
trochanter

Ischial tuberosity

(b) Posterior view

Coxal (Hip) Joint—Anterior and Posterior Views
Figure 9.18a,b

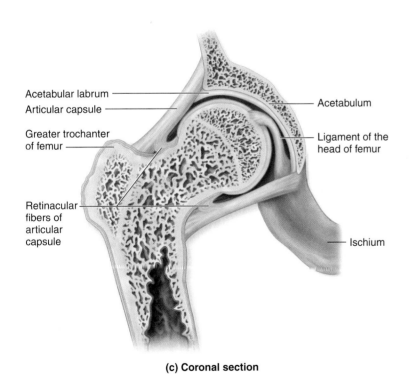

Acetabular labrum

Articular capsule

Greater trochanter
of femur

Retinacular
fibers of
articular
capsule

Acetabulum

Ligament of the
head of femur

Ischium

(c) Coronal section

Coxal (Hip) Joint—Coronal Section
Figure 9.18c

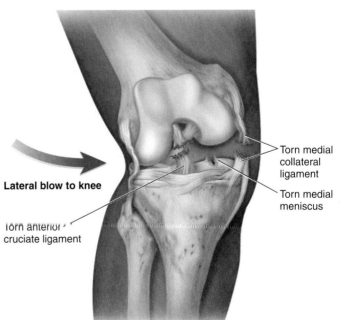

Lateral blow to knee

Torn medial
collateral
ligament

Torn medial
meniscus

Torn anterior
cruciate ligament

Unhappy Triad of Injuries to the Right Knee
Clinical View p. 279

137

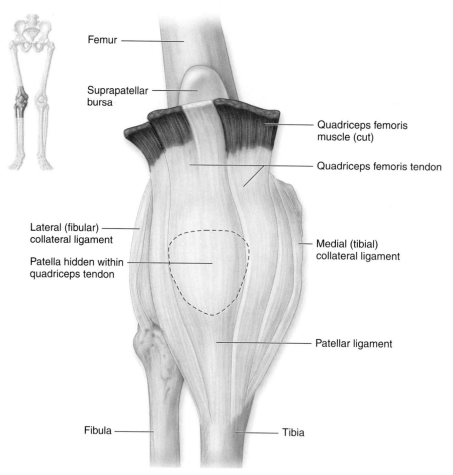

Femur

Suprapatellar
bursa

Quadriceps femoris
muscle (cut)

Quadriceps femoris tendon

Lateral (fibular)
collateral ligament

Medial (tibial)
collateral ligament

Patella hidden within
quadriceps tendon

Patellar ligament

Fibula

Tibia

(a) Anterior superficial view

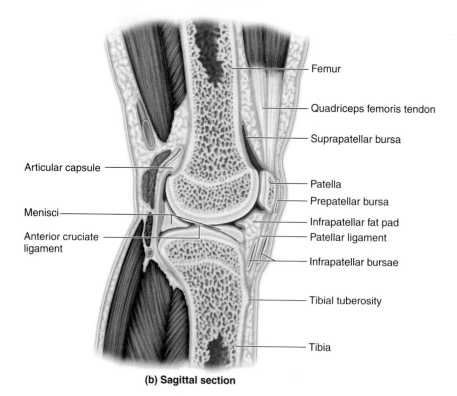

Femur

Quadriceps femoris tendon

Suprapatellar bursa

Articular capsule

Patella

Prepatellar bursa

Menisci

Infrapatellar fat pad

Anterior cruciate
ligament

Patellar ligament

Infrapatellar bursae

Tibial tuberosity

Tibia

(b) Sagittal section

Knee Joint—Anterior Superficial View and Sagittal Section
Figure 9.19a, b

138

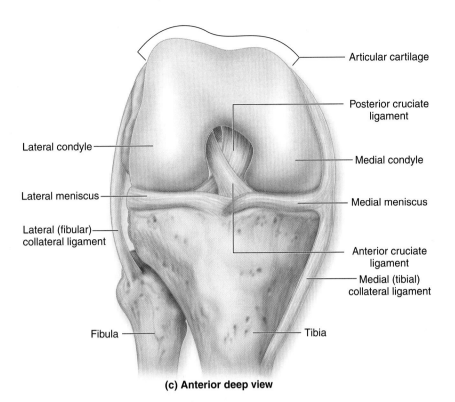

(c) Anterior deep view

Knee Joint—Anterior Deep View
Figure 9.19c

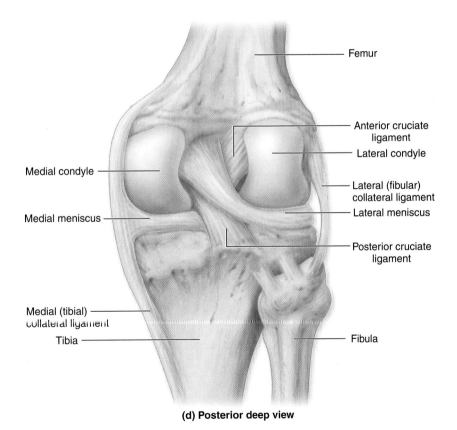

(d) Posterior deep view

Knee Joint—Posterior Deep View
Figure 9.19d

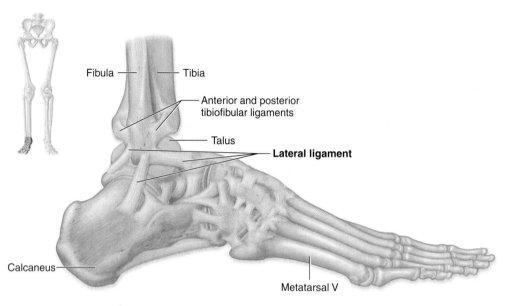

Fibula — Tibia

Anterior and posterior tibiofibular ligaments

Talus

Lateral ligament

Calcaneus

Metatarsal V

(a) Lateral view

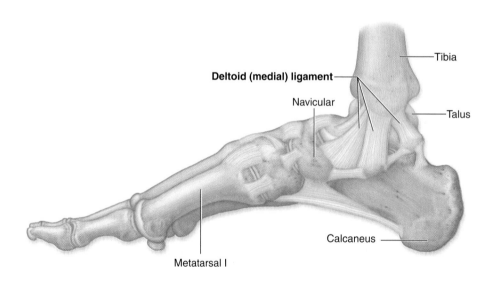

Tibia

Deltoid (medial) ligament

Navicular

Talus

Metatarsal I

Calcaneus

(c) Medial view

Talocrural (Ankle) Joint—Lateral and Medial Views
Figure 9.20a, c

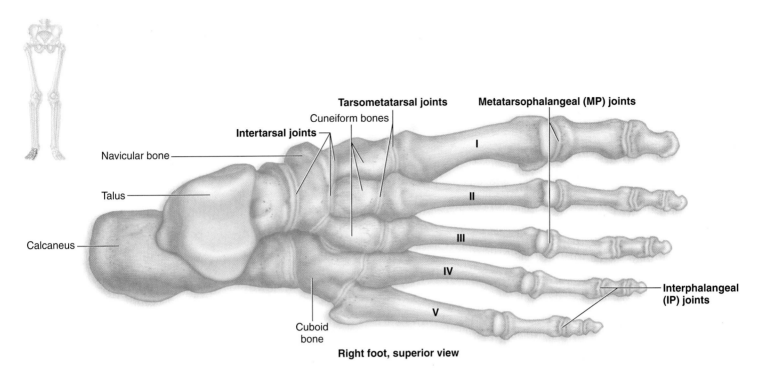

Tarsometatarsal joints
Cuneiform bones
Metatarsophalangeal (MP) joints
Intertarsal joints
Navicular bone
Talus
Calcaneus
I
II
III
IV
V
Interphalangeal (IP) joints
Cuboid bone

Right foot, superior view

Joints of the Foot
Figure 9.21

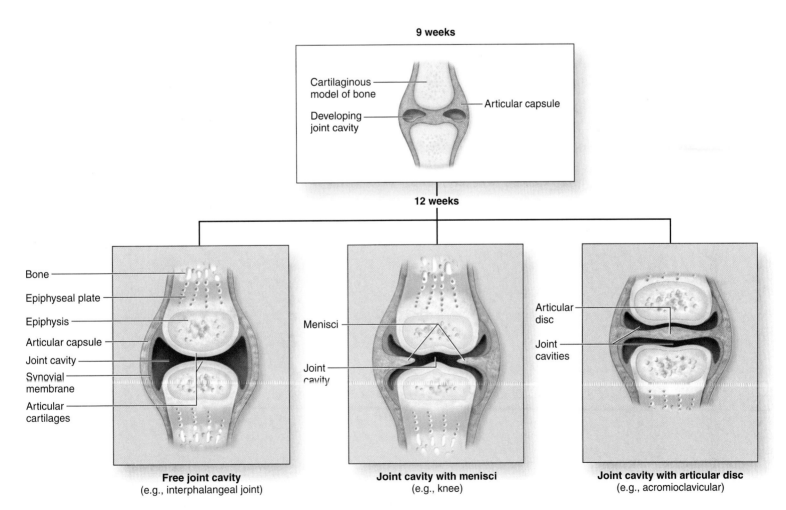

9 weeks

Cartilaginous model of bone
Developing joint cavity
Articular capsule

12 weeks

Bone
Epiphyseal plate
Epiphysis
Articular capsule
Joint cavity
Synovial membrane
Articular cartilages

Free joint cavity
(e.g., interphalangeal joint)

Menisci
Joint cavity

Joint cavity with menisci
(e.g., knee)

Articular disc
Joint cavities

Joint cavity with articular disc
(e.g., acromioclavicular)

Development of Synovial Joints
Figure 9.22

141

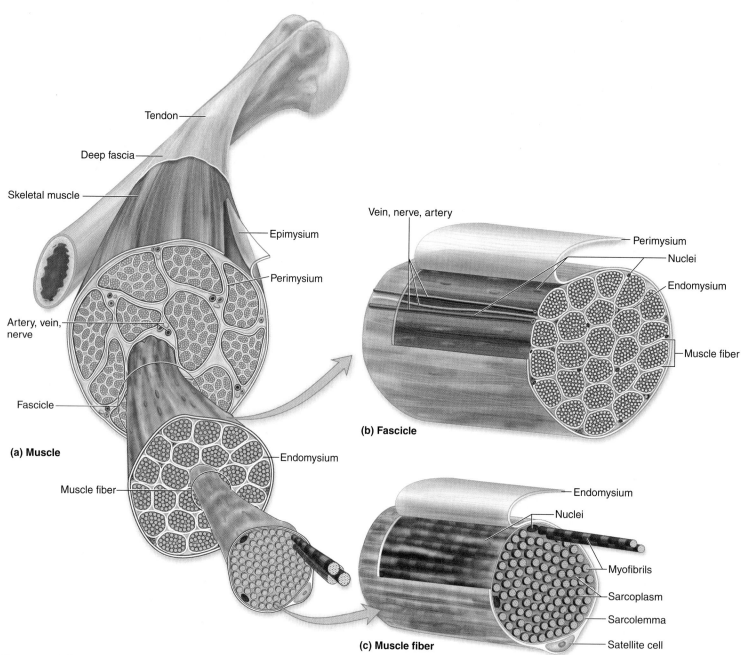

Tendon

Deep fascia

Skeletal muscle

Epimysium

Perimysium

Artery, vein, nerve

Fascicle

(a) Muscle

Endomysium

Muscle fiber

Vein, nerve, artery

Perimysium

Nuclei

Endomysium

Muscle fiber

(b) Fascicle

Endomysium

Nuclei

Myofibrils

Sarcoplasm

Sarcolemma

Satellite cell

(c) Muscle fiber

Structural Organization of Skeletal Muscle
Figure 10.1

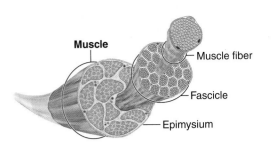

Muscle
- Muscle fiber
- Fascicle
- Epimysium

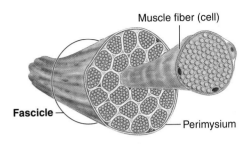

Muscle fiber (cell)

Fascicle
- Perimysium

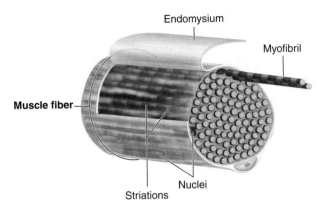

Endomysium

Myofibril

Muscle fiber

Striations

Nuclei

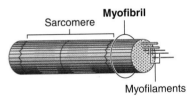

Sarcomere

Myofibril

Myofilaments

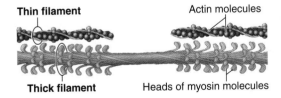

Thin filament

Actin molecules

Thick filament

Heads of myosin molecules

Organizational Levels of Skeletal Muscle
Table 10.1

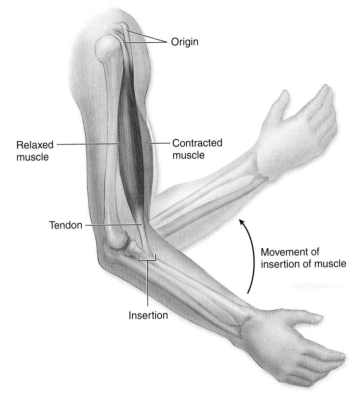

Origin

Relaxed muscle

Contracted muscle

Tendon

Insertion

Movement of insertion of muscle

Muscle Origin and Insertion
Figure 10.2

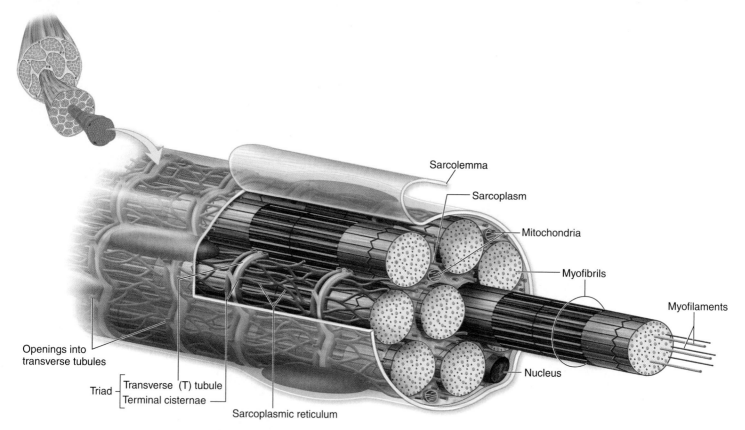

Sarcolemma

Sarcoplasm

Mitochondria

Myofibrils

Myofilaments

Nucleus

Openings into
transverse tubules

Triad — [Transverse (T) tubule
         Terminal cisternae]

Sarcoplasmic reticulum

Formation, Structure, and Organization of a Skeletal Muscle Fiber
Figure 10.3

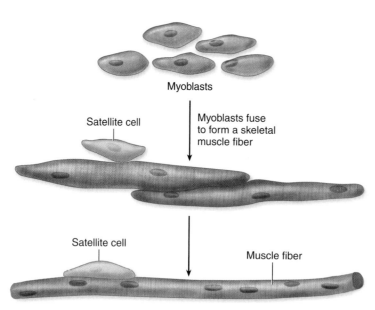

Myoblasts

Satellite cell

Myoblasts fuse
to form a skeletal
muscle fiber

Satellite cell

Muscle fiber

Development of Skeletal Muscle
Figure 10.4

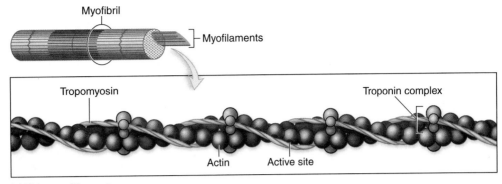

Myofibril

Myofilaments

Tropomyosin

Troponin complex

Actin   Active site

(a) Thin myofilament

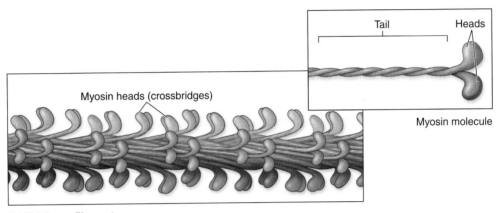

Tail   Heads

Myosin heads (crossbridges)

Myosin molecule

(b) Thick myofilament

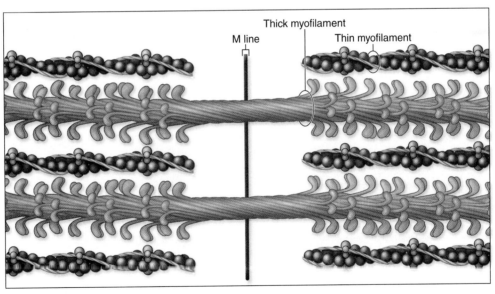

M line

Thick myofilament

Thin myofilament

(c) Comparison of thick and thin myofilaments

Molecular Structure of Thin and Thick Filaments
Figure 10.5

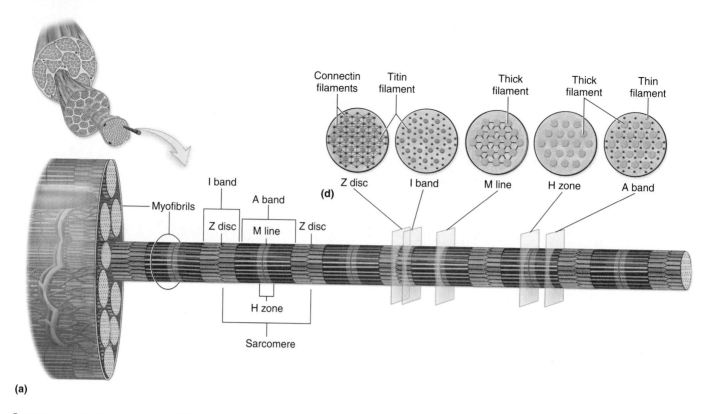

(a)

Structure of a Sarcomere—Microscopic Arrangement of Bands and Zones; Myofibril Cross Sections
Figure 10.6a,d

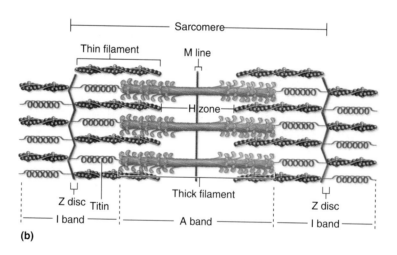

(b)

Sarcomere Organization—Diagram
Figure 10.6b

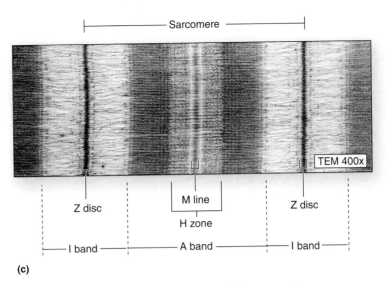

Sarcomere Organization—Electron Micrograph
Figure 10.6c
Figure 10.6c: © James Dennis/Phototake

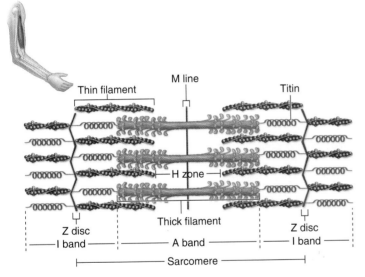

(a) Relaxed muscle
Sarcomere, I band, and H zone at an expanded/relaxed length
Note: The thick and thin filaments do not change length when the muscle contracts.

Sliding Filament Model of Contraction—Relaxed Muscle
Figure 10.7a

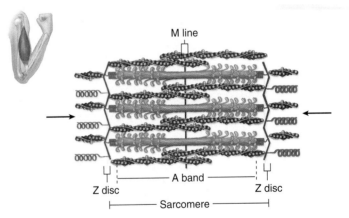

(c) Fully contracted muscle
The H zone and I band disappear, and the sarcomere is at its shortest length. Note: The length of the thick and thin filaments does not change.

Sliding Filament Model of Contraction—Fully Contracted Muscle
Figure 10.7c

147

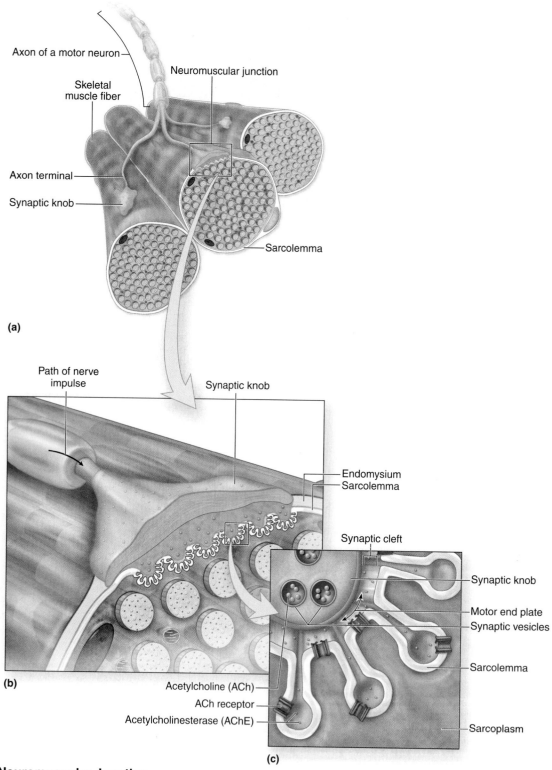

(a)

Axon of a motor neuron

Neuromuscular junction

Skeletal
muscle fiber

Axon terminal

Synaptic knob

Sarcolemma

(b)

Path of nerve
impulse

Synaptic knob

Endomysium
Sarcolemma

Synaptic cleft

Synaptic knob

Motor end plate
Synaptic vesicles

Sarcolemma

Acetylcholine (ACh)

ACh receptor

Acetylcholinesterase (AChE)

Sarcoplasm

(c)

Neuromuscular Junction
Figure 10.8

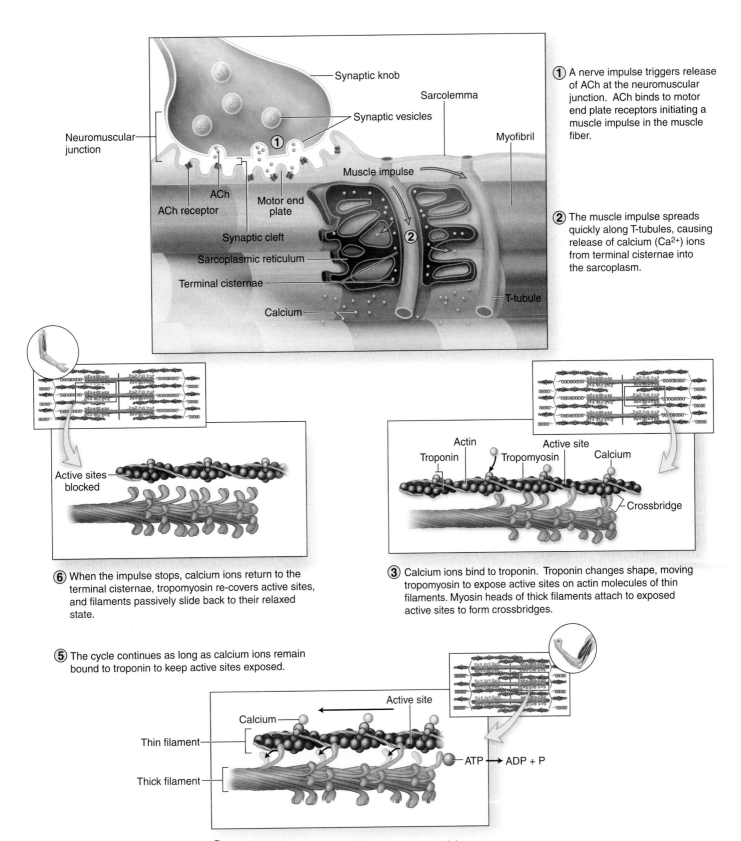

① A nerve impulse triggers release of ACh at the neuromuscular junction. ACh binds to motor end plate receptors initiating a muscle impulse in the muscle fiber.

② The muscle impulse spreads quickly along T-tubules, causing release of calcium (Ca^{2+}) ions from terminal cisternae into the sarcoplasm.

⑥ When the impulse stops, calcium ions return to the terminal cisternae, tropomyosin re-covers active sites, and filaments passively slide back to their relaxed state.

③ Calcium ions bind to troponin. Troponin changes shape, moving tropomyosin to expose active sites on actin molecules of thin filaments. Myosin heads of thick filaments attach to exposed active sites to form crossbridges.

⑤ The cycle continues as long as calcium ions remain bound to troponin to keep active sites exposed.

④ Myosin heads pivot, moving thin filaments toward the sarcomere center. ATP binds myosin heads, which detach from thin filaments and return to their pre-pivot position. The repeating cycle of *attach–pivot–detach–return* slides thick and thin filaments past one another. The sarcomere shortens and the muscle contracts.

Events in Muscle Contraction
Figure 10.9

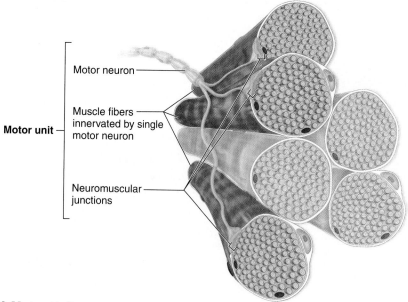

Motor unit {
  Motor neuron
  Muscle fibers innervated by single motor neuron
  Neuromuscular junctions
}

A Motor Unit
Figure 10.10

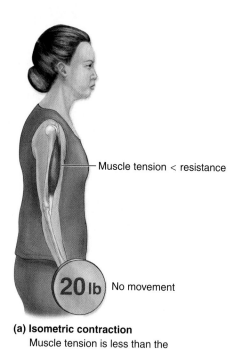

Muscle tension < resistance

20 lb No movement

(a) Isometric contraction
Muscle tension is less than the resistance; muscle does not shorten, and no movement occurs.

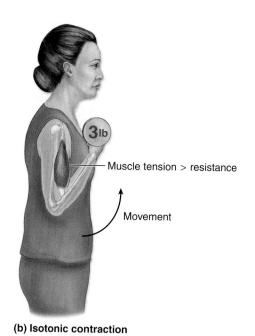

3 lb

Muscle tension > resistance

Movement

(b) Isotonic contraction
Muscle tension is greater than the resistance; muscle shortens, and movement occurs.

Isometric Versus Isotonic Contraction
Figure 10.11

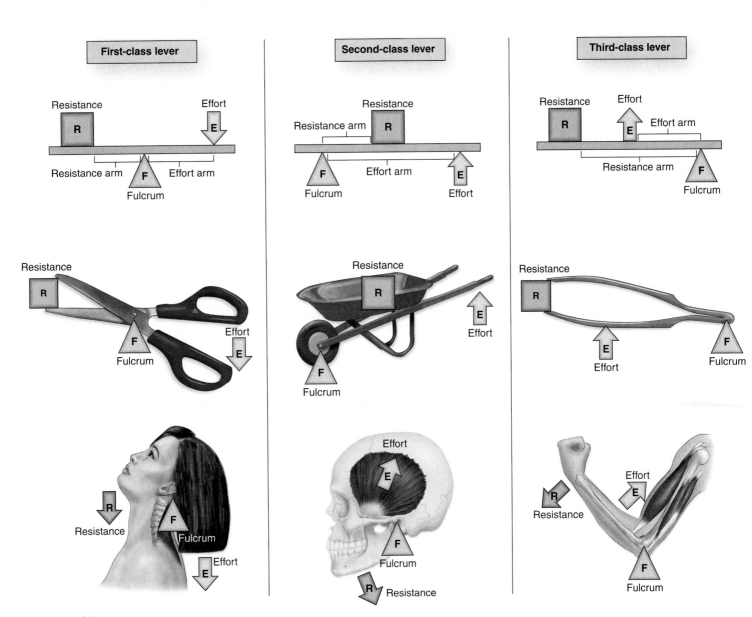

Classes of Levers
Figure 10.13

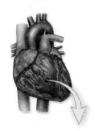

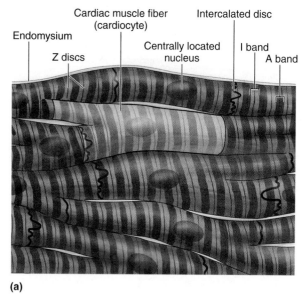

Endomysium

Z discs

Cardiac muscle fiber
(cardiocyte)

Centrally located
nucleus

I band

A band

Intercalated disc

(a)

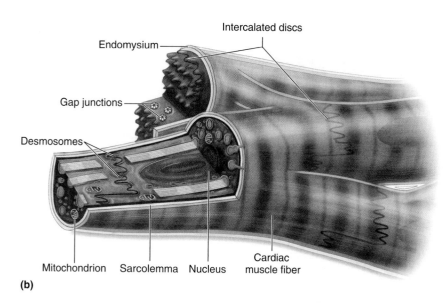

Endomysium

Intercalated discs

Gap junctions

Desmosomes

Mitochondrion   Sarcolemma   Nucleus

Cardiac
muscle fiber

(b)

Cardiac Muscle
Figure 10.15

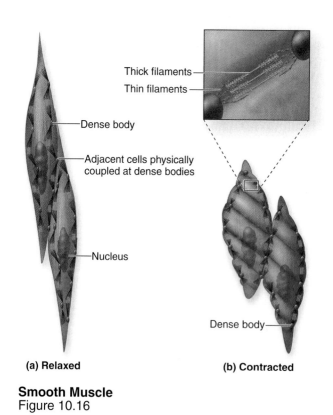

Thick filaments

Thin filaments

Dense body

Adjacent cells physically
coupled at dense bodies

Nucleus

Dense body

(a) Relaxed

(b) Contracted

Smooth Muscle
Figure 10.16

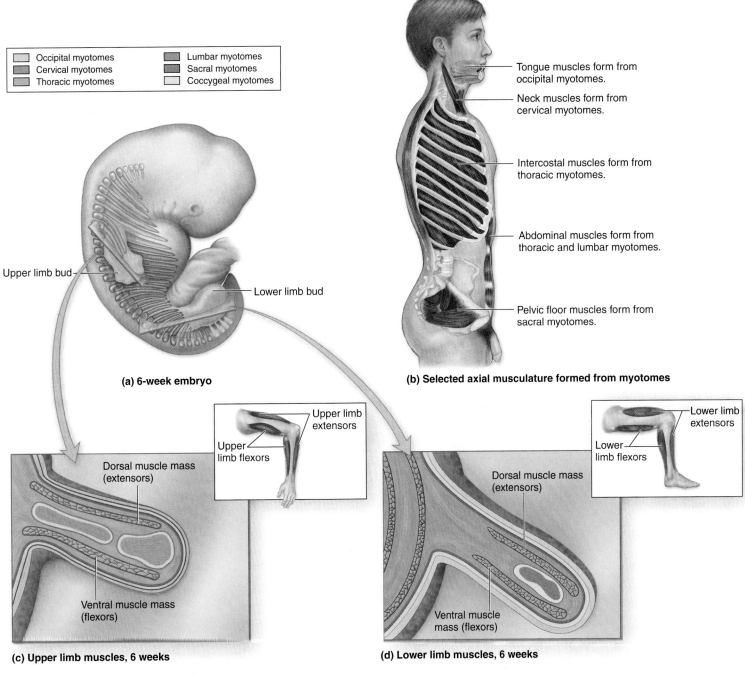

Legend:
- Occipital myotomes
- Cervical myotomes
- Thoracic myotomes
- Lumbar myotomes
- Sacral myotomes
- Coccygeal myotomes

Upper limb bud

Lower limb bud

(a) 6-week embryo

Tongue muscles form from occipital myotomes.

Neck muscles form from cervical myotomes.

Intercostal muscles form from thoracic myotomes.

Abdominal muscles form from thoracic and lumbar myotomes.

Pelvic floor muscles form from sacral myotomes.

(b) Selected axial musculature formed from myotomes

Upper limb extensors

Upper limb flexors

Dorsal muscle mass (extensors)

Ventral muscle mass (flexors)

(c) Upper limb muscles, 6 weeks

Lower limb extensors

Lower limb flexors

Dorsal muscle mass (extensors)

Ventral muscle mass (flexors)

(d) Lower limb muscles, 6 weeks

Development of Skeletal Muscles
Figure 10.17

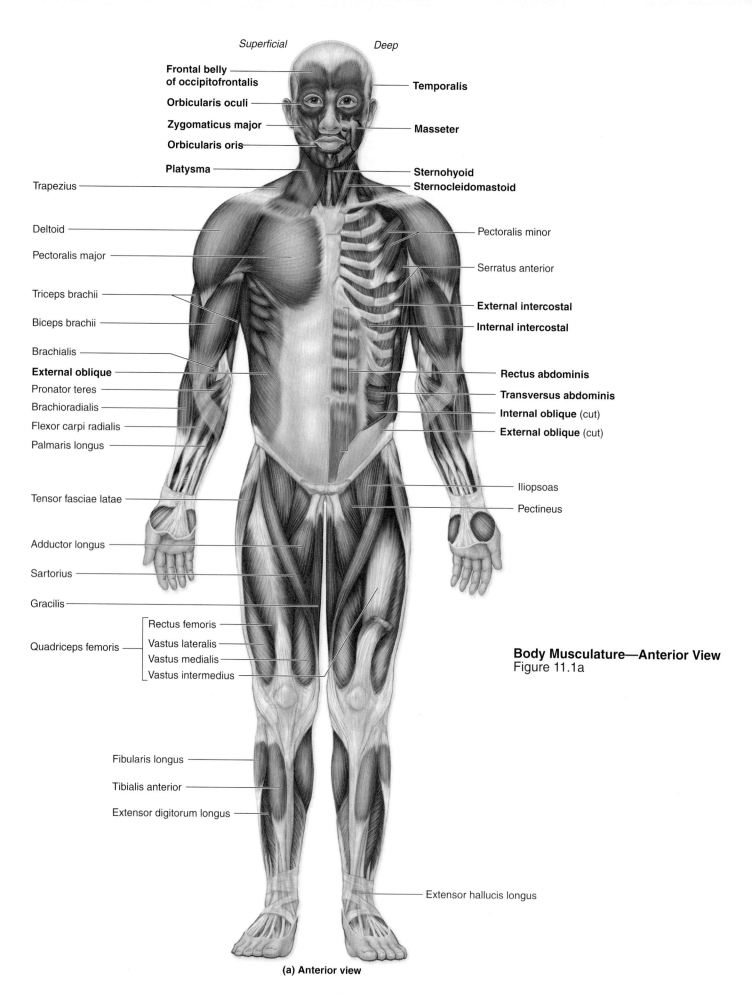

Superficial   Deep

**Frontal belly
of occipitofrontalis**

Orbicularis oculi

Zygomaticus major

Orbicularis oris

Platysma

Trapezius

Deltoid

Pectoralis major

Triceps brachii

Biceps brachii

Brachialis

External oblique

Pronator teres

Brachioradialis

Flexor carpi radialis

Palmaris longus

Tensor fasciae latae

Adductor longus

Sartorius

Gracilis

Quadriceps femoris — Rectus femoris
Vastus lateralis
Vastus medialis
Vastus intermedius

Fibularis longus

Tibialis anterior

Extensor digitorum longus

Temporalis

Masseter

Sternohyoid
Sternocleidomastoid

Pectoralis minor

Serratus anterior

External intercostal

Internal intercostal

Rectus abdominis

Transversus abdominis

Internal oblique (cut)

External oblique (cut)

Iliopsoas

Pectineus

Body Musculature—Anterior View
Figure 11.1a

Extensor hallucis longus

(a) Anterior view

154

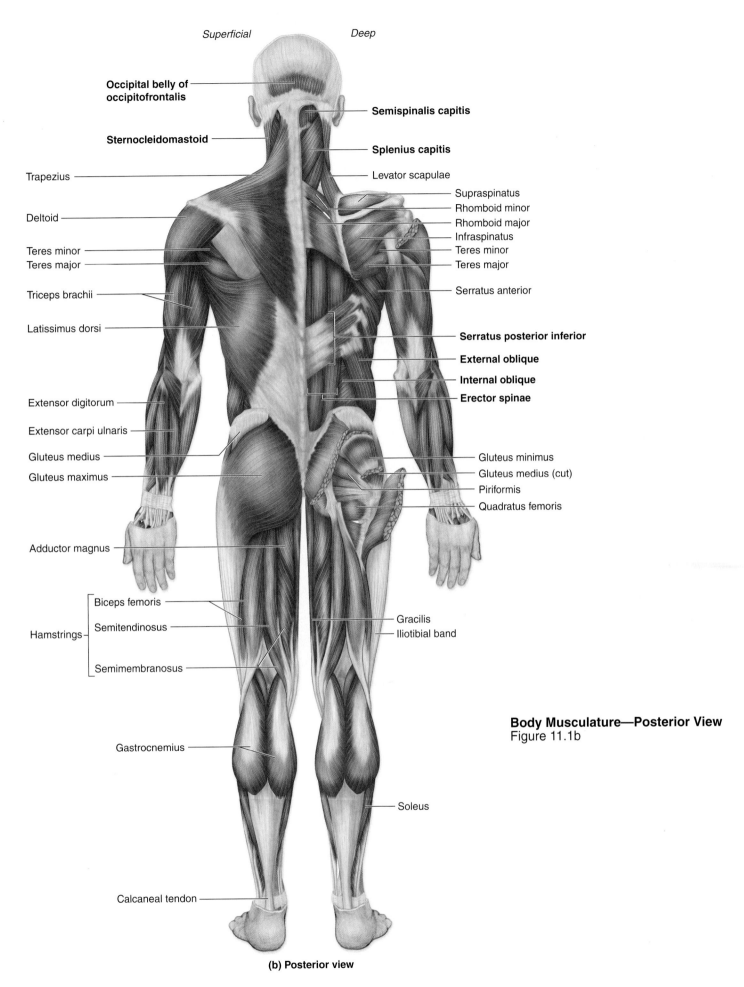

Superficial   _Deep_

Occipital belly of occipitofrontalis

Semispinalis capitis

Sternocleidomastoid

Splenius capitis

Trapezius

Levator scapulae

Supraspinatus

Rhomboid minor

Deltoid

Rhomboid major

Infraspinatus

Teres minor

Teres minor

Teres major

Teres major

Triceps brachii

Serratus anterior

Latissimus dorsi

Serratus posterior inferior

External oblique

Internal oblique

Erector spinae

Extensor digitorum

Extensor carpi ulnaris

Gluteus minimus

Gluteus medius

Gluteus medius (cut)

Gluteus maximus

Piriformis

Quadratus femoris

Adductor magnus

Biceps femoris

Gracilis

Semitendinosus

Iliotibial band

Hamstrings

Semimembranosus

Body Musculature—Posterior View
Figure 11.1b

Gastrocnemius

Soleus

Calcaneal tendon

(b) Posterior view

155

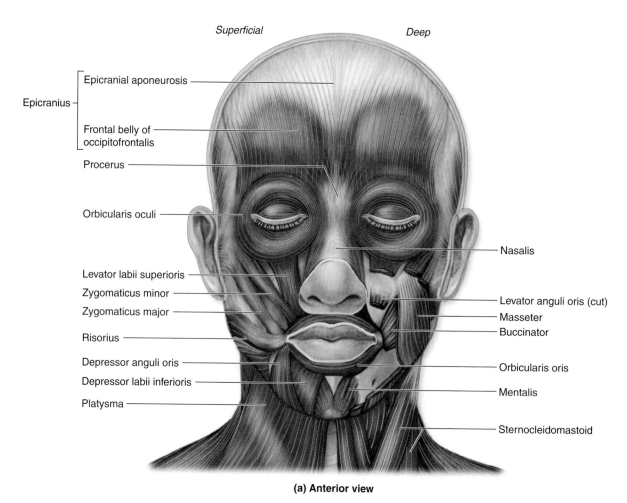

Superficial   *Deep*

Epicranius — {
Epicranial aponeurosis —

Frontal belly of —
occipitofrontalis

Procerus —

Orbicularis oculi —

Levator labii superioris —
Zygomaticus minor —
Zygomaticus major —

Risorius —

Depressor anguli oris —

Depressor labii inferioris —

Platysma —

Nasalis —

Levator anguli oris (cut) —
Masseter —
Buccinator —

Orbicularis oris —

Mentalis —

Sternocleidomastoid —

(a) Anterior view

Muscles of Facial Expression—Anterior View
Figure 11.2a

156

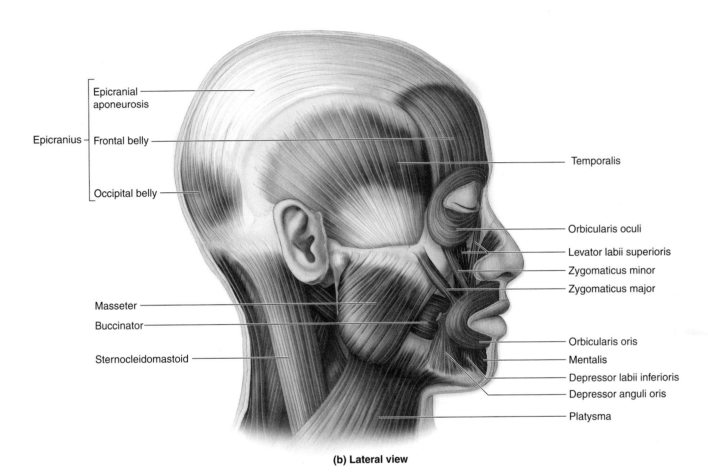

Epicranial aponeurosis

Epicranius {
  Frontal belly

  Occipital belly
}

Temporalis

Orbicularis oculi

Levator labii superioris

Zygomaticus minor

Zygomaticus major

Masseter

Buccinator

Sternocleidomastoid

Orbicularis oris

Mentalis

Depressor labii inferioris

Depressor anguli oris

Platysma

(b) Lateral view

Muscles of Facial Expression—Lateral View
Figure 11.2b

Depressor anguli oris
(frown)

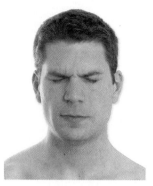

Orbicularis oculi
(blink/close eyes)

Zygomaticus major
(smile)

Orbicularis oris
(close mouth/kiss)

Frontal belly of occipitofrontalis
(wrinkle forehead, raise eyebrows)

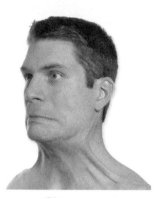

Platysma
(tense skin of neck)

Surface Anatomy of Some Muscles of Facial Expression
Figure 11.3
Figure 11.3: © The McGraw-Hill Companies, Inc./Photo by Jw Ramsey

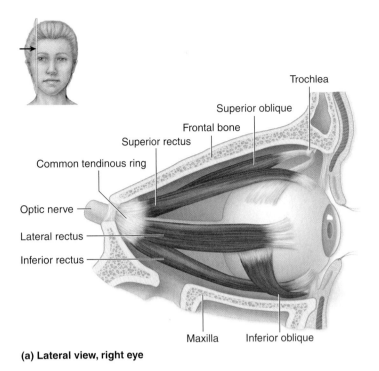

(a) Lateral view, right eye

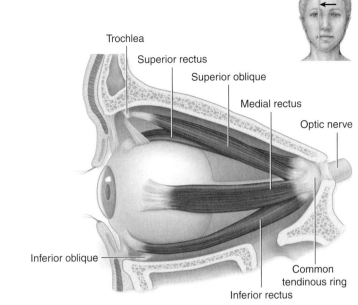

(b) Medial view, right eye

Extrinsic Muscle of the Eye—Lateral and Medial Views
Figure 11.4a, b

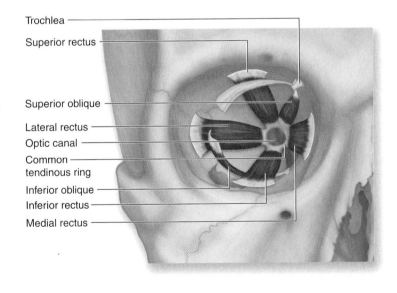

(c) Anterior view of right orbit, eye removed

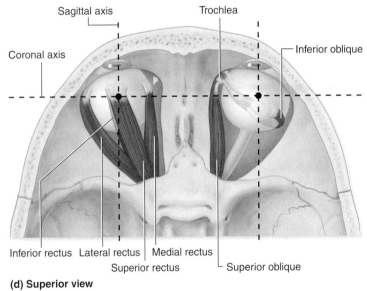

(d) Superior view

Extrinsic Muscle of the Eye—Anterior and Superior Views
Figure 11.4c, d

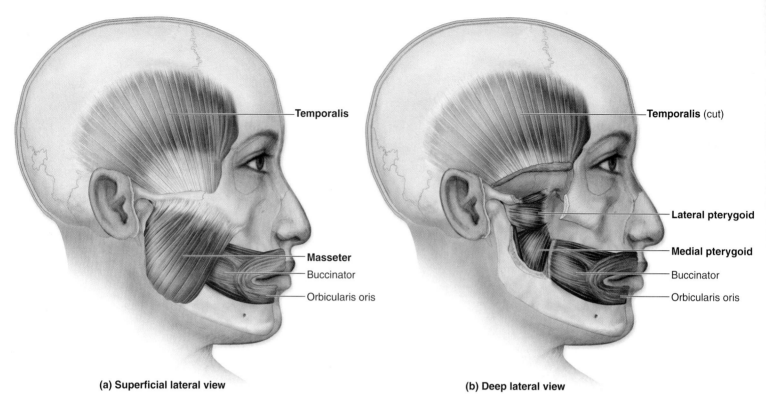

(a) Superficial lateral view

(b) Deep lateral view

Muscles of Mastication
Figure 11.5

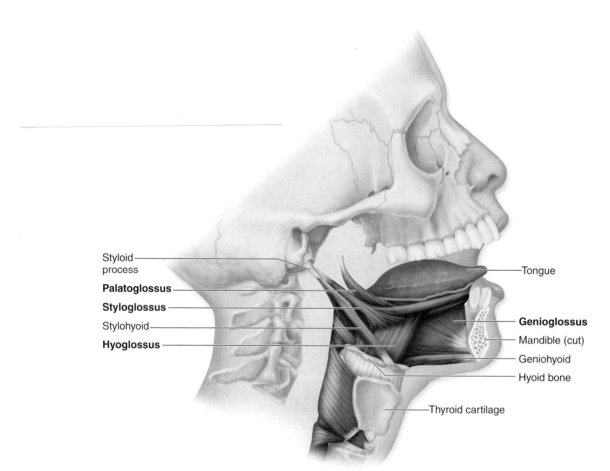

Right lateral view

Muscle That Move the Tongue
Figure 11.6

160

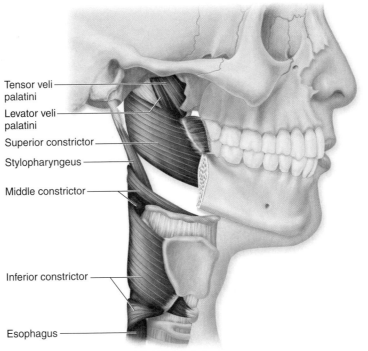

Tensor veli palatini

Levator veli palatini

Superior constrictor

Stylopharyngeus

Middle constrictor

Inferior constrictor

Esophagus

Right lateral view

Muscles of the Pharynx
Figure 11.7

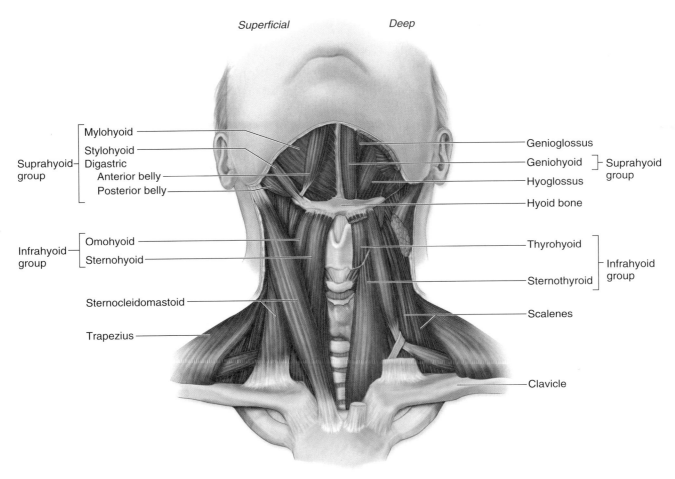

Superficial   *Deep*

Suprahyoid group
- Mylohyoid
- Stylohyoid
- Digastric
  - Anterior belly
  - Posterior belly

Infrahyoid group
- Omohyoid
- Sternohyoid

Sternocleidomastoid

Trapezius

Genioglossus

Geniohyoid ⎤ Suprahyoid group

Hyoglossus

Hyoid bone

Thyrohyoid ⎤ Infrahyoid group

Sternothyroid

Scalenes

Clavicle

Muscles of the Anterior Neck
Figure 11.8

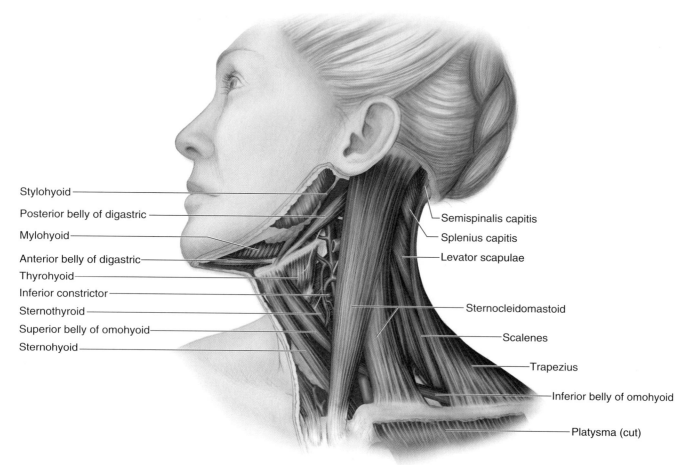

Stylohyoid

Posterior belly of digastric

Mylohyoid

Anterior belly of digastric

Thyrohyoid

Inferior constrictor

Sternothyroid

Superior belly of omohyoid

Sternohyoid

Semispinalis capitis

Splenius capitis

Levator scapulae

Sternocleidomastoid

Scalenes

Trapezius

Inferior belly of omohyoid

Platysma (cut)

Anterolateral view

Anterolateral Muscles That Move the Head and Neck
Figure 11.9

Deep   *Deeper*

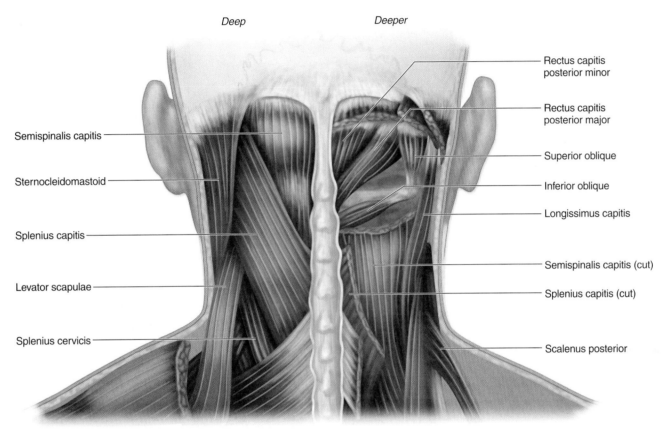

Semispinalis capitis

Sternocleidomastoid

Splenius capitis

Levator scapulae

Splenius cervicis

Rectus capitis posterior minor

Rectus capitis posterior major

Superior oblique

Inferior oblique

Longissimus capitis

Semispinalis capitis (cut)

Splenius capitis (cut)

Scalenus posterior

Posterior view

Posterior Neck Muscles
Figure 11.10

163

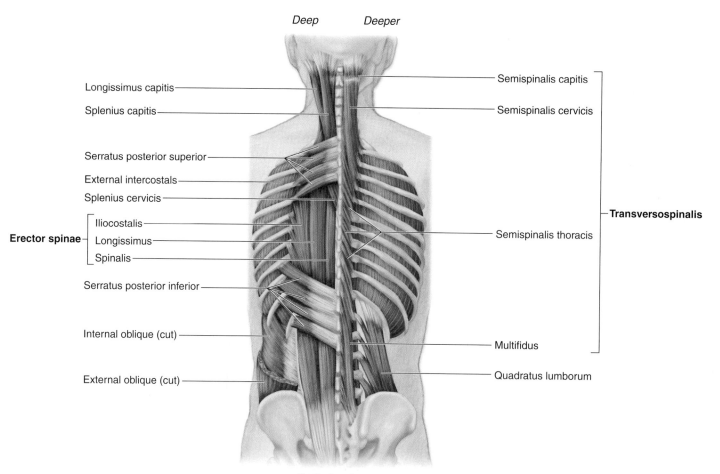

Deep   *Deeper*

Longissimus capitis

Splenius capitis

Serratus posterior superior

External intercostals

Splenius cervicis

Erector spinae
- Iliocostalis
- Longissimus
- Spinalis

Serratus posterior inferior

Internal oblique (cut)

External oblique (cut)

Semispinalis capitis

Semispinalis cervicis

Transversospinalis

Semispinalis thoracis

Multifidus

Quadratus lumborum

Posterior view

Superficial and Deep Muscles of the Vertebral Column
Figure 11.11

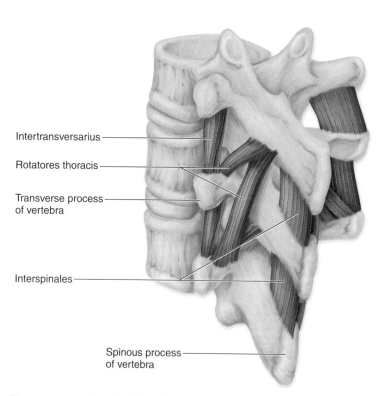

Intertransversarius

Rotatores thoracis

Transverse process
of vertebra

Interspinales

Spinous process
of vertebra

Transversospinalis Muscles
Figure 11.12

164

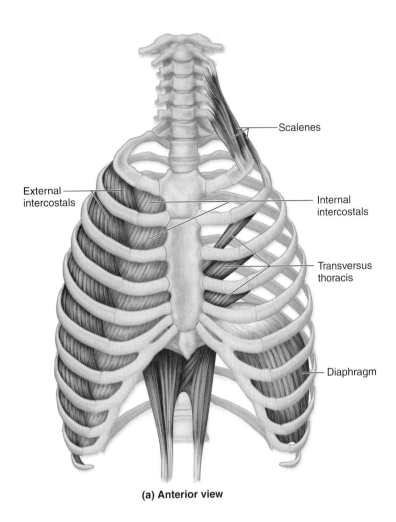

(a) Anterior view

Scalenes

External intercostals

Internal intercostals

Transversus thoracis

Diaphragm

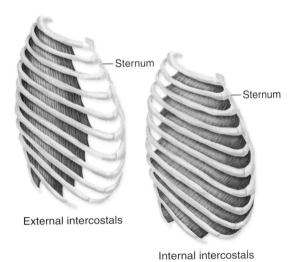

Sternum

Sternum

External intercostals

Internal intercostals

(c) Lateral view

Muscles of Respiration—Anterior and Lateral Views
Figure 11.13a,c

165

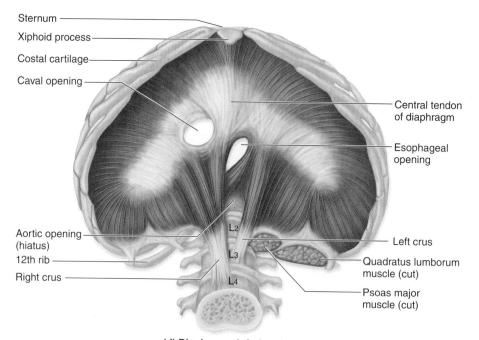

Sternum
Xiphoid process
Costal cartilage
Caval opening

Central tendon
of diaphragm

Esophageal
opening

L2
L3
L4

Aortic opening
(hiatus)
12th rib
Right crus

Left crus

Quadratus lumborum
muscle (cut)

Psoas major
muscle (cut)

(d) Diaphragm, inferior view

Muscles of Respiration—Diaphragm, Inferior View
Figure 11.13d

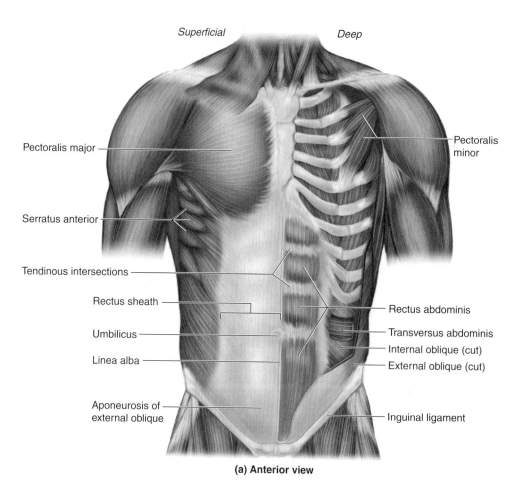

Superficial   *Deep*

Pectoralis major

Pectoralis
minor

Serratus anterior

Tendinous intersections

Rectus sheath

Umbilicus

Linea alba

Aponeurosis of
external oblique

Rectus abdominis

Transversus abdominis

Internal oblique (cut)

External oblique (cut)

Inguinal ligament

(a) Anterior view

Muscle of the Abdominal Wall—Anterior View
Figure 11.14a

166

External oblique

Internal oblique
and
rectus abdominis

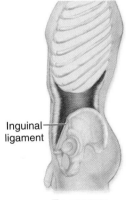

Inguinal
ligament

Transversus
abdominis

(c)

**Muscle of the
Abdominal Wall—
Individual Muscles**
Figure 11.14c

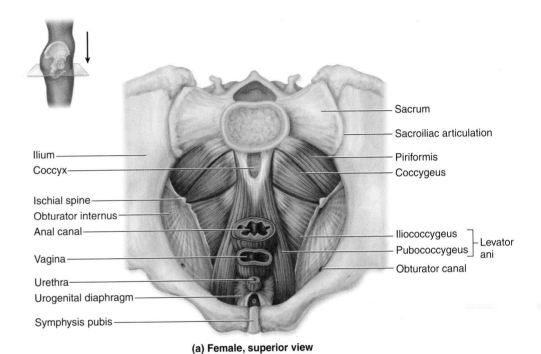

Ilium
Coccyx
Ischial spine
Obturator internus
Anal canal
Vagina
Urethra
Urogenital diaphragm
Symphysis pubis

Sacrum
Sacroiliac articulation
Piriformis
Coccygeus
Iliococcygeus ⎤ Levator
Pubococcygeus ⎦ ani
Obturator canal

(a) Female, superior view

Muscles of the Pelvic Floor—Female, Superior View
Figure 11.15a

167

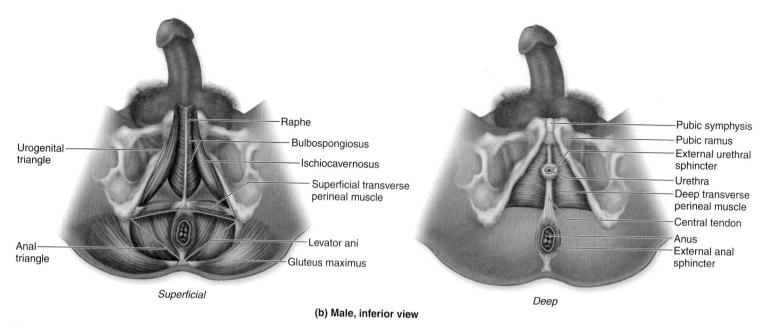

(b) Male, inferior view

Raphe
Bulbospongiosus
Ischiocavernosus
Superficial transverse perineal muscle
Levator ani
Gluteus maximus
Urogenital triangle
Anal triangle
Superficial

Pubic symphysis
Pubic ramus
External urethral sphincter
Urethra
Deep transverse perineal muscle
Central tendon
Anus
External anal sphincter
Deep

Muscles of the Pelvic Floor—Male, Inferior View
Figure 11.15b

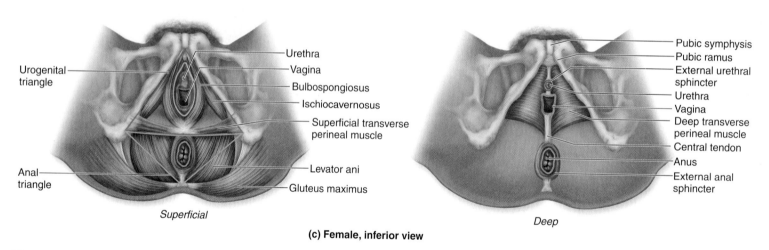

(c) Female, inferior view

Urethra
Vagina
Bulbospongiosus
Ischiocavernosus
Superficial transverse perineal muscle
Levator ani
Gluteus maximus
Urogenital triangle
Anal triangle
Superficial

Pubic symphysis
Pubic ramus
External urethral sphincter
Urethra
Vagina
Deep transverse perineal muscle
Central tendon
Anus
External anal sphincter
Deep

Muscles of the Pelvic Floor—Female, Inferior View
Figure 11.15c

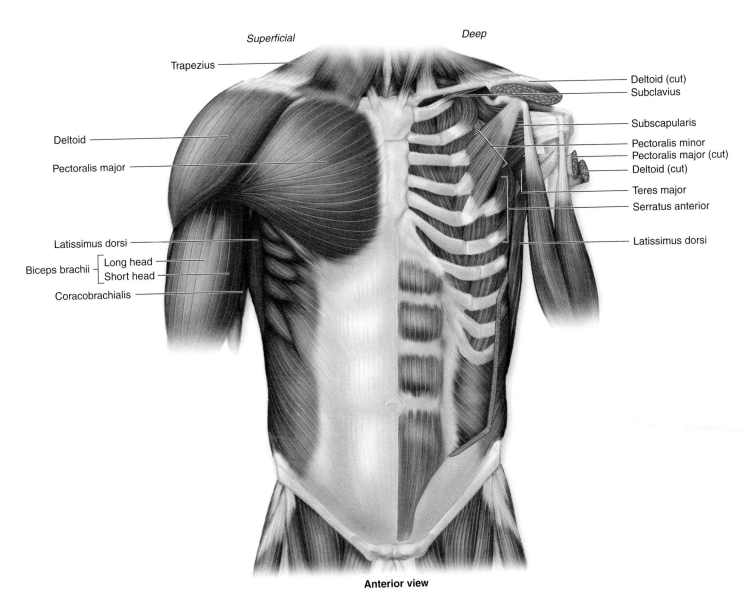

Trapezius

Deltoid (cut)
Subclavius

Subscapularis

Deltoid

Pectoralis minor
Pectoralis major (cut)
Deltoid (cut)

Pectoralis major

Teres major
Serratus anterior

Latissimus dorsi

Biceps brachii ⎡ Long head
⎣ Short head

Latissimus dorsi

Coracobrachialis

Anterior view

Muscles of the Anterior Trunk
Figure 12.1

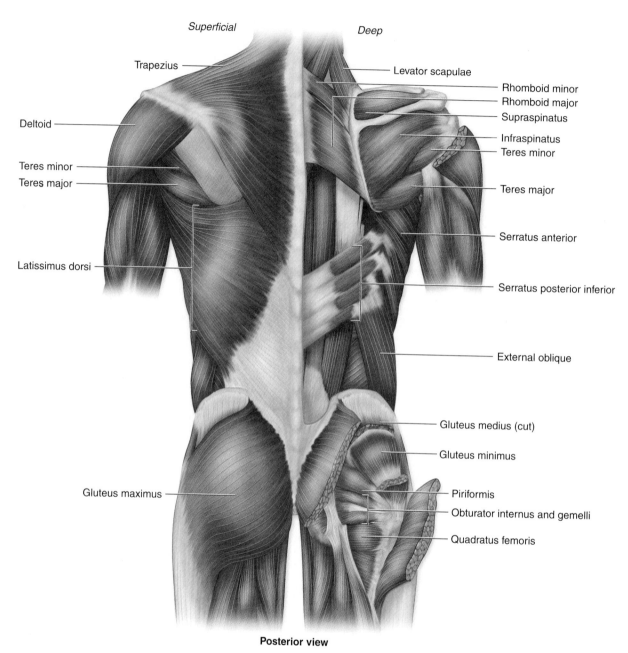

Superficial *Deep*

Trapezius — Levator scapulae

Rhomboid minor
Rhomboid major
Supraspinatus

Deltoid — Infraspinatus
Teres minor

Teres minor —
Teres major — Teres major

Serratus anterior

Latissimus dorsi — Serratus posterior inferior

External oblique

Gluteus medius (cut)

Gluteus minimus

Gluteus maximus — Piriformis
Obturator internus and gemelli
Quadratus femoris

Posterior view

Muscles of the Posterior Trunk
Figure 12.2

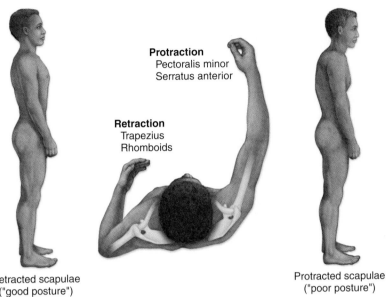

Protraction
Pectoralis minor
Serratus anterior

Retraction
Trapezius
Rhomboids

Retracted scapulae
("good posture")

Protracted scapulae
("poor posture")

(a)

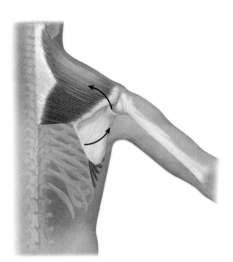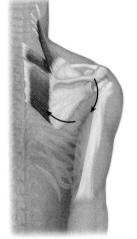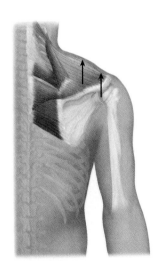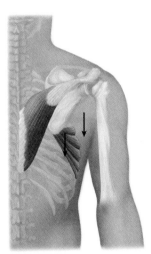

Superior rotators
Serratus anterior
Trapezius (superior part)

(b)

Inferior rotators
Rhomboid major
Rhomboid minor
Levator scapulae

Elevators
Rhomboid major
Rhomboid minor
Levator scapulae
Trapezius (superior part)

Depressors
Trapezius (inferior part)
Serratus anterior
Pectoralis minor (not shown)

Actions of Some Thoracic Muscles on the Scapula
Figure 12.3

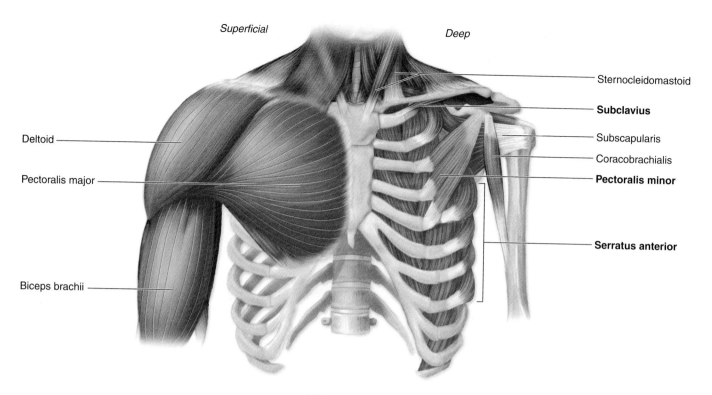

Superficial   Deep

Deltoid

Pectoralis major

Biceps brachii

Sternocleidomastoid

Subclavius

Subscapularis

Coracobrachialis

Pectoralis minor

Serratus anterior

(a) Anterior view

Muscles That Move the Pectoral Girdle and Glenohumeral Joint—Anterior View
Figure 12.4a

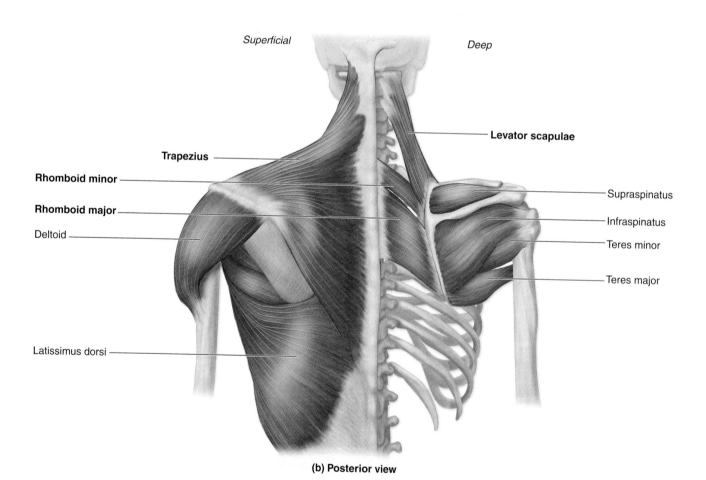

Superficial   Deep

Trapezius

Rhomboid minor

Rhomboid major

Deltoid

Latissimus dorsi

Levator scapulae

Supraspinatus

Infraspinatus

Teres minor

Teres major

(b) Posterior view

Muscles That Move the Pectoral Girdle and Glenohumeral Joint—Posterior View
Figure 12.4b

172

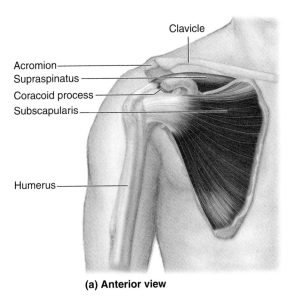

Clavicle

Acromion

Supraspinatus

Coracoid process

Subscapularis

Humerus

(a) Anterior view

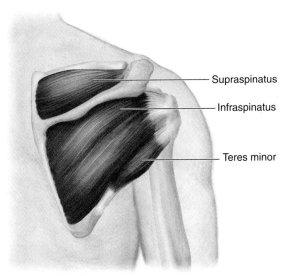

Supraspinatus

Infraspinatus

Teres minor

(b) Posterior view

Rotator Cuff Muscles
Figure 12.5a,b

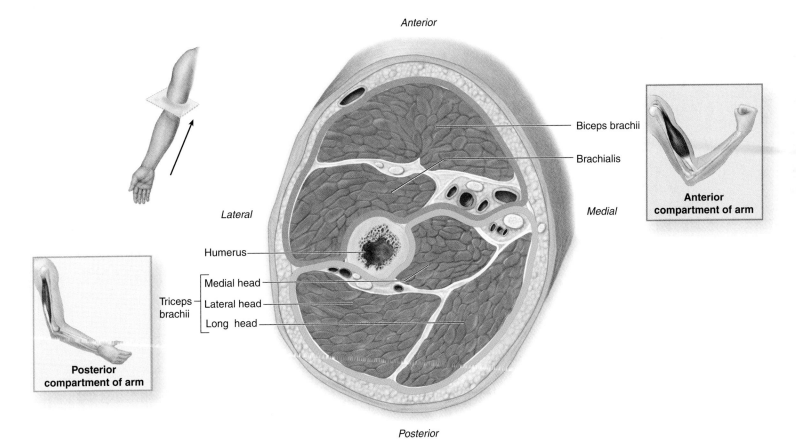

Anterior

Lateral

Medial

Posterior

Biceps brachii

Brachialis

Humerus

Triceps brachii
- Medial head
- Lateral head
- Long head

Anterior compartment of arm

Posterior compartment of arm

Action of Arm Muscles on the Forearm
Figure 12.6

173

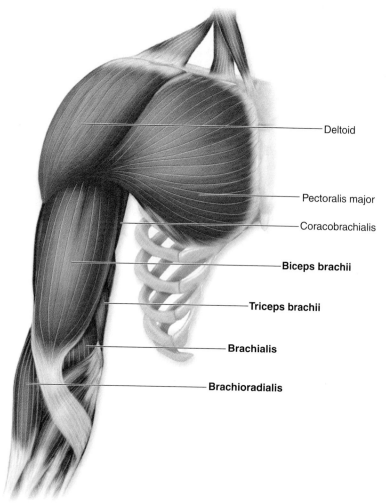

Deltoid

Pectoralis major

Coracobrachialis

Biceps brachii

Triceps brachii

Brachialis

Brachioradialis

(a) Anterior view

Anterior Muscles That Move the Elbow Joint/Forearm
Figure 12.7a

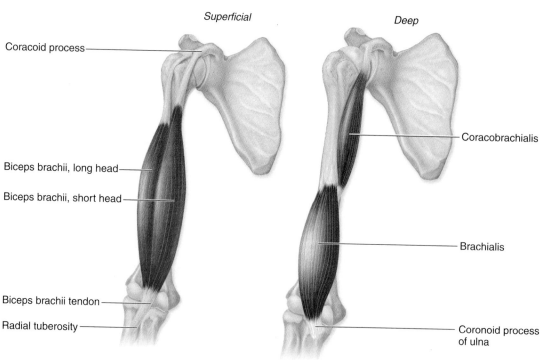

Superficial

Deep

Coracoid process

Biceps brachii, long head

Biceps brachii, short head

Biceps brachii tendon

Radial tuberosity

Coracobrachialis

Brachialis

Coronoid process
of ulna

(b) Anterior muscles

Anterior Muscles That Move the Elbow Joint/Forearm—Individual Muscles
Figure 12.7b

174

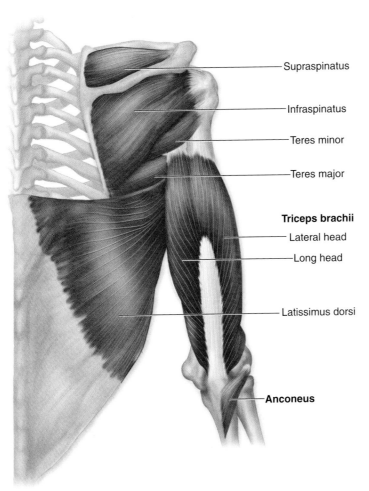

- Supraspinatus
- Infraspinatus
- Teres minor
- Teres major

Triceps brachii
- Lateral head
- Long head

- Latissimus dorsi

- **Anconeus**

(a) Posterior view

Posterior Muscles That Move the Elbow Joint/Forearm
Figure 12.8a

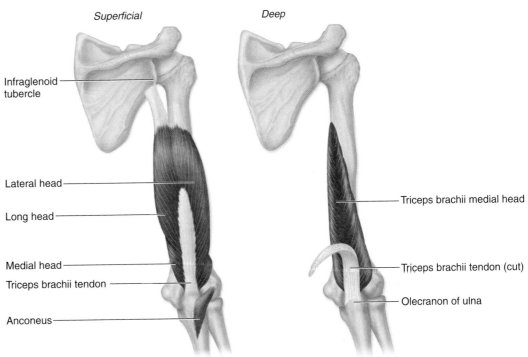

Superficial   *Deep*

Infraglenoid tubercle

Lateral head

Long head

Medial head

Triceps brachii tendon

Anconeus

Triceps brachii medial head

Triceps brachii tendon (cut)

Olecranon of ulna

(b) Posterior muscles

Posterior Muscles That Move the Elbow Joint/Forearm—Individual Muscles
Figure 12.8b

175

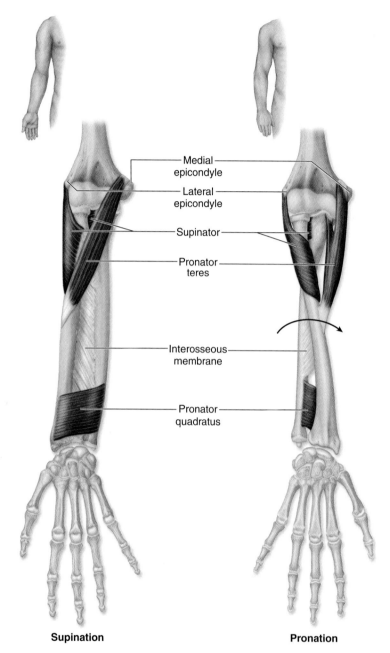

Medial epicondyle

Lateral epicondyle

Supinator

Pronator teres

Interosseous membrane

Pronator quadratus

Supination

Pronation

Forearm Muscles That Supinate and Pronate
Figure 12.9

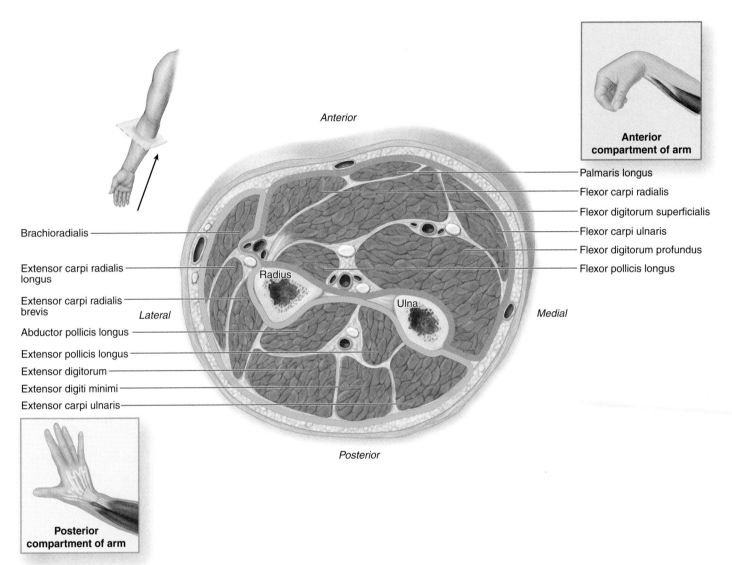

Anterior

Palmaris longus

Flexor carpi radialis

Flexor digitorum superficialis

Flexor carpi ulnaris

Flexor digitorum profundus

Flexor pollicis longus

Brachioradialis

Extensor carpi radialis longus

Extensor carpi radialis brevis

Radius

Ulna

Lateral

Medial

Abductor pollicis longus

Extensor pollicis longus

Extensor digitorum

Extensor digiti minimi

Extensor carpi ulnaris

Posterior

Anterior compartment of arm

Posterior compartment of arm

Actions of the Muscles of the Forearm
Figure 12.10

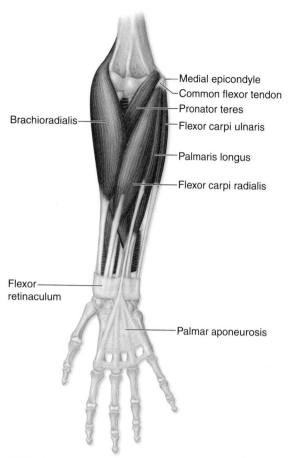

Medial epicondyle
Common flexor tendon
Pronator teres
Flexor carpi ulnaris
Brachioradialis
Palmaris longus
Flexor carpi radialis

Flexor retinaculum

Palmar aponeurosis

(a) Right anterior forearm, superficial view

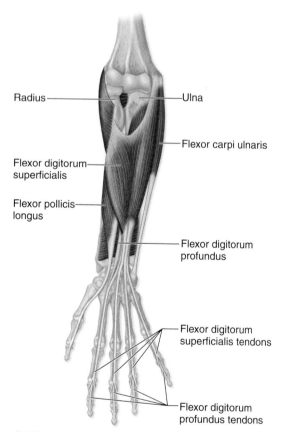

Radius
Ulna
Flexor carpi ulnaris
Flexor digitorum superficialis
Flexor pollicis longus
Flexor digitorum profundus

Flexor digitorum superficialis tendons

Flexor digitorum profundus tendons

(b) Right anterior forearm, intermediate view

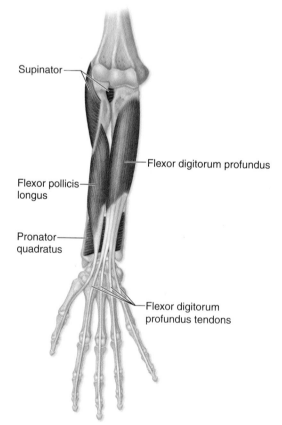

Supinator
Flexor digitorum profundus
Flexor pollicis longus
Pronator quadratus
Flexor digitorum profundus tendons

(c) Right anterior forearm, deep view

Anterior Forearm Muscles
Figure 12.11

178

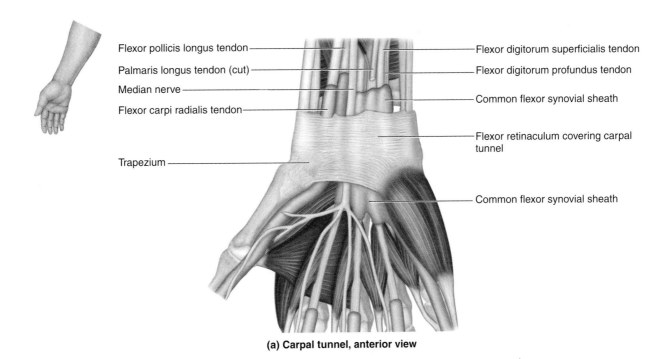

Flexor pollicis longus tendon

Palmaris longus tendon (cut)

Median nerve

Flexor carpi radialis tendon

Trapezium

Flexor digitorum superficialis tendon

Flexor digitorum profundus tendon

Common flexor synovial sheath

Flexor retinaculum covering carpal tunnel

Common flexor synovial sheath

(a) Carpal tunnel, anterior view

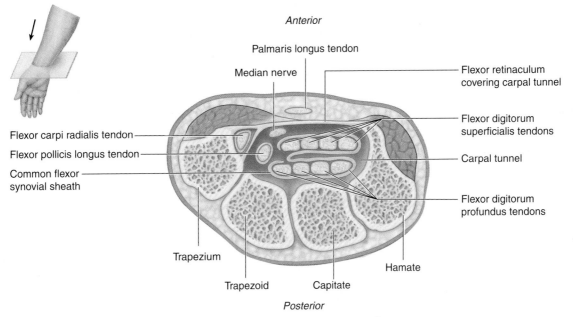

Anterior

Palmaris longus tendon

Median nerve

Flexor carpi radialis tendon

Flexor pollicis longus tendon

Common flexor synovial sheath

Flexor retinaculum covering carpal tunnel

Flexor digitorum superficialis tendons

Carpal tunnel

Flexor digitorum profundus tendons

Trapezium

Trapezoid

Capitate

Hamate

Posterior

(b) Carpal tunnel, transverse section

Carpal Tunnel
Clinical View p. 371

Clinical View p. 371

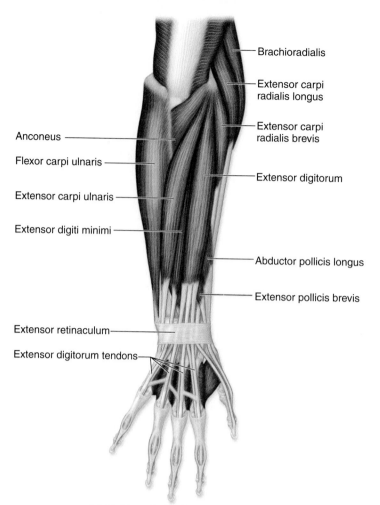

Brachioradialis

Extensor carpi radialis longus

Extensor carpi radialis brevis

Anconeus

Flexor carpi ulnaris

Extensor carpi ulnaris

Extensor digiti minimi

Extensor digitorum

Abductor pollicis longus

Extensor pollicis brevis

Extensor retinaculum

Extensor digitorum tendons

(a) Right posterior forearm, superficial view

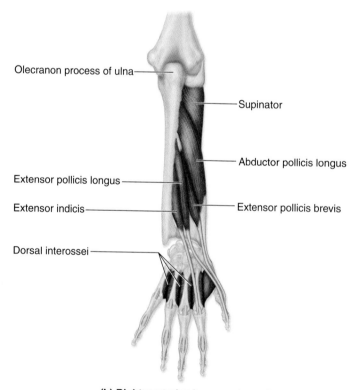

Olecranon process of ulna

Supinator

Abductor pollicis longus

Extensor pollicis longus

Extensor indicis

Extensor pollicis brevis

Dorsal interossei

(b) Right posterior forearm, deep view

Posterior Forearm Muscles
Figure 12.13

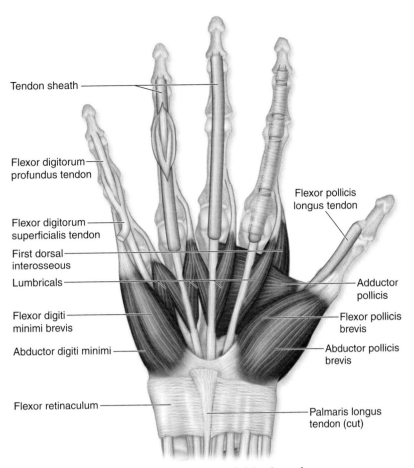

Tendon sheath

Flexor digitorum profundus tendon

Flexor digitorum superficialis tendon

First dorsal interosseous

Lumbricals

Flexor digiti minimi brevis

Abductor digiti minimi

Flexor retinaculum

Flexor pollicis longus tendon

Adductor pollicis

Flexor pollicis brevis

Abductor pollicis brevis

Palmaris longus tendon (cut)

(a) Right hand, superficial palmar view

Intrinsic Muscles of the Hand—Superficial Palmar View
Figure 12.14a

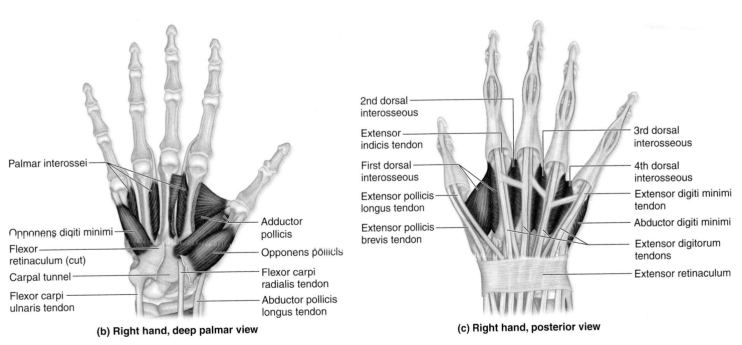

Palmar interossei

Opponens digiti minimi

Flexor retinaculum (cut)

Carpal tunnel

Flexor carpi ulnaris tendon

Adductor pollicis

Opponens pollicis

Flexor carpi radialis tendon

Abductor pollicis longus tendon

(b) Right hand, deep palmar view

2nd dorsal interosseous

Extensor indicis tendon

First dorsal interosseous

Extensor pollicis longus tendon

Extensor pollicis brevis tendon

3rd dorsal interosseous

4th dorsal interosseous

Extensor digiti minimi tendon

Abductor digiti minimi

Extensor digitorum tendons

Extensor retinaculum

(c) Right hand, posterior view

Intrinsic Muscles of the Hand—Deep Palmar and Posterior Views
Figure 12.14b,c

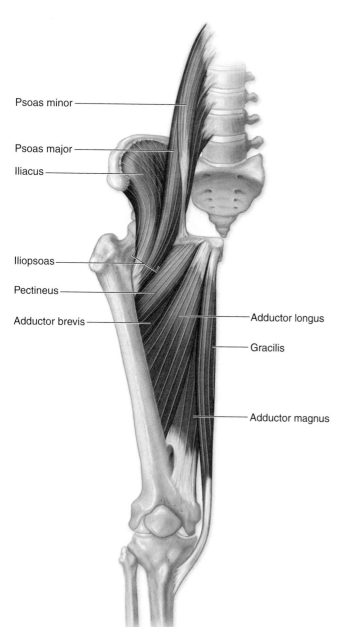

Psoas minor

Psoas major

Iliacus

Iliopsoas

Pectineus

Adductor brevis

Adductor longus

Gracilis

Adductor magnus

(a) Right thigh, anterior view

Muscles That Act on the Hip and Thigh—Anterior View
Figure 12.15a

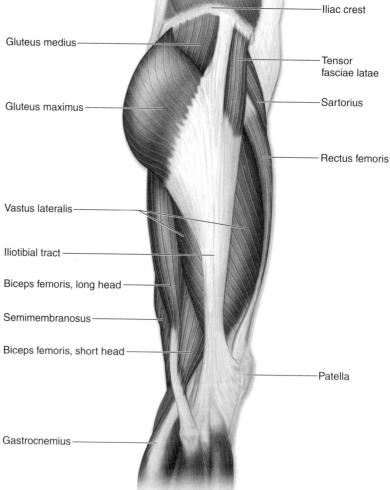

Iliac crest

Gluteus medius

Tensor fasciae latae

Gluteus maximus

Sartorius

Rectus femoris

Vastus lateralis

Iliotibial tract

Biceps femoris, long head

Semimembranosus

Biceps femoris, short head

Patella

Gastrocnemius

(b) Right thigh, lateral view

Muscles That Act on the Hip and Thigh—Lateral View
Figure 12.15b

182

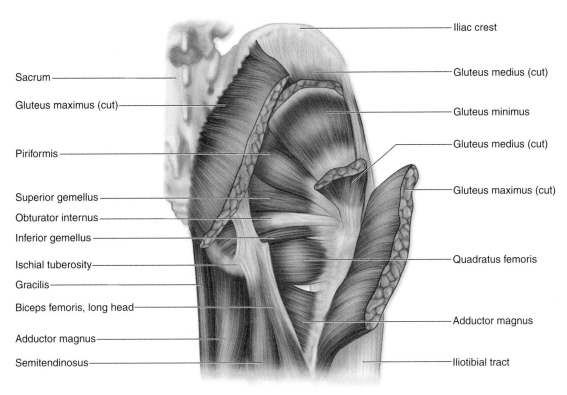

Iliac crest

Sacrum

Gluteus maximus (cut)

Piriformis

Superior gemellus

Obturator internus

Inferior gemellus

Ischial tuberosity

Gracilis

Biceps femoris, long head

Adductor magnus

Semitendinosus

Gluteus medius (cut)

Gluteus minimus

Gluteus medius (cut)

Gluteus maximus (cut)

Quadratus femoris

Adductor magnus

Iliotibial tract

(c) Right thigh, deep posterior view

Muscles That Act on the Hip and Thigh—Deep Posterior View
Figure 12.15c

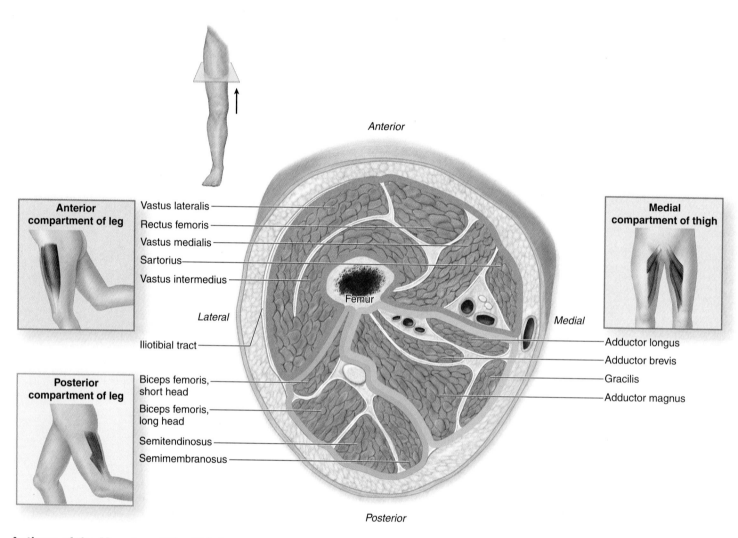

Anterior

Anterior compartment of leg

Vastus lateralis
Rectus femoris
Vastus medialis
Sartorius
Vastus intermedius

Femur

Lateral

Iliotibial tract

Medial

Medial compartment of thigh

Adductor longus
Adductor brevis
Gracilis
Adductor magnus

Posterior compartment of leg

Biceps femoris, short head
Biceps femoris, long head
Semitendinosus
Semimembranosus

Posterior

Actions of the Muscles of the Thigh
Figure 12.16

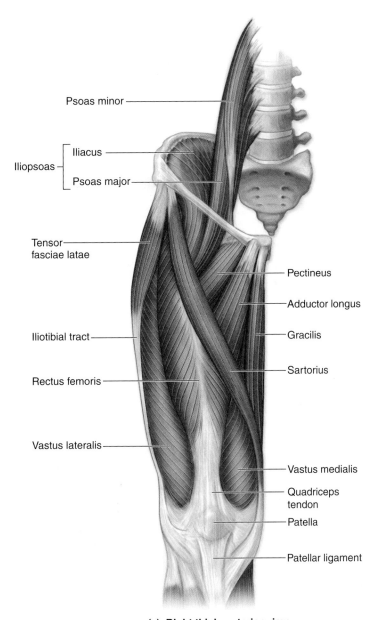

Psoas minor

Iliopsoas
Iliacus
Psoas major

Tensor fasciae latae

Pectineus

Adductor longus

Iliotibial tract

Gracilis

Sartorius

Rectus femoris

Vastus lateralis

Vastus medialis

Quadriceps tendon

Patella

Patellar ligament

(a) Right thigh, anterior view

Muscles of the Right Anterior Thigh
Figure 12.17a

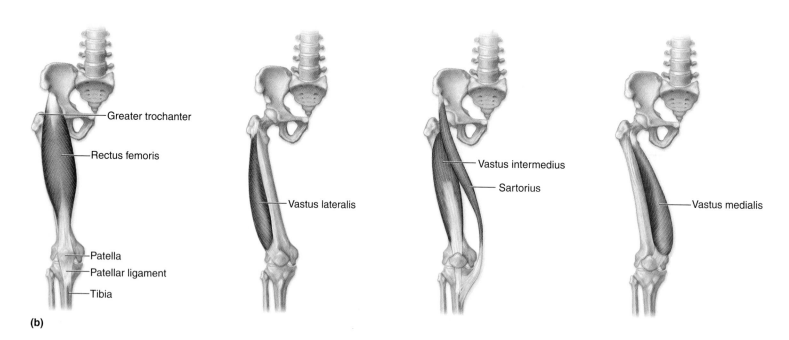

(b)

Muscles of the Right Anterior Thigh—Individual Muscles
Figure 12.17b

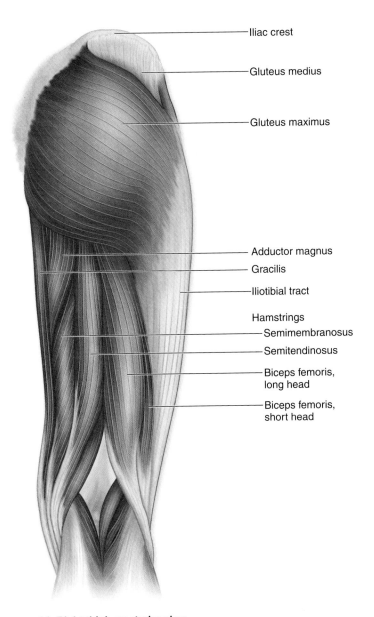

Iliac crest

Gluteus medius

Gluteus maximus

Adductor magnus

Gracilis

Iliotibial tract

Hamstrings
Semimembranosus

Semitendinosus

Biceps femoris,
long head

Biceps femoris,
short head

(a) Right thigh, posterior view

Muscles of the Right Posterior Thigh
Figure 12.18a

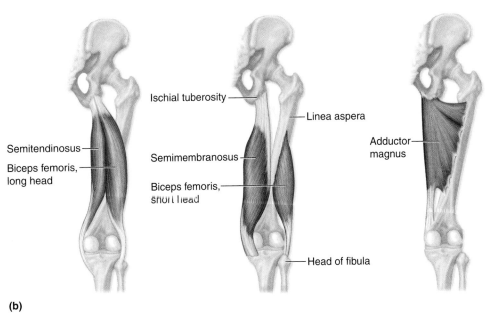

Semitendinosus

Biceps femoris,
long head

Ischial tuberosity

Linea aspera

Semimembranosus

Biceps femoris,
short head

Adductor
magnus

Head of fibula

(b)

Muscles of the Right Posterior Thigh—Individual Muscles
Figure 12.18b

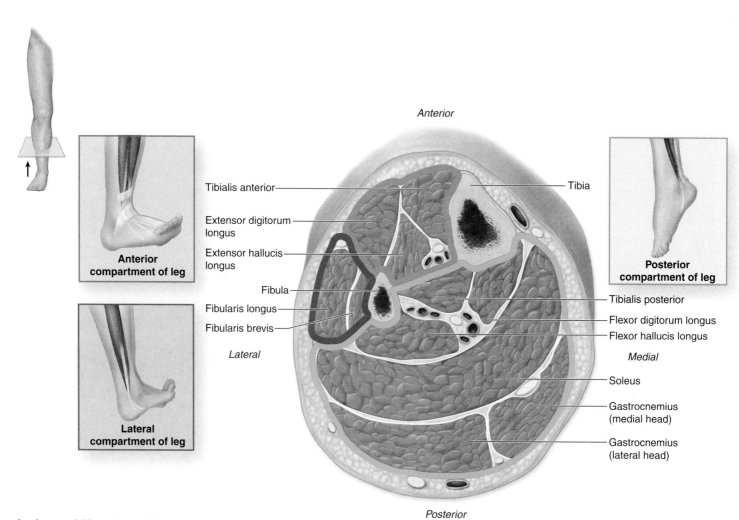

Anterior

Tibialis anterior

Extensor digitorum longus

Extensor hallucis longus

Fibula

Fibularis longus

Fibularis brevis

Lateral

Anterior compartment of leg

Lateral compartment of leg

Tibia

Posterior compartment of leg

Tibialis posterior

Flexor digitorum longus

Flexor hallucis longus

Medial

Soleus

Gastrocnemius (medial head)

Gastrocnemius (lateral head)

Posterior

Actions of Muscles of the Leg
Figure 12.19

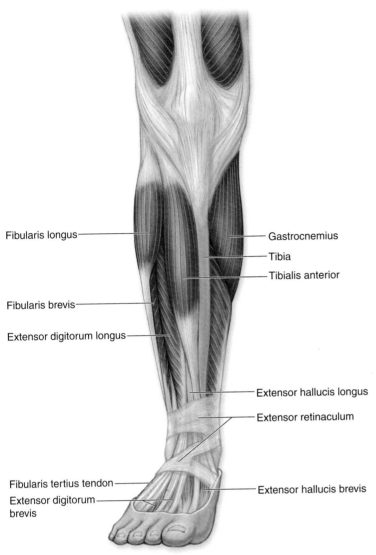

Fibularis longus

Fibularis brevis

Extensor digitorum longus

Gastrocnemius

Tibia

Tibialis anterior

Extensor hallucis longus

Extensor retinaculum

Fibularis tertius tendon

Extensor digitorum brevis

Extensor hallucis brevis

(a) Right leg, anterior view

Muscles of the Right Anterior Leg
Figure 12.20a

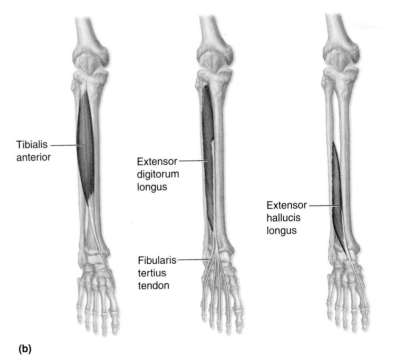

Tibialis anterior

Extensor digitorum longus

Fibularis tertius tendon

Extensor hallucis longus

(b)

Muscles of the Right Anterior Leg—Individual Muscles
Figure 12.20b

189

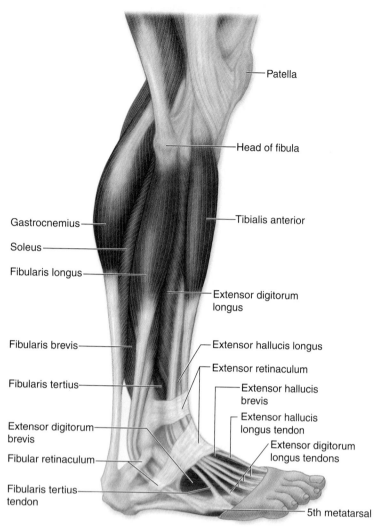

Patella

Head of fibula

Tibialis anterior

Gastrocnemius

Soleus

Fibularis longus

Extensor digitorum longus

Fibularis brevis

Extensor hallucis longus

Extensor retinaculum

Fibularis tertius

Extensor hallucis brevis

Extensor hallucis longus tendon

Extensor digitorum brevis

Extensor digitorum longus tendons

Fibular retinaculum

Fibularis tertius tendon

5th metatarsal

(a) Right leg, lateral view

Muscles of the Right Lateral Leg
Figure 12.21a

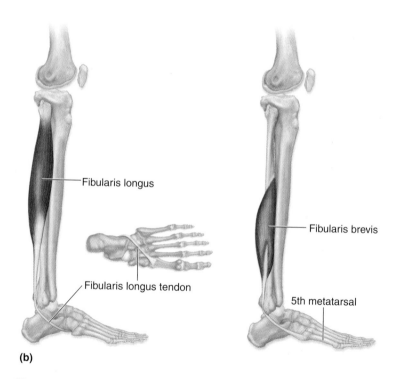

Fibularis longus

Fibularis brevis

Fibularis longus tendon

5th metatarsal

(b)

Muscles of the Right Lateral Leg—Individual Muscles
Figure 12.21b

190

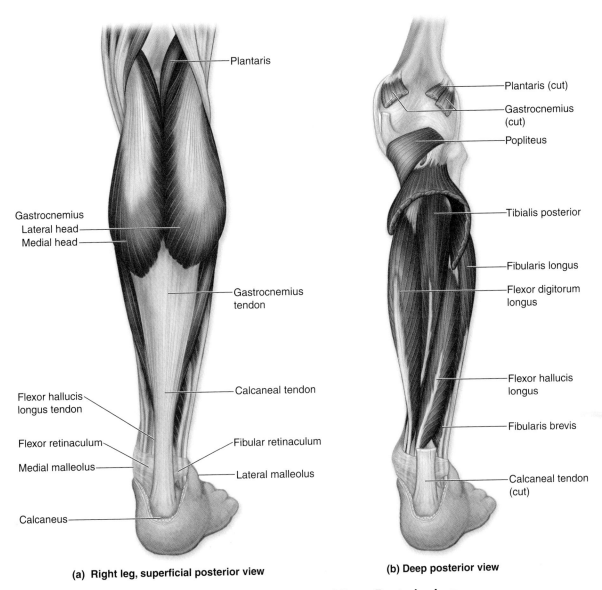

Plantaris

Plantaris (cut)

Gastrocnemius
(cut)

Popliteus

Gastrocnemius
Lateral head
Medial head

Tibialis posterior

Gastrocnemius
tendon

Fibularis longus

Flexor digitorum
longus

Calcaneal tendon

Flexor hallucis
longus tendon

Flexor hallucis
longus

Flexor retinaculum

Fibularis brevis

Fibular retinaculum

Medial malleolus

Lateral malleolus

Calcaneal tendon
(cut)

Calcaneus

(a) Right leg, superficial posterior view

(b) Deep posterior view

Muscles of the Right Posterior Leg—Superficial and Deep Posterior Leg
Figure 12.22a, b

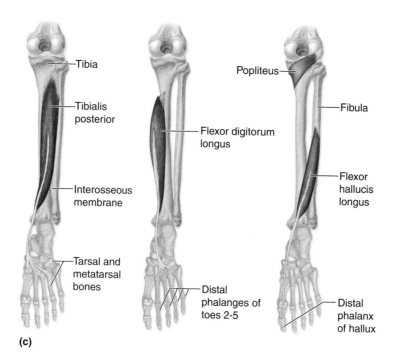

(c)

Muscles of the Right Posterior Leg—Individual Muscles
Figure 12.22c

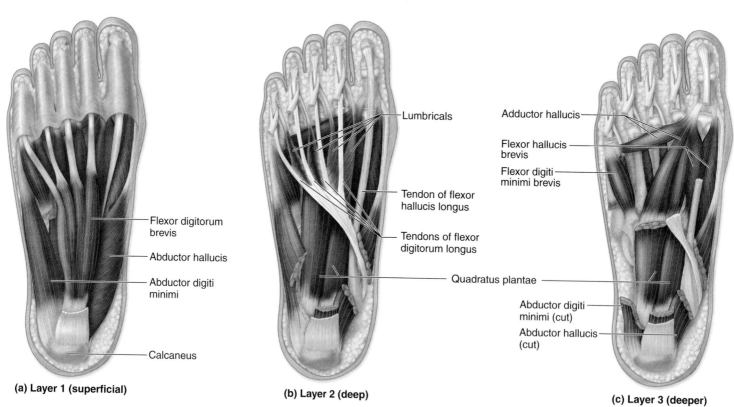

Plantar Intrinsic Muscles of the Right Foot—Layers 1-3
Figure 12.23a,b,c

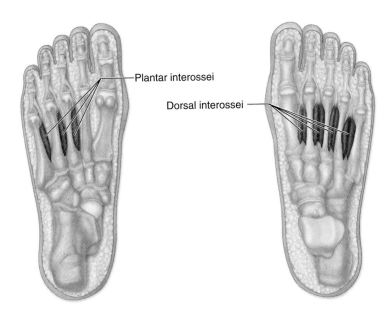

Plantar interossei

Dorsal interossei

(d) Layer 4 (deepest), plantar view

(e) Layer 4 (deepest), dorsal view

Plantar Intrinsic Muscles of the Right Foot—Layer 4, Plantar and Dorsal Views
Figure 12.23d,e

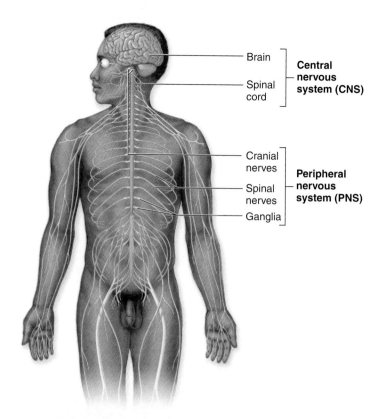

Organization of the Nervous System
Figure 14.1

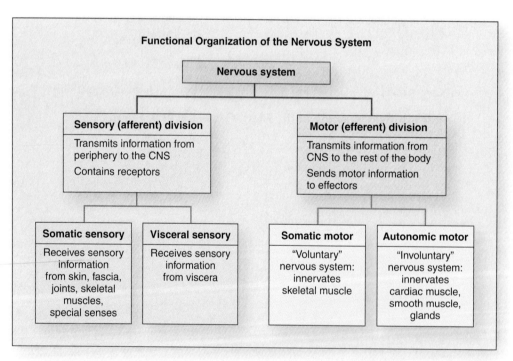

Functional Organization of the Nervous System
Figure 14.2

194

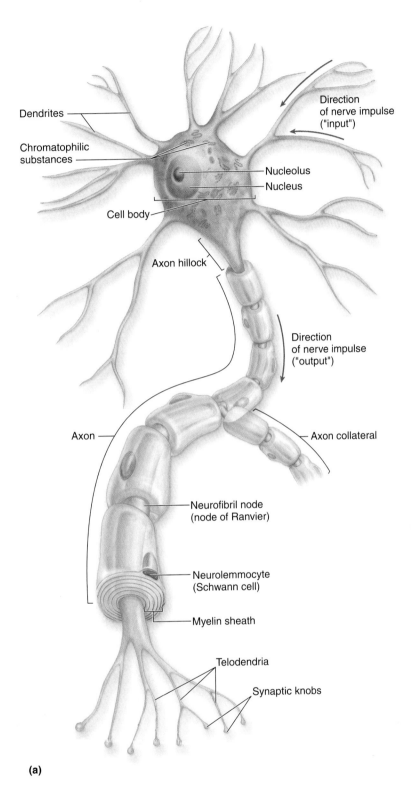

Dendrites

Chromatophilic
substances

Nucleolus

Nucleus

Cell body

Direction
of nerve impulse
("input")

Axon hillock

Direction
of nerve impulse
("output")

Axon

Axon collateral

Neurofibril node
(node of Ranvier)

Neurolemmocyte
(Schwann cell)

Myelin sheath

Telodendria

Synaptic knobs

(a)

Structure of a Typical Neuron
Figure 14.3a

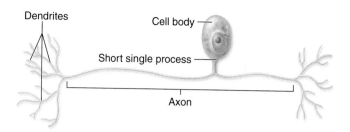

Dendrites
Cell body
Short single process
Axon

(a) Unipolar neuron

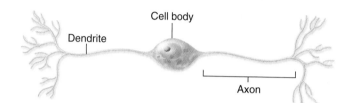

Cell body
Dendrite
Axon

(b) Bipolar neuron

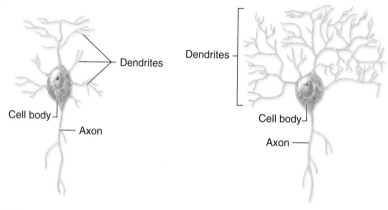

Dendrites
Cell body
Axon

Dendrites
Cell body
Axon

(c) Multipolar neurons

Structural Classification of Neurons
Figure 14.4

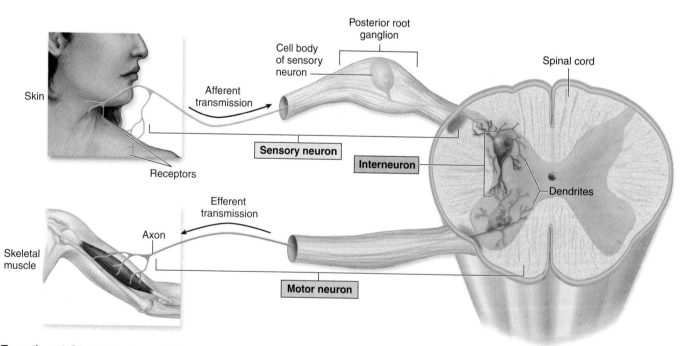

Skin

Receptors

Afferent transmission

Cell body of sensory neuron

Posterior root ganglion

Spinal cord

Sensory neuron

Interneuron

Dendrites

Efferent transmission

Axon

Skeletal muscle

Motor neuron

Functional Classification of Neurons
Figure 14.5

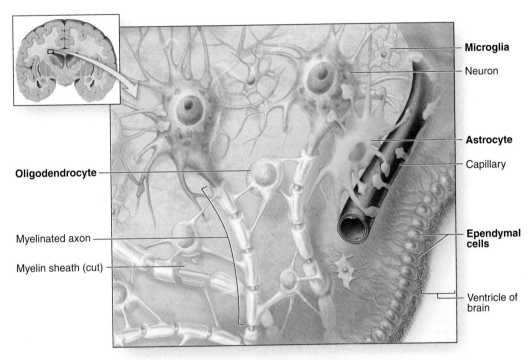

Microglia

Neuron

Astrocyte

Capillary

Oligodendrocyte

Ependymal
cells

Myelinated axon

Myelin sheath (cut)

Ventricle of
brain

Cellular Organization of Neural Tissue Within the CNS
Figure 14.6

CNS Glial Cells

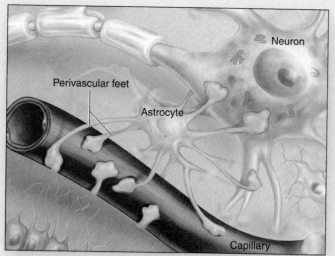

(a) Astrocyte

Neuron
Perivascular feet
Astrocyte
Capillary

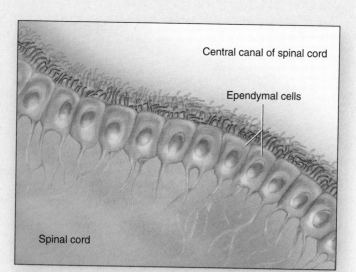

(b) Ependymal cells

Central canal of spinal cord
Ependymal cells
Spinal cord

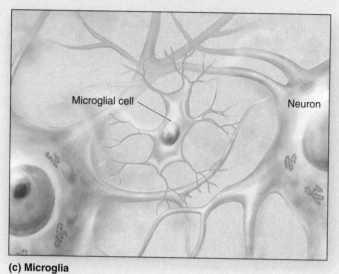

(c) Microglia

Microglial cell
Neuron

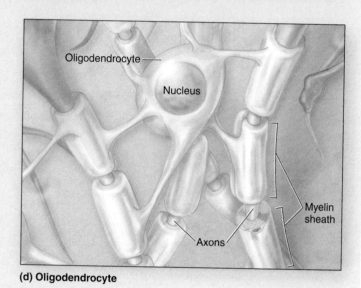

(d) Oligodendrocyte

Oligodendrocyte
Nucleus
Myelin sheath
Axons

PNS Glial Cells

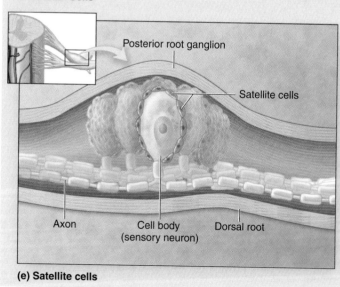

(e) Satellite cells

Posterior root ganglion
Satellite cells
Axon
Cell body (sensory neuron)
Dorsal root

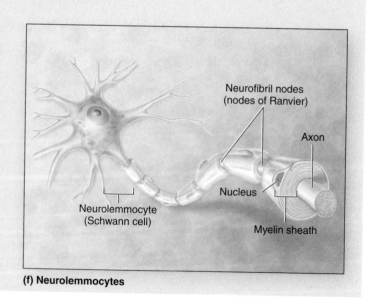

(f) Neurolemmocytes

Neurofibril nodes (nodes of Ranvier)
Axon
Neurolemmocyte (Schwann cell)
Nucleus
Myelin sheath

Glial Cells
Figure 14.7

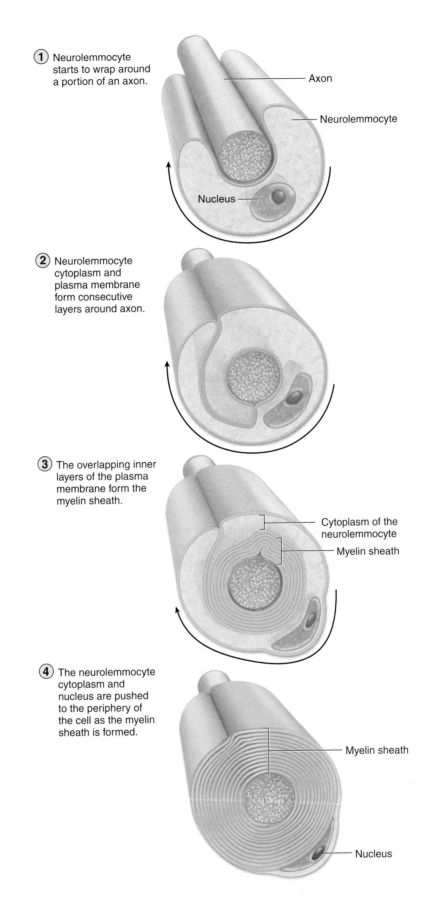

1. Neurolemmocyte starts to wrap around a portion of an axon.

Axon

Neurolemmocyte

Nucleus

2. Neurolemmocyte cytoplasm and plasma membrane form consecutive layers around axon.

3. The overlapping inner layers of the plasma membrane form the myelin sheath.

Cytoplasm of the neurolemmocyte

Myelin sheath

4. The neurolemmocyte cytoplasm and nucleus are pushed to the periphery of the cell as the myelin sheath is formed.

Myelin sheath

Nucleus

Myelination of PNS Axons
Figure 14.8

CNS

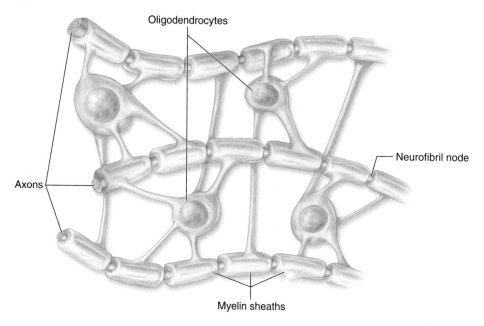

Oligodendrocytes

Neurofibril node

Axons

Myelin sheaths

PNS

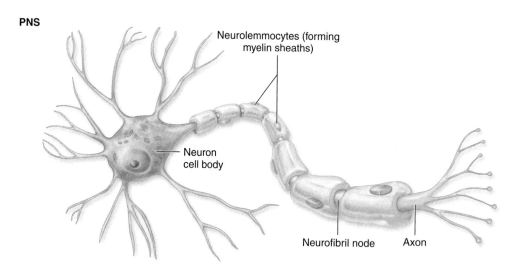

Neurolemmocytes (forming myelin sheaths)

Neuron cell body

Neurofibril node

Axon

Myelination Sheaths in the CNS and PNS
Figure 14.9

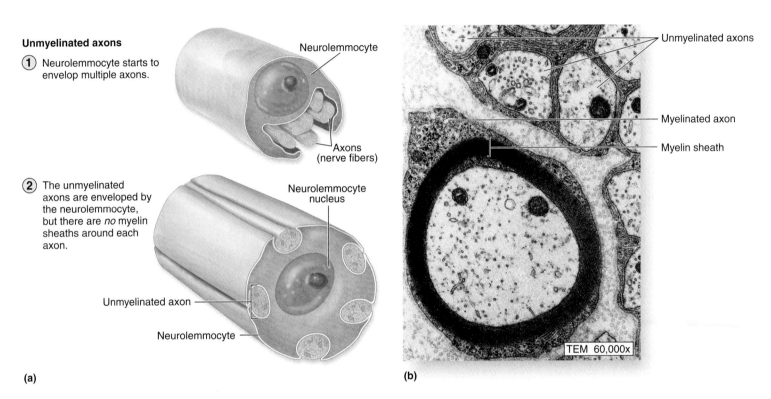

Unmyelinated axons

(1) Neurolemmocyte starts to envelop multiple axons.

Neurolemmocyte

Axons (nerve fibers)

(2) The unmyelinated axons are enveloped by the neurolemmocyte, but there are *no* myelin sheaths around each axon.

Neurolemmocyte nucleus

Unmyelinated axon

Neurolemmocyte

(a)

Unmyelinated axons

Myelinated axon

Myelin sheath

TEM 60,000x

(b)

Comparison of Unmyelinated and Myelinated Axons
Figure 14.10
Figure 14.10: © Dr. D. W. Fawcett /Visuals Unlimited

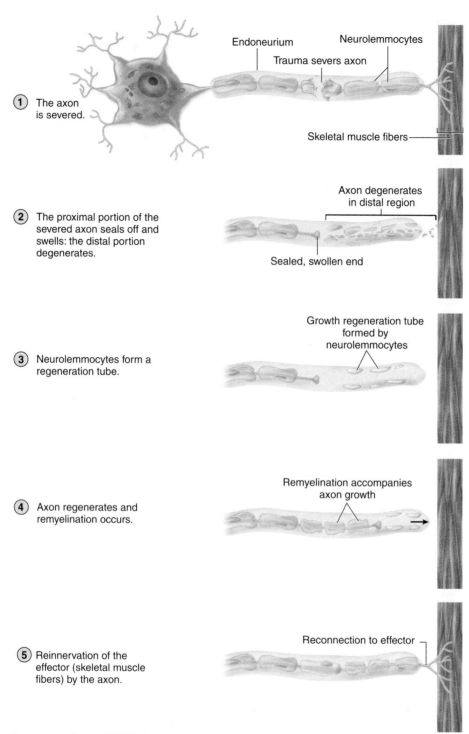

1 The axon is severed.

Endoneurium
Trauma severs axon
Neurolemmocytes

Skeletal muscle fibers

2 The proximal portion of the severed axon seals off and swells: the distal portion degenerates.

Axon degenerates in distal region

Sealed, swollen end

3 Neurolemmocytes form a regeneration tube.

Growth regeneration tube formed by neurolemmocytes

4 Axon regenerates and remyelination occurs.

Remyelination accompanies axon growth

5 Reinnervation of the effector (skeletal muscle fibers) by the axon.

Reconnection to effector

Regeneration of PNS Axons
Figure 14.11

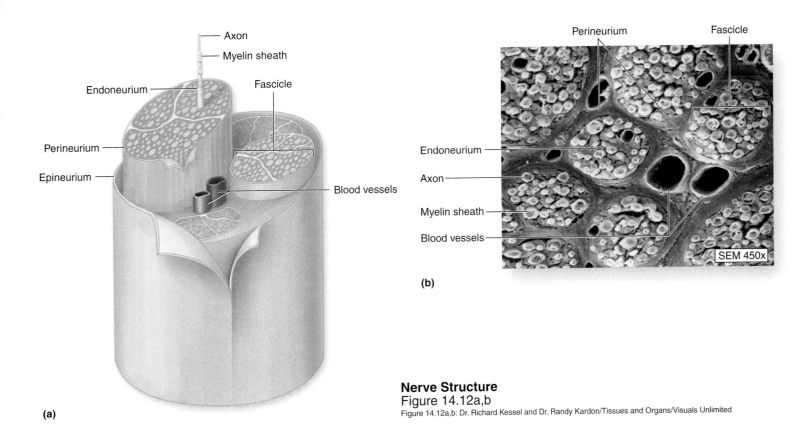

(a)

(b)

Nerve Structure
Figure 14.12a,b
Figure 14.12a,b: Dr. Richard Kessel and Dr. Randy Kardon/Tissues and Organs/Visuals Unlimited

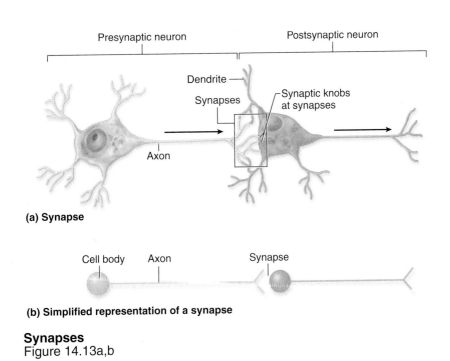

(a) Synapse

(b) Simplified representation of a synapse

Synapses
Figure 14.13a,b

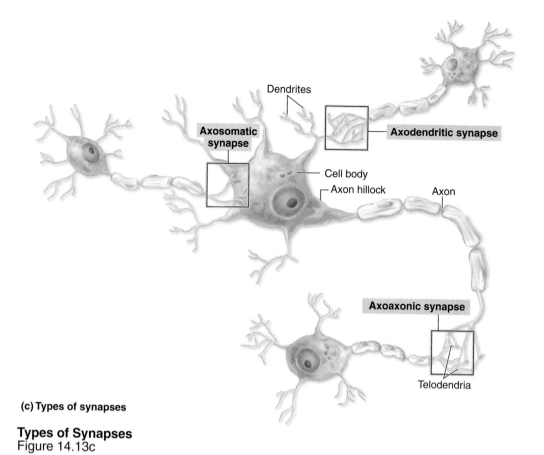

Dendrites

Axosomatic synapse

Axodendritic synapse

Cell body

Axon hillock

Axon

Axoaxonic synapse

Telodendria

(c) Types of synapses

Types of Synapses
Figure 14.13c

Electrical synapse

Smooth muscle cells

Presynaptic cell

Postsynaptic cell

Gap junction

Local current

Positively charged ions

Connexons

Plasma membrane

Inner surface of plasma membrane

(a) Electrical synapse

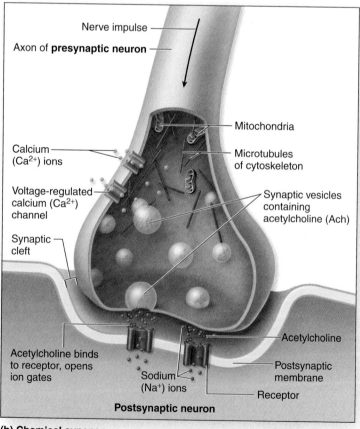

Nerve impulse

Axon of **presynaptic neuron**

Mitochondria

Calcium (Ca^{2+}) ions

Microtubules of cytoskeleton

Voltage-regulated calcium (Ca^{2+}) channel

Synaptic vesicles containing acetylcholine (Ach)

Synaptic cleft

Acetylcholine binds to receptor, opens ion gates

Sodium (Na$^+$) ions

Acetylcholine

Postsynaptic membrane

Receptor

Postsynaptic neuron

(b) Chemical synapse

Electrical and Chemical Synapses
Figure 14.14

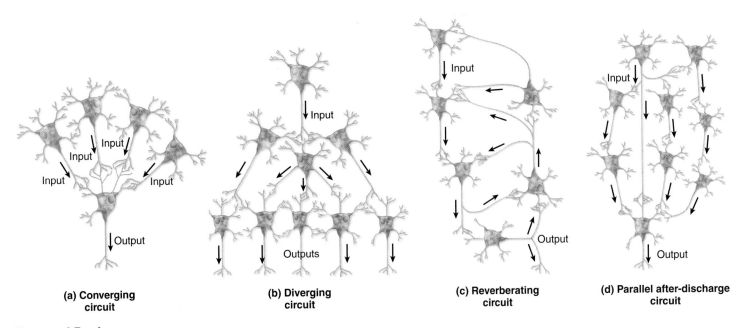

(a) **Converging circuit**

(b) **Diverging circuit**

(c) **Reverberating circuit**

(d) **Parallel after-discharge circuit**

Input

Input

Input

Input

Input

Input

Input

Output

Outputs

Output

Output

Neuronal Pools
Figure 14.15

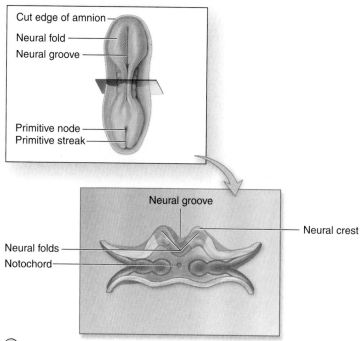

Cut edge of amnion
Neural fold
Neural groove

Primitive node
Primitive streak

Neural groove
Neural crest
Neural folds
Notochord

(1) Neural folds and neural groove form from the neural plate.

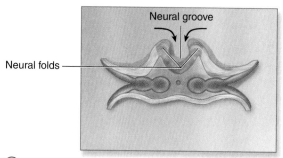

Neural groove
Neural folds

(2) Neural folds elevate and approach one another.

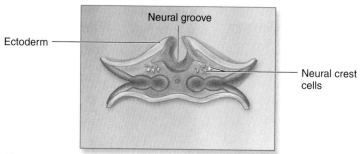

Neural groove
Ectoderm
Neural crest cells

(3) Neural crest cells begin to "pinch off" of the neural folds and form other structures.

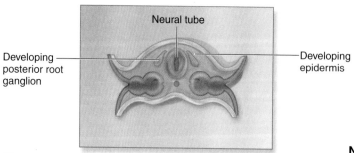

Neural tube
Developing posterior root ganglion
Developing epidermis

(4) Neural folds fuse to form the neural tube.

Nervous System Development
Figure 14.16

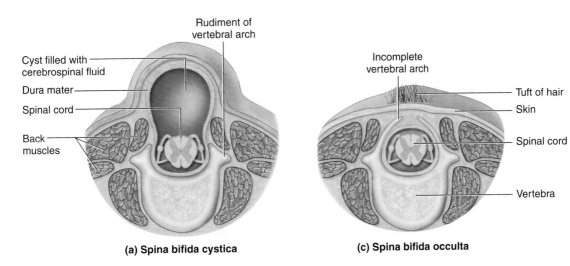

Rudiment of
vertebral arch

Cyst filled with
cerebrospinal fluid

Dura mater

Spinal cord

Back
muscles

(a) Spina bifida cystica

Incomplete
vertebral arch

Tuft of hair

Skin

Spinal cord

Vertebra

(c) Spina bifida occulta

Spina Bifida
Clinical View p. 437

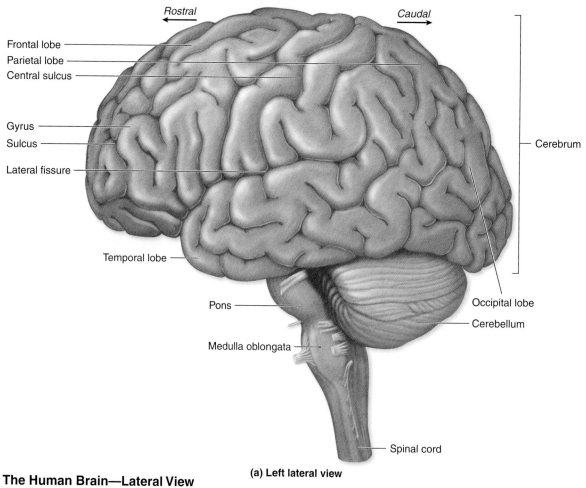

Rostral ←   Caudal →

Frontal lobe

Parietal lobe

Central sulcus

Gyrus

Sulcus

Lateral fissure

Temporal lobe

Pons

Medulla oblongata

Cerebrum

Occipital lobe

Cerebellum

Spinal cord

(a) Left lateral view

The Human Brain—Lateral View
Figure 15.1a

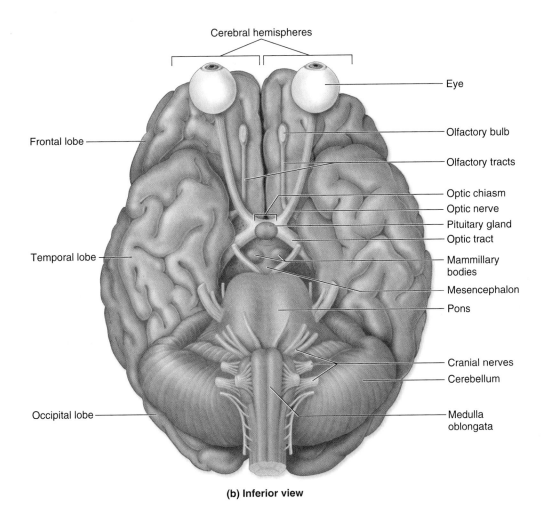

Cerebral hemispheres

Eye

Olfactory bulb

Olfactory tracts

Frontal lobe

Optic chiasm
Optic nerve
Pituitary gland
Optic tract

Temporal lobe

Mammillary bodies

Mesencephalon

Pons

Cranial nerves
Cerebellum

Occipital lobe

Medulla oblongata

(b) Inferior view

The Human Brain—Inferior View
Figure 15.1b

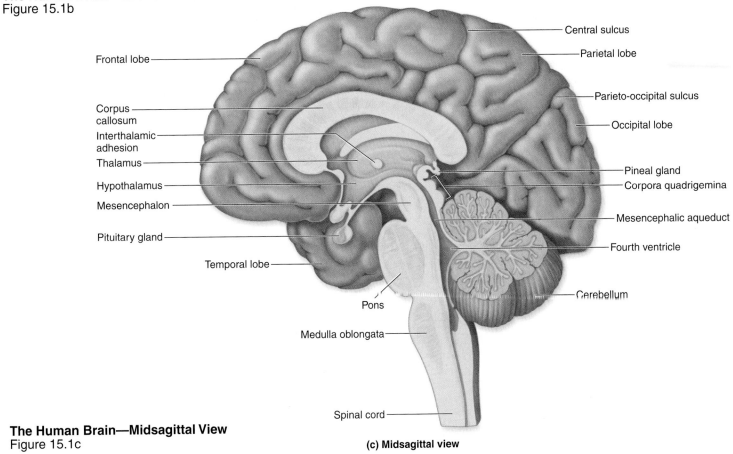

Central sulcus

Parietal lobe

Frontal lobe

Parieto-occipital sulcus

Corpus callosum

Occipital lobe

Interthalamic adhesion

Thalamus

Hypothalamus

Pineal gland
Corpora quadrigemina

Mesencephalon

Mesencephalic aqueduct

Pituitary gland

Fourth ventricle

Temporal lobe

Cerebellum

Pons

Medulla oblongata

Spinal cord

The Human Brain—Midsagittal View
Figure 15.1c

(c) Midsagittal view

209

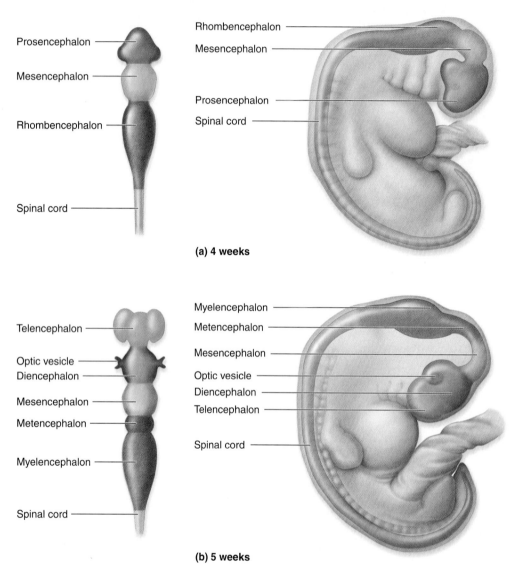

Prosencephalon

Mesencephalon

Rhombencephalon

Spinal cord

Rhombencephalon

Mesencephalon

Prosencephalon

Spinal cord

(a) 4 weeks

Telencephalon

Optic vesicle

Diencephalon

Mesencephalon

Metencephalon

Myelencephalon

Spinal cord

Myelencephalon

Metencephalon

Mesencephalon

Optic vesicle

Diencephalon

Telencephalon

Spinal cord

(b) 5 weeks

Structural Changes in the Developing Brain—4 Weeks, 5 Weeks
Figure 15.2a,b

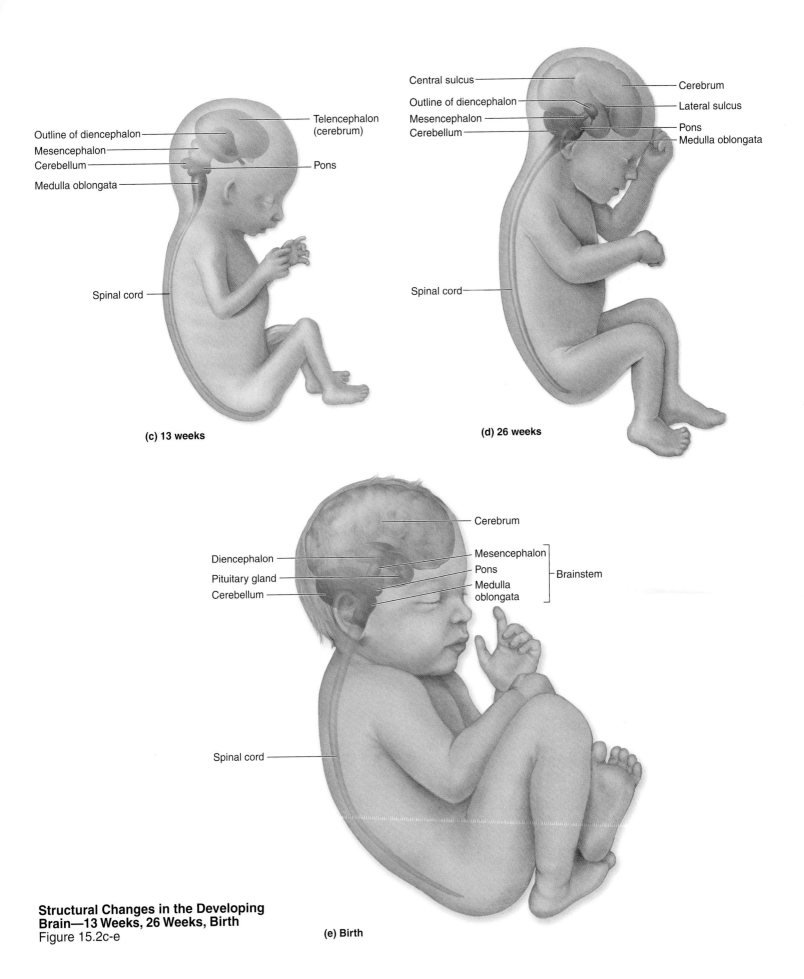

(c) 13 weeks

Outline of diencephalon
Mesencephalon
Cerebellum
Medulla oblongata
Telencephalon (cerebrum)
Pons
Spinal cord

(d) 26 weeks

Central sulcus
Outline of diencephalon
Mesencephalon
Cerebellum
Cerebrum
Lateral sulcus
Pons
Medulla oblongata
Spinal cord

(e) Birth

Diencephalon
Pituitary gland
Cerebellum
Cerebrum
Mesencephalon
Pons
Medulla oblongata
Brainstem
Spinal cord

Structural Changes in the Developing Brain—13 Weeks, 26 Weeks, Birth
Figure 15.2c-e

Table 15.1	Major Brain Structures: Embryonic Through Adult			
EMBRYONIC DEVELOPMENT				**ADULT STRUCTURE**
Neural Tube	Primary Brain Vesicles	Secondary Brain Vesicles (future adult brain regions)[1]	Neural Canal Derivative[2]	Structures Within Brain Region
		Telencephalon	Lateral ventricles	Cerebrum
	Prosencephalor (forebrain)	Diencephalon	Third ventricle	Epithalamus Thalamus Hypothalamus
	Mesencephalon (midbrain)	Mesencephalon	Mesencephalic (cerebral) aqueduct	Cerebral peduncles Superior colliculi Inferior colliculi
		Metencephalon	Rostral part of fourth ventricle	Pons Cerebellum
	Rhombencephalon (hindbrain)	Myelencephalon	Caudal part of fourth ventricle; central canal caudally	Medulla oblongata

Rostral

Caudal   Neural canal

Neural canal

Major Brain Structures: Embryonic Through Adult
Table 15.1

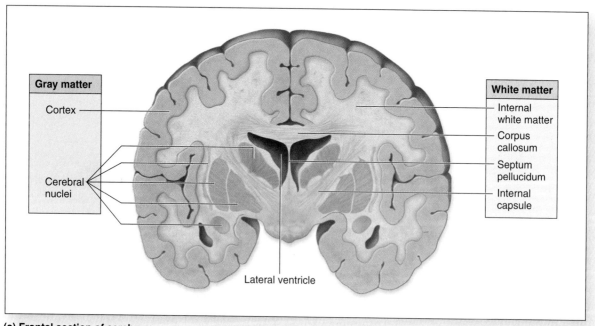

Gray matter
- Cortex
- Cerebral nuclei

White matter
- Internal white matter
- Corpus callosum
- Septum pellucidum
- Internal capsule

Lateral ventricle

(a) Frontal section of cerebrum

Gray and White Matter in the CNS
Figure 15.3a

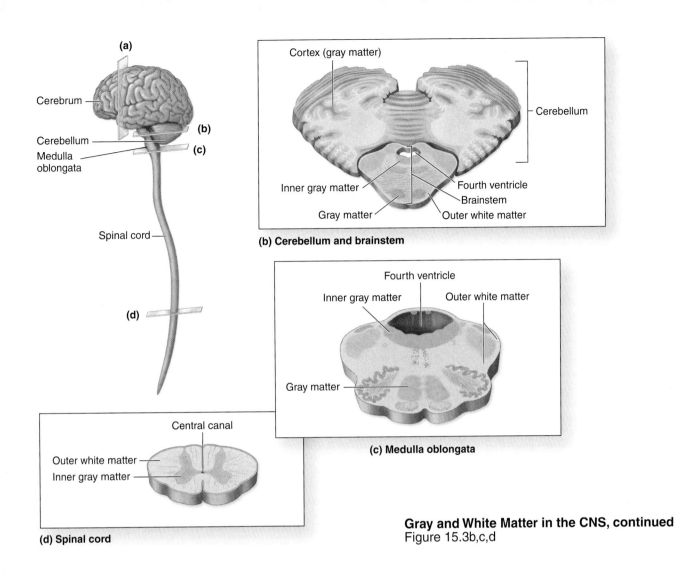

(a)

Cerebrum

Cerebellum

Medulla
oblongata

Spinal cord

(b)

(c)

(d)

(b) Cerebellum and brainstem

Cortex (gray matter)

Cerebellum

Inner gray matter

Gray matter

Fourth ventricle

Brainstem

Outer white matter

(c) Medulla oblongata

Fourth ventricle

Inner gray matter

Outer white matter

Gray matter

Central canal

Outer white matter

Inner gray matter

(d) Spinal cord

Gray and White Matter in the CNS, continued
Figure 15.3b,c,d

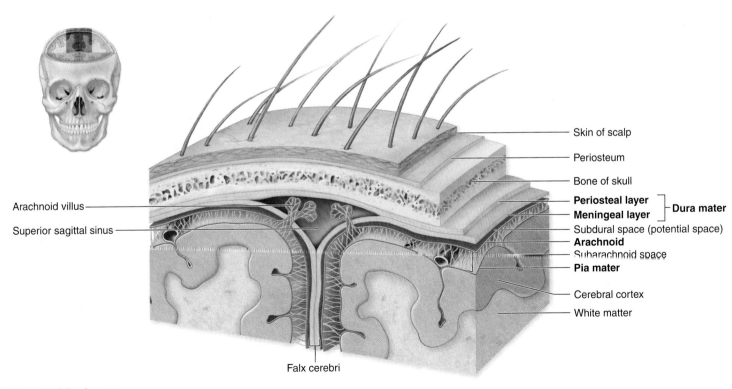

Skin of scalp

Periosteum

Bone of skull

Periosteal layer

Meningeal layer

┐ Dura mater

Subdural space (potential space)

Arachnoid

Subarachnoid space

Pia mater

Cerebral cortex

White matter

Arachnoid villus

Superior sagittal sinus

Falx cerebri

Cranial Meninges
Figure 15.4

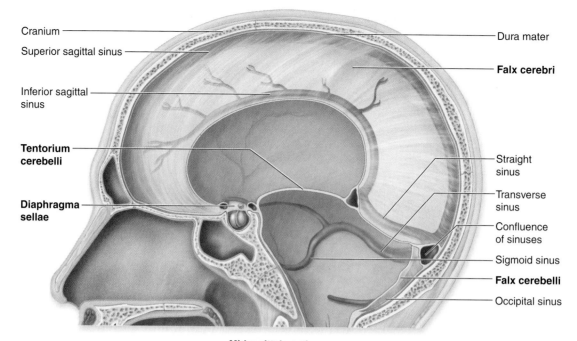

Midsagittal section

Cranial Dural Septa
Figure 15.5

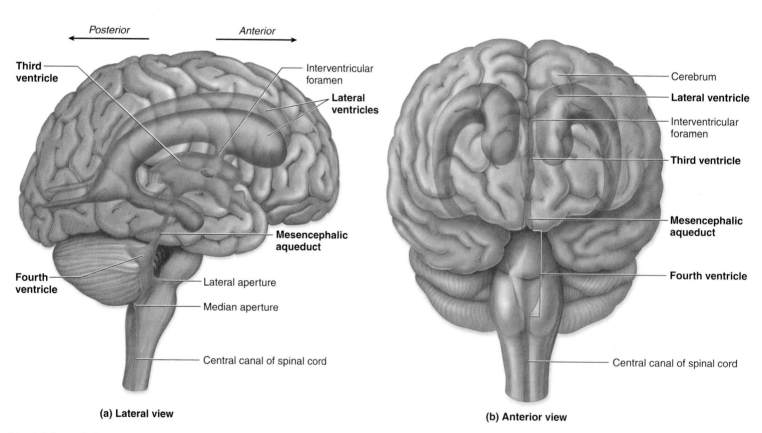

(a) Lateral view

(b) Anterior view

Ventricles of the Brain
Figure 15.6

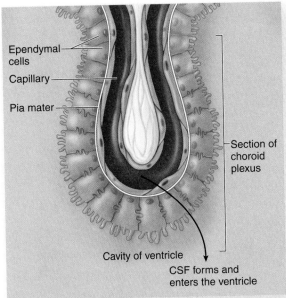

Ependymal cells

Capillary

Pia mater

Section of choroid plexus

Cavity of ventricle

CSF forms and enters the ventricle

(a)

Choroid Plexus
Figure 15.7a

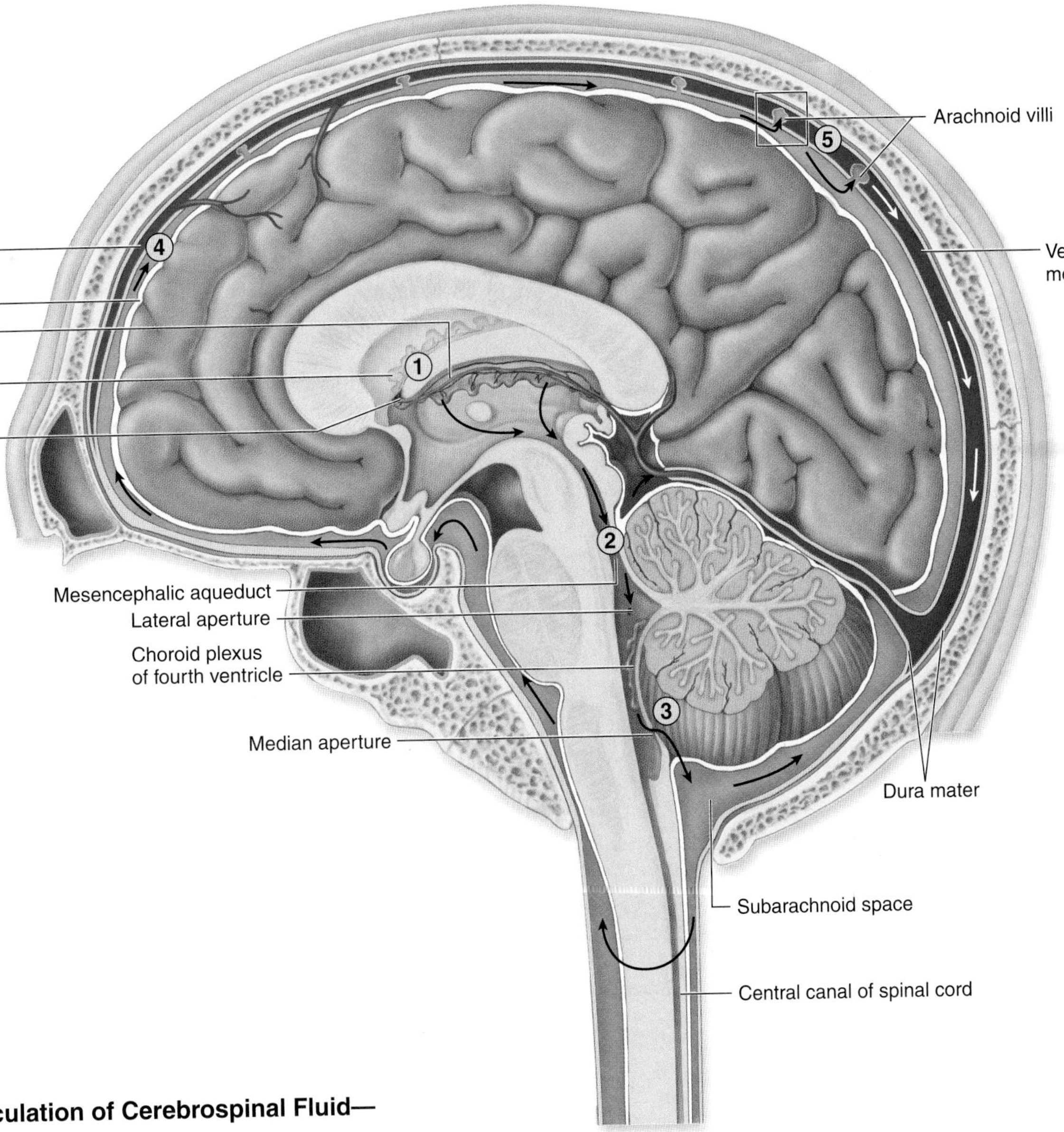

Superior sagittal sinus (dural venous sinus)

Pia mater

Choroid plexus of third ventricle

Choroid plexus of lateral ventricle

Interventricular foramen

Mesencephalic aqueduct

Lateral aperture

Choroid plexus of fourth ventricle

Median aperture

Arachnoid villi

Venous fluid movement

Dura mater

Subarachnoid space

Central canal of spinal cord

Production and Circulation of Cerebrospinal Fluid—Midsagittal Section
Figure 15.8a

(a) Midsagittal section

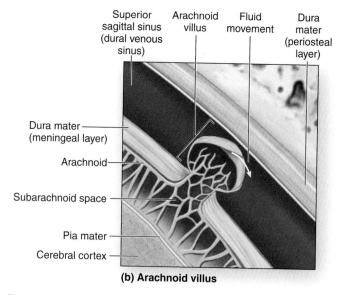

Superior sagittal sinus (dural venous sinus)

Arachnoid villus

Fluid movement

Dura mater (periosteal layer)

Dura mater (meningeal layer)

Arachnoid

Subarachnoid space

Pia mater

Cerebral cortex

(b) Arachnoid villus

Production and Circulation of Cerebrospinal Fluid— Arachnoid Villus
Figure 15.8b

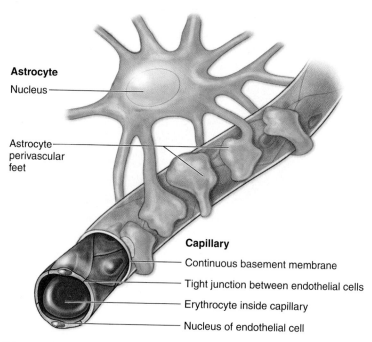

Astrocyte

Nucleus

Astrocyte perivascular feet

Capillary

Continuous basement membrane

Tight junction between endothelial cells

Erythrocyte inside capillary

Nucleus of endothelial cell

Blood-Brain Barrier
Figure 15.9

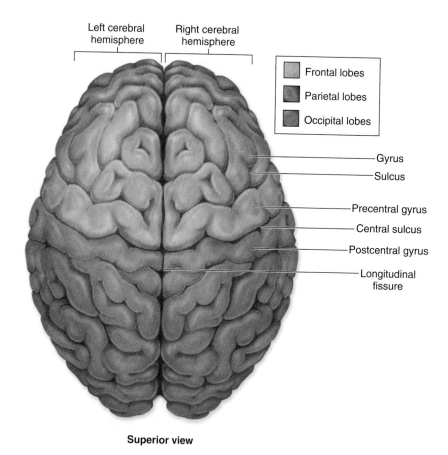

Left cerebral hemisphere

Right cerebral hemisphere

Frontal lobes

Parietal lobes

Occipital lobes

Gyrus

Sulcus

Precentral gyrus

Central sulcus

Postcentral gyrus

Longitudinal fissure

Superior view

Cerebral Hemispheres
Figure 15.10

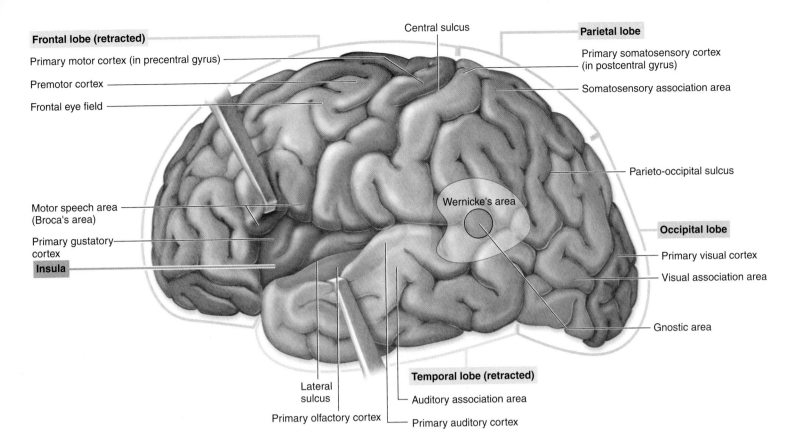

Central sulcus

Frontal lobe (retracted)

Primary motor cortex (in precentral gyrus)

Premotor cortex

Frontal eye field

Motor speech area (Broca's area)

Primary gustatory cortex

Insula

Parietal lobe

Primary somatosensory cortex (in postcentral gyrus)

Somatosensory association area

Parieto-occipital sulcus

Wernicke's area

Occipital lobe

Primary visual cortex

Visual association area

Gnostic area

Temporal lobe (retracted)

Auditory association area

Primary auditory cortex

Lateral sulcus

Primary olfactory cortex

Cerebral Lobes
Figure 15.11

Left lateral view

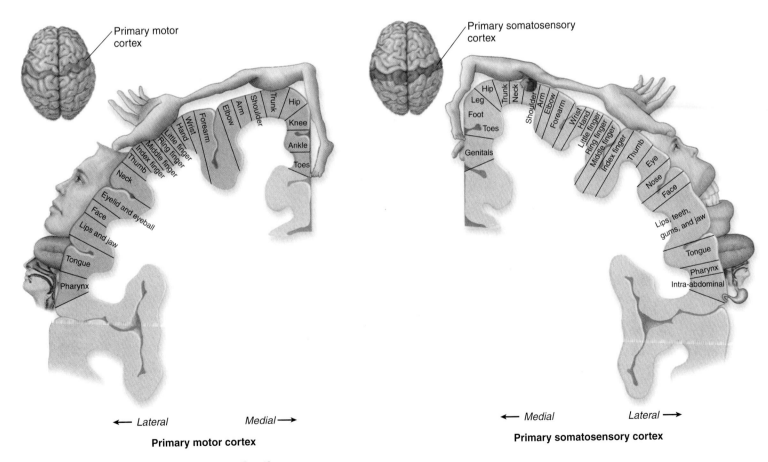

Primary motor cortex

Trunk
Hip
Knee
Ankle
Toes
Shoulder
Elbow
Arm
Forearm
Wrist
Hand
Little finger
Ring finger
Middle finger
Index finger
Thumb
Neck
Eyelid and eyeball
Face
Lips and jaw
Tongue
Pharynx

← *Lateral*   *Medial* →

Primary motor cortex

Primary somatosensory cortex

Hip
Leg
Foot
Toes
Genitals
Trunk
Neck
Shoulder
Arm
Elbow
Forearm
Wrist
Hand
Little finger
Ring finger
Middle finger
Index finger
Thumb
Eye
Nose
Face
Lips, teeth, gums, and jaw
Tongue
Pharynx
Intra-abdominal

← *Medial*   *Lateral* →

Primary somatosensory cortex

Primary Motor and Somatosensory Cortices
Figure 15.12

217

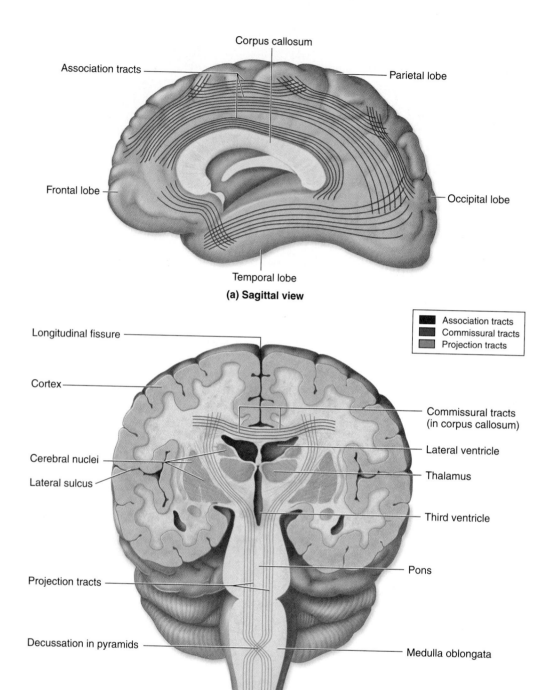

(a) Sagittal view

Corpus callosum

Association tracts

Parietal lobe

Frontal lobe

Occipital lobe

Temporal lobe

Longitudinal fissure

Cortex

Cerebral nuclei

Lateral sulcus

Projection tracts

Decussation in pyramids

	Association tracts
	Commissural tracts
	Projection tracts

Commissural tracts
(in corpus callosum)

Lateral ventricle

Thalamus

Third ventricle

Pons

Medulla oblongata

(b) Coronal section

Cerebral White Matter
Figure 15.13

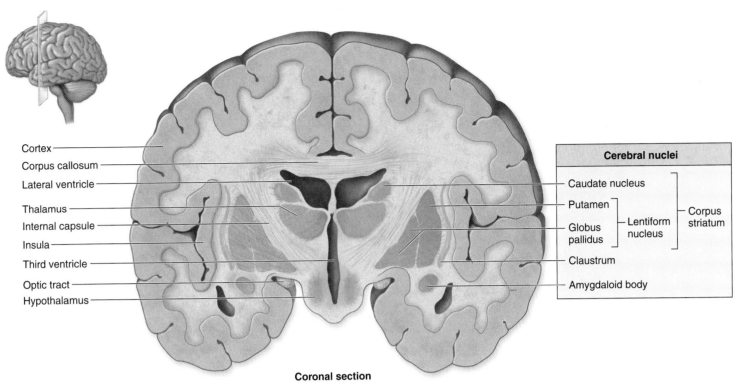

Cortex

Corpus callosum

Lateral ventricle

Thalamus

Internal capsule

Insula

Third ventricle

Optic tract

Hypothalamus

Cerebral nuclei

Caudate nucleus

Putamen

Globus pallidus — Lentiform nucleus — Corpus striatum

Claustrum

Amygdaloid body

Coronal section

Cerebral Nuclei
Figure 15.14

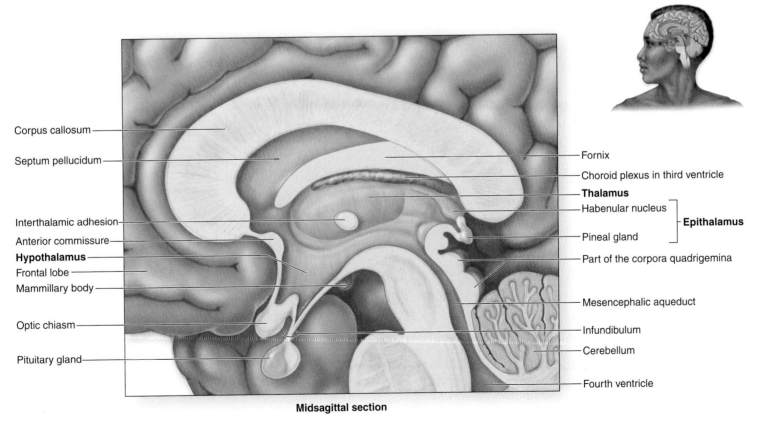

Corpus callosum

Septum pellucidum

Interthalamic adhesion

Anterior commissure

Hypothalamus

Frontal lobe

Mammillary body

Optic chiasm

Pituitary gland

Fornix

Choroid plexus in third ventricle

Thalamus

Habenular nucleus — **Epithalamus**

Pineal gland

Part of the corpora quadrigemina

Mesencephalic aqueduct

Infundibulum

Cerebellum

Fourth ventricle

Midsagittal section

Diencephalon
Figure 15.15

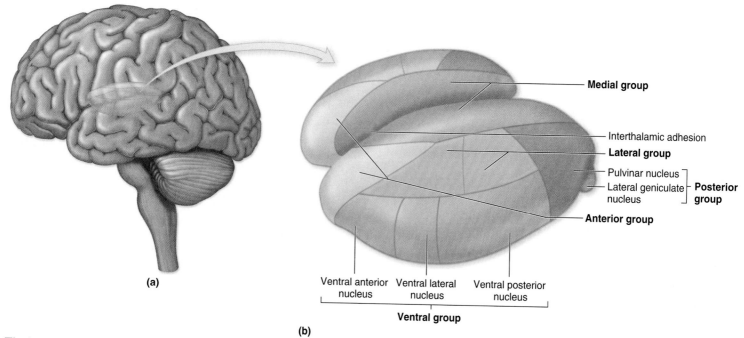

Medial group

Interthalamic adhesion

Lateral group

Pulvinar nucleus ⎤
Lateral geniculate ⎬ **Posterior group**
nucleus ⎦

Anterior group

(a)

Ventral anterior nucleus   Ventral lateral nucleus   Ventral posterior nucleus

Ventral group

(b)

Thalamus
Figure 15.16

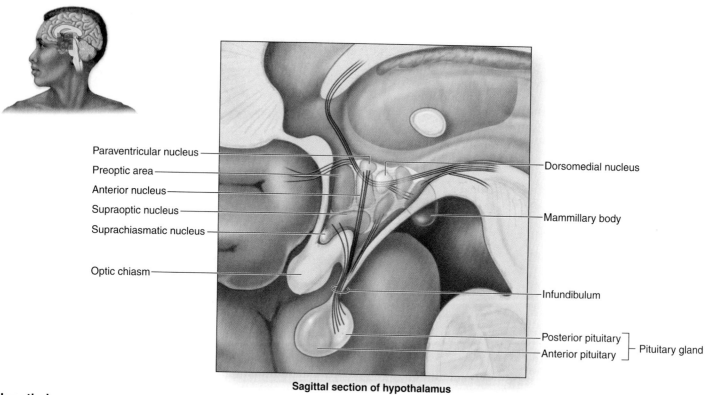

Paraventricular nucleus

Preoptic area

Anterior nucleus

Supraoptic nucleus

Suprachiasmatic nucleus

Optic chiasm

Dorsomedial nucleus

Mammillary body

Infundibulum

Posterior pituitary ⎤
⎬ Pituitary gland
Anterior pituitary ⎦

Sagittal section of hypothalamus

Hypothalamus
Figure 15.17

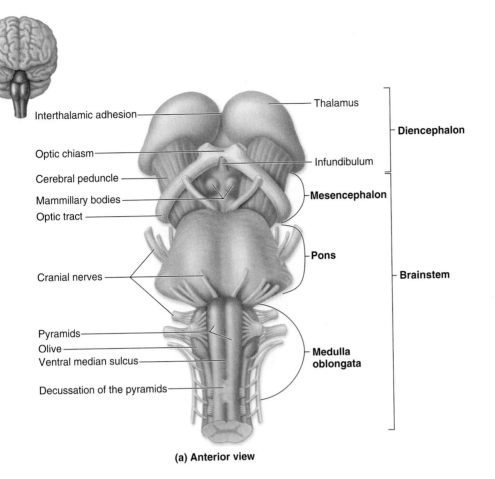

Interthalamic adhesion

Optic chiasm

Cerebral peduncle

Mammillary bodies

Optic tract

Cranial nerves

Pyramids

Olive

Ventral median sulcus

Decussation of the pyramids

Thalamus

Infundibulum

Diencephalon

Mesencephalon

Pons

Brainstem

Medulla
oblongata

(a) Anterior view

Brainstem—Anterior View
Figure 15.18a

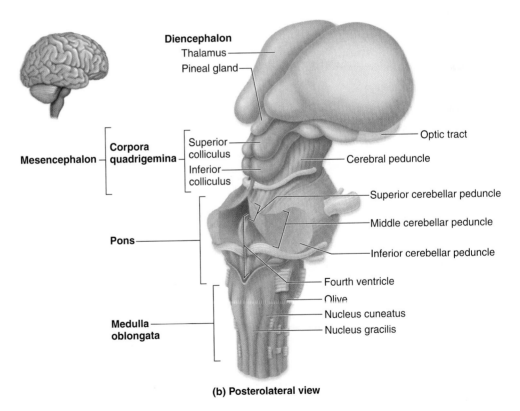

Diencephalon
Thalamus

Pineal gland

Mesencephalon

Corpora
quadrigemina

Superior
colliculus

Inferior
colliculus

Optic tract

Cerebral peduncle

Superior cerebellar peduncle

Middle cerebellar peduncle

Inferior cerebellar peduncle

Fourth ventricle

Olive

Nucleus cuneatus

Nucleus gracilis

Pons

Medulla
oblongata

(b) Posterolateral view

Brainstem—Posterolateral View
Figure 15.18b

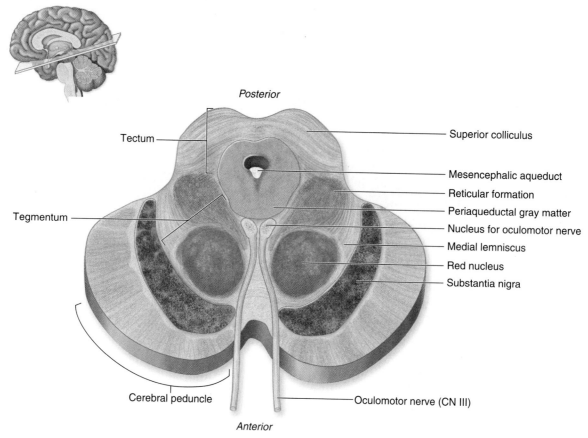

Posterior

Tectum

Superior colliculus

Mesencephalic aqueduct

Reticular formation

Periaqueductal gray matter

Tegmentum

Nucleus for oculomotor nerve

Medial lemniscus

Red nucleus

Substantia nigra

Cerebral peduncle

Oculomotor nerve (CN III)

Anterior

Cross-sectional view of mesencephalon

Mesencephalon
Figure 15.19

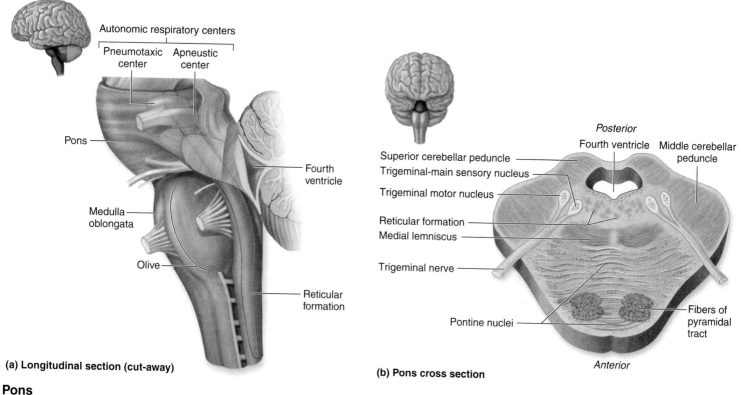

Autonomic respiratory centers

Pneumotaxic center   Apneustic center

Pons

Medulla oblongata

Olive

Fourth ventricle

Reticular formation

Posterior

Fourth ventricle   Middle cerebellar peduncle

Superior cerebellar peduncle

Trigeminal-main sensory nucleus

Trigeminal motor nucleus

Reticular formation

Medial lemniscus

Trigeminal nerve

Pontine nuclei

Fibers of pyramidal tract

Anterior

(a) Longitudinal section (cut-away)

(b) Pons cross section

Pons
Figure 15.20

222

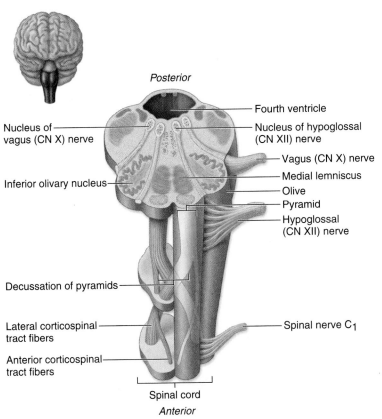

Posterior

Nucleus of vagus (CN X) nerve

Fourth ventricle

Nucleus of hypoglossal (CN XII) nerve

Inferior olivary nucleus

Vagus (CN X) nerve

Medial lemniscus

Olive

Pyramid

Hypoglossal (CN XII) nerve

Decussation of pyramids

Lateral corticospinal tract fibers

Spinal nerve C$_1$

Anterior corticospinal tract fibers

Spinal cord

Anterior

(a) Medulla oblongata, cross-sectional view

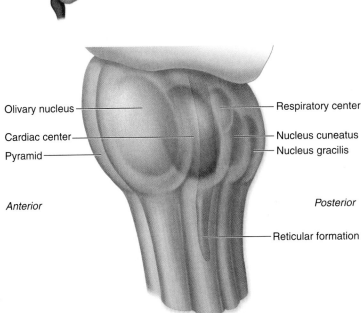

Olivary nucleus

Respiratory center

Cardiac center

Nucleus cuneatus

Pyramid

Nucleus gracilis

Anterior

Posterior

Reticular formation

(b) Medulla oblongata, lateral view

Medulla Oblongata
Figure 15.21

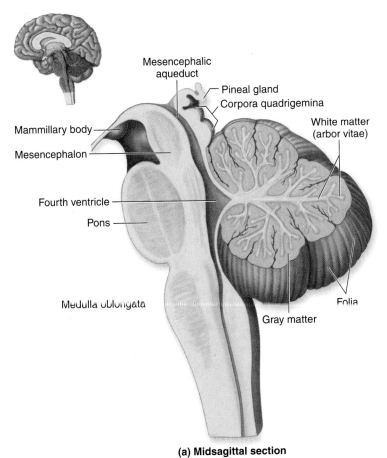

Mesencephalic aqueduct

Pineal gland

Corpora quadrigemina

Mammillary body

White matter (arbor vitae)

Mesencephalon

Fourth ventricle

Pons

Medulla oblongata

Folia

Gray matter

(a) Midsagittal section

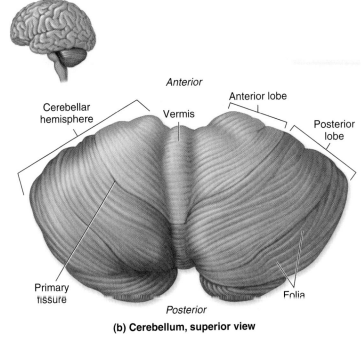

Anterior

Cerebellar hemisphere

Vermis

Anterior lobe

Posterior lobe

Primary fissure

Folia

Posterior

(b) Cerebellum, superior view

Cerebellum
Figure 15.22

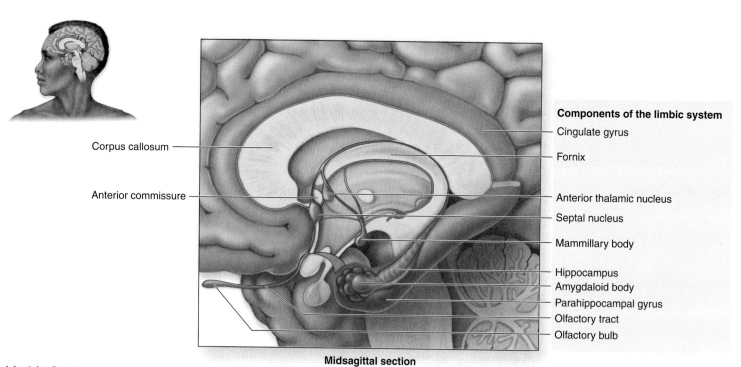

Corpus callosum —

Anterior commissure —

Components of the limbic system
— Cingulate gyrus

— Fornix

— Anterior thalamic nucleus
— Septal nucleus
— Mammillary body

— Hippocampus
— Amygdaloid body
— Parahippocampal gyrus
— Olfactory tract
— Olfactory bulb

Midsagittal section

Limbic System
Figure 15.23

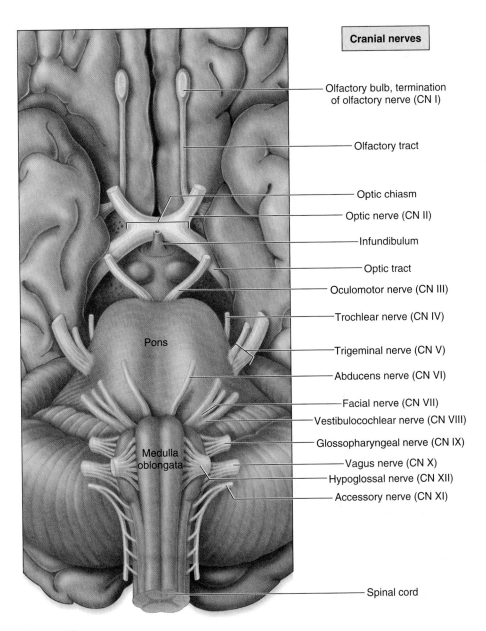

Cranial nerves

- Olfactory bulb, termination of olfactory nerve (CN I)
- Olfactory tract
- Optic chiasm
- Optic nerve (CN II)
- Infundibulum
- Optic tract
- Oculomotor nerve (CN III)
- Trochlear nerve (CN IV)
- Trigeminal nerve (CN V)
- Abducens nerve (CN VI)
- Facial nerve (CN VII)
- Vestibulocochlear nerve (CN VIII)
- Glossopharyngeal nerve (CN IX)
- Vagus nerve (CN X)
- Hypoglossal nerve (CN XII)
- Accessory nerve (CN XI)
- Spinal cord

Pons

Medulla oblongata

Cranial Nerves
Figure 15.24

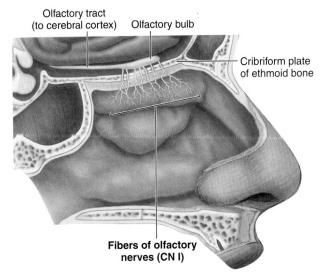

Olfactory tract (to cerebral cortex)   Olfactory bulb

Cribriform plate of ethmoid bone

Fibers of olfactory nerves (CN I)

Olfactory Nerve
Table 15.7 I

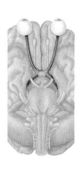

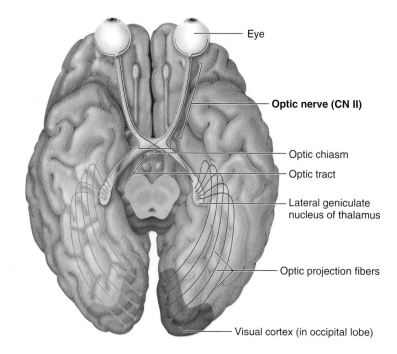

Eye

Optic nerve (CN II)

Optic chiasm

Optic tract

Lateral geniculate nucleus of thalamus

Optic projection fibers

Visual cortex (in occipital lobe)

Optic Nerve
Table 15.7 II

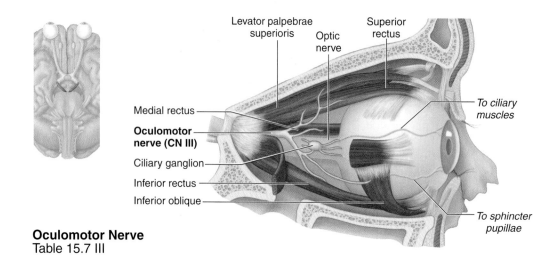

Levator palpebrae superioris

Optic nerve

Superior rectus

Medial rectus

Oculomotor nerve (CN III)

Ciliary ganglion

Inferior rectus

Inferior oblique

To ciliary muscles

To sphincter pupillae

Oculomotor Nerve
Table 15.7 III

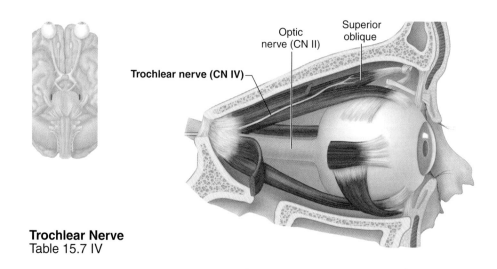

Optic nerve (CN II)

Superior oblique

Trochlear nerve (CN IV)

Trochlear Nerve
Table 15.7 IV

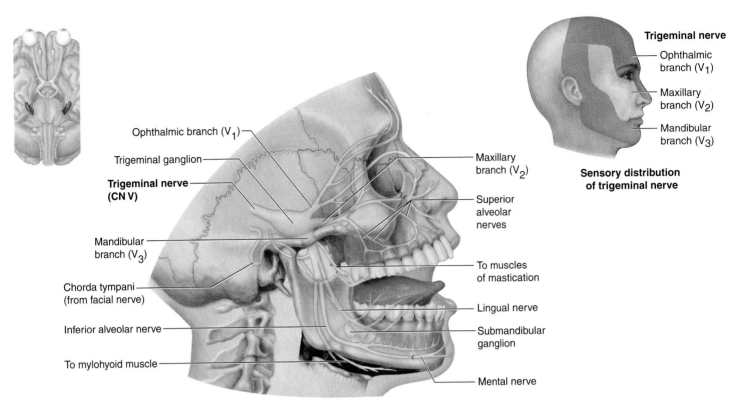

Trigeminal nerve

Ophthalmic branch (V₁) → V_1
Maxillary branch (V₂) → V_2
Mandibular branch (V₃) → V_3

Sensory distribution of trigeminal nerve

Ophthalmic branch (V₁)
Trigeminal ganglion
Trigeminal nerve (CN V)
Mandibular branch (V₃)
Chorda tympani (from facial nerve)
Inferior alveolar nerve
To mylohyoid muscle

Maxillary branch (V₂)
Superior alveolar nerves
To muscles of mastication
Lingual nerve
Submandibular ganglion
Mental nerve

Trigeminal Nerve
Table 15.7 V

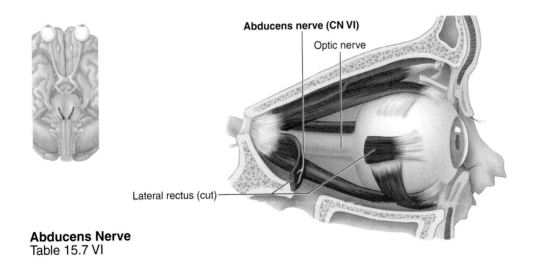

Abducens nerve (CN VI)
Optic nerve
Lateral rectus (cut)

Abducens Nerve
Table 15.7 VI

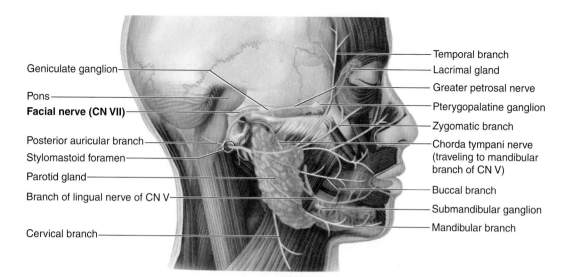

Facial Nerve
Table 15.7 VII

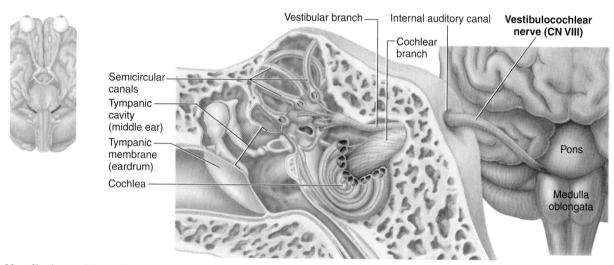

Vestibulocochlear Nerve
Table 15.7 VIII

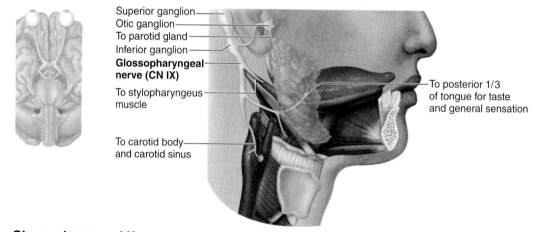

Glossopharyngeal Nerve
Table 15.7 IX

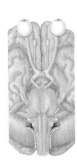

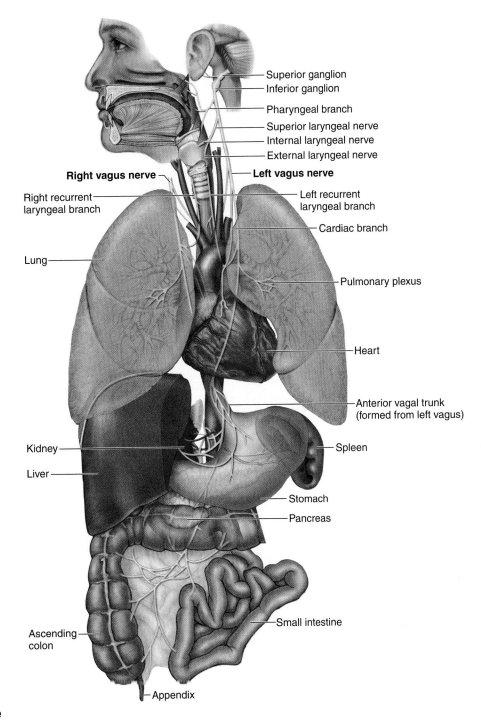

Superior ganglion
Inferior ganglion
Pharyngeal branch
Superior laryngeal nerve
Internal laryngeal nerve
External laryngeal nerve

Right vagus nerve

Left vagus nerve

Right recurrent
laryngeal branch

Left recurrent
laryngeal branch

Cardiac branch

Lung

Pulmonary plexus

Heart

Anterior vagal trunk
(formed from left vagus)

Kidney

Spleen

Liver

Stomach

Pancreas

Ascending
colon

Small intestine

Appendix

Vagus Nerve
Table 15.7 X

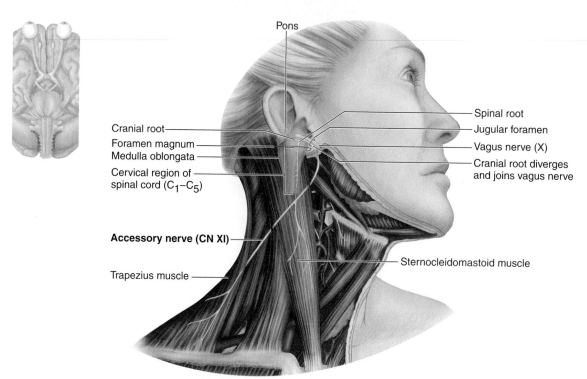

Pons

Cranial root

Foramen magnum

Medulla oblongata

Cervical region of
spinal cord (C$_1$–C$_5$)

Spinal root

Jugular foramen

Vagus nerve (X)

Cranial root diverges
and joins vagus nerve

Accessory nerve (CN XI)

Sternocleidomastoid muscle

Trapezius muscle

Accessory Nerve
Table 15.7 XI

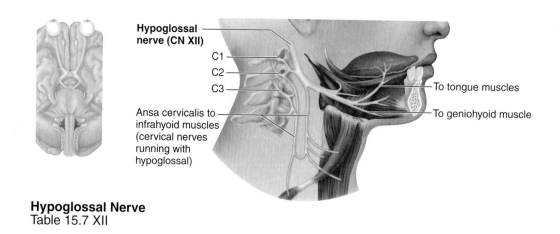

**Hypoglossal
nerve (CN XII)**

C1

C2

C3

Ansa cervicalis to
infrahyoid muscles
(cervical nerves
running with
hypoglossal)

To tongue muscles

To geniohyoid muscle

Hypoglossal Nerve
Table 15.7 XII

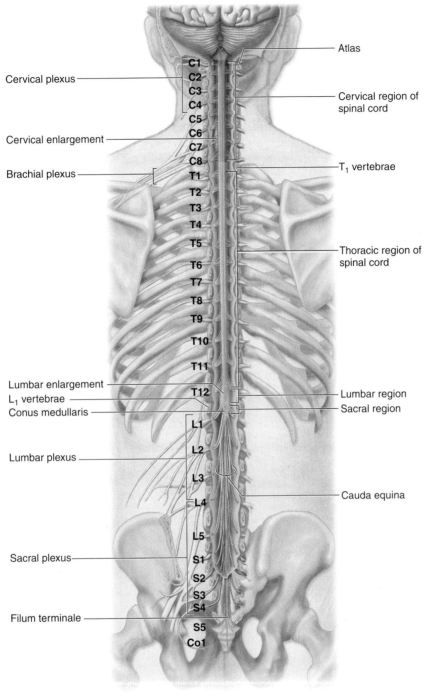

Cervical plexus

Cervical enlargement

Brachial plexus

Lumbar enlargement

L_1 vertebrae

Conus medullaris

Lumbar plexus

Sacral plexus

Filum terminale

Atlas

Cervical region of spinal cord

T_1 vertebrae

Thoracic region of spinal cord

Lumbar region

Sacral region

Cauda equina

C1
C2
C3
C4
C5
C6
C7
C8
T1
T2
T3
T4
T5
T6
T7
T8
T9
T10
T11
T12
L1
L2
L3
L4
L5
S1
S2
S3
S4
S5
Co1

(a) Posterior view

Gross Anatomy of the Spinal Cord
Figure 16.1a

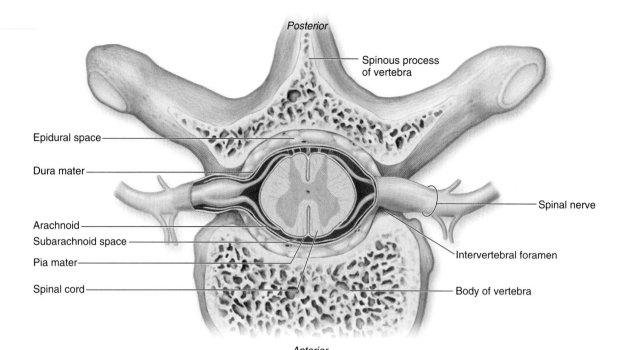

(a) Cross section of vertebra and spinal cord

Spinal Meninges and Structure of the Spinal Cord—Cross Section
Figure 16.2a

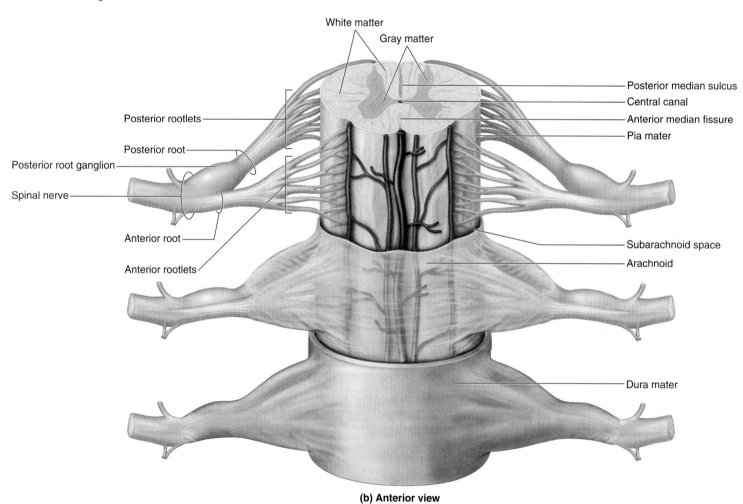

(b) Anterior view

Spinal Meninges and Structure of the Spinal Cord—Anterior View
Figure 16.2b

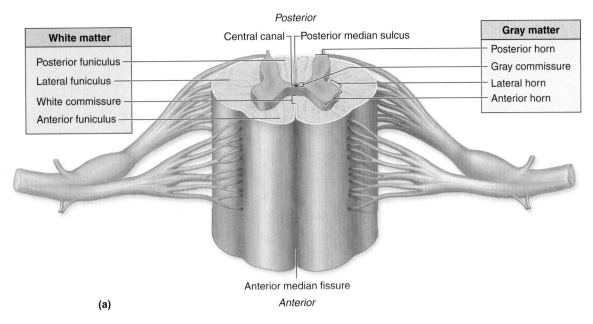

Gray Matter and White Matter Organization in the Spinal Cord
Figure 16.3a

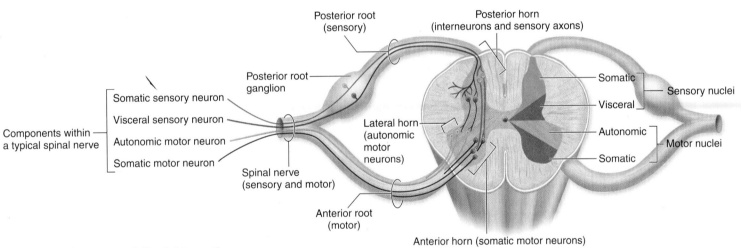

Neuron Pathways and Nuclei Locations
Figure 16.4

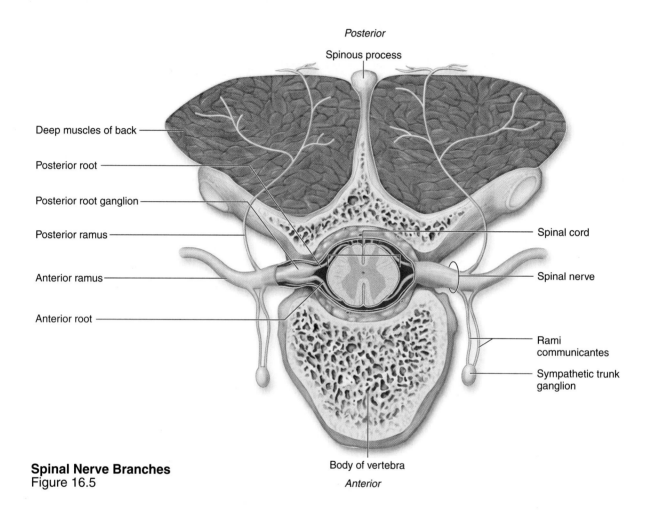

Posterior

Spinous process

Deep muscles of back

Posterior root

Posterior root ganglion

Posterior ramus

Anterior ramus

Anterior root

Spinal cord

Spinal nerve

Rami communicantes

Sympathetic trunk ganglion

Body of vertebra

Anterior

Spinal Nerve Branches
Figure 16.5

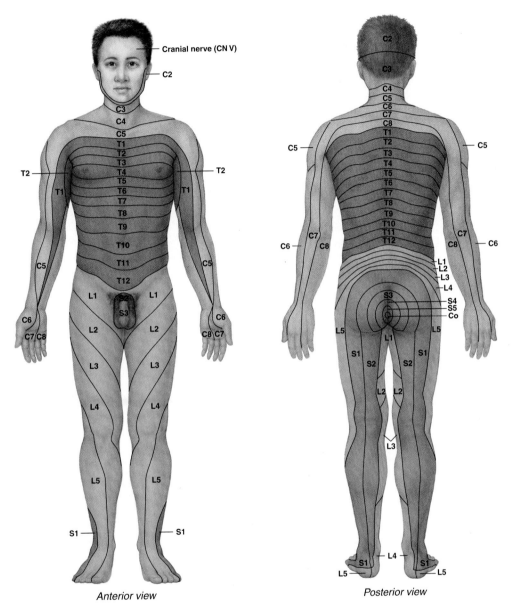

Anterior view

Posterior view

Dermatome Maps
Figure 16.6

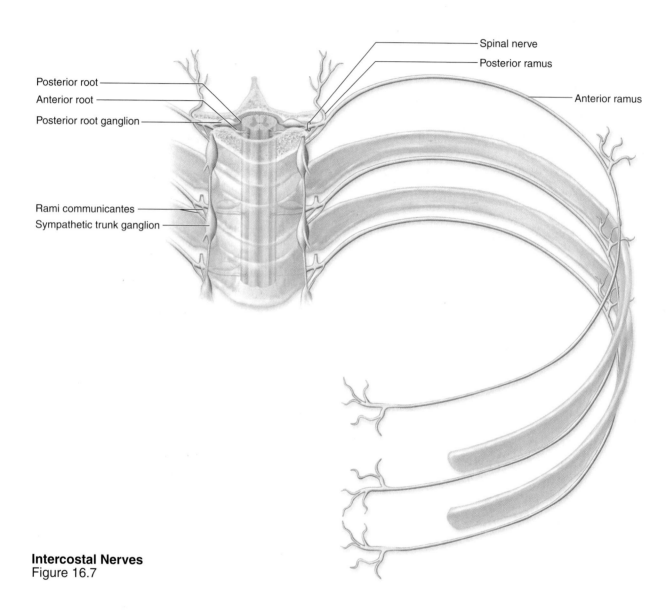

Posterior root

Anterior root

Posterior root ganglion

Rami communicantes

Sympathetic trunk ganglion

Spinal nerve

Posterior ramus

Anterior ramus

Intercostal Nerves
Figure 16.7

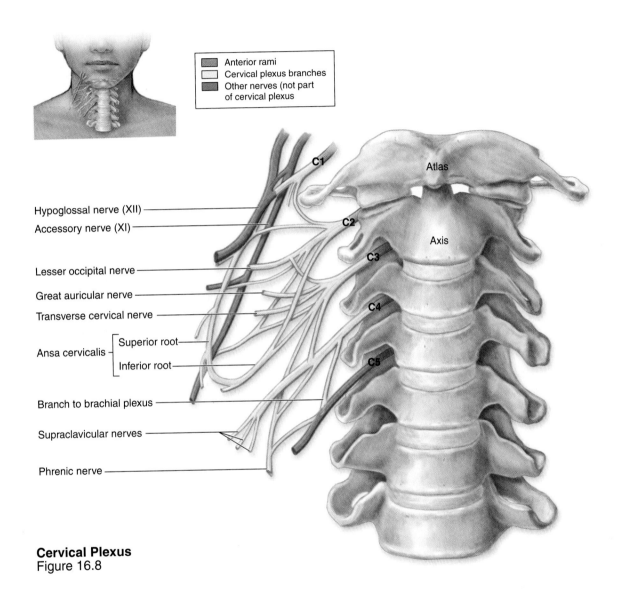

Anterior rami
Cervical plexus branches
Other nerves (not part of cervical plexus

C1
Atlas
Hypoglossal nerve (XII)
Accessory nerve (XI)
C2
Axis
C3
Lesser occipital nerve
Great auricular nerve
C4
Transverse cervical nerve
Superior root
Ansa cervicalis
C5
Inferior root
Branch to brachial plexus
Supraclavicular nerves
Phrenic nerve

Cervical Plexus
Figure 16.8

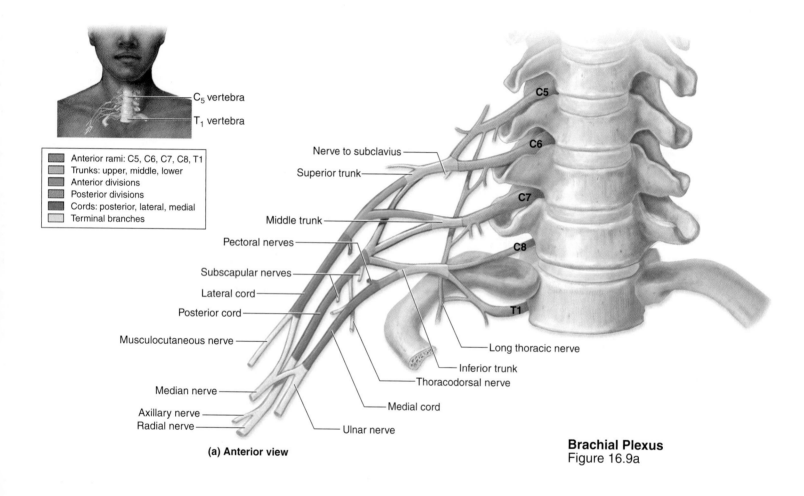

C_5 vertebra

T_1 vertebra

Anterior rami: C5, C6, C7, C8, T1
Trunks: upper, middle, lower
Anterior divisions
Posterior divisions
Cords: posterior, lateral, medial
Terminal branches

Nerve to subclavius

Superior trunk

C5

C6

C7

Middle trunk

C8

Pectoral nerves

Subscapular nerves

T1

Lateral cord

Posterior cord

Long thoracic nerve

Musculocutaneous nerve

Inferior trunk

Thoracodorsal nerve

Median nerve

Medial cord

Axillary nerve

Radial nerve

Ulnar nerve

(a) Anterior view

Brachial Plexus
Figure 16.9a

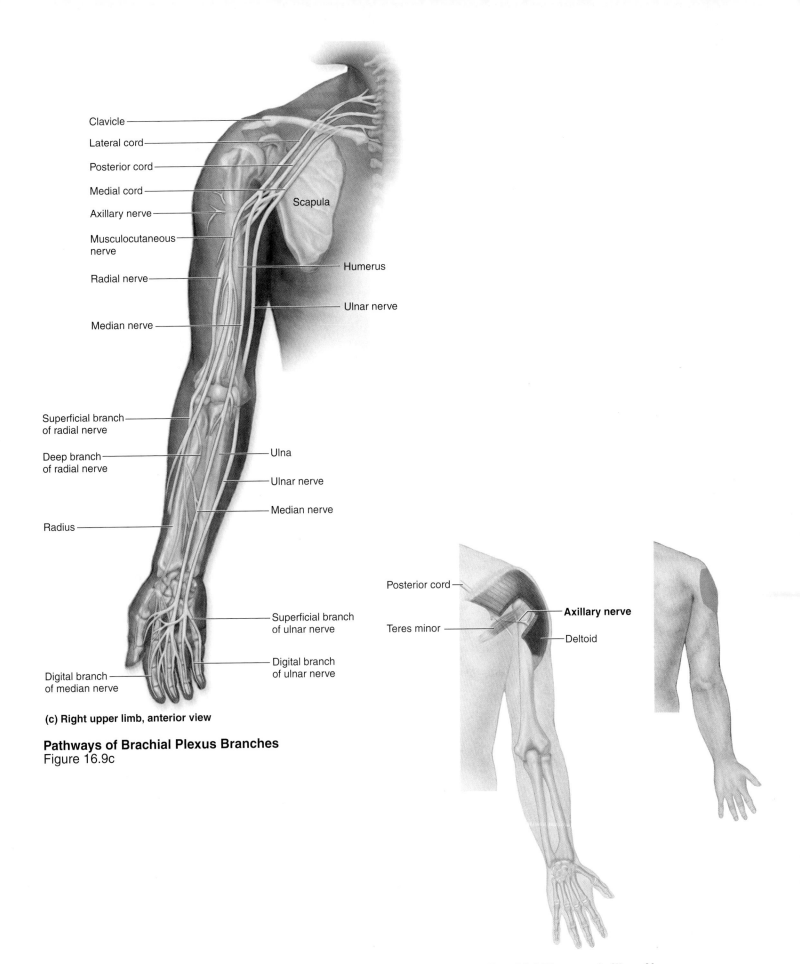

Clavicle

Lateral cord

Posterior cord

Medial cord

Axillary nerve

Musculocutaneous
nerve

Radial nerve

Median nerve

Superficial branch
of radial nerve

Deep branch
of radial nerve

Radius

Scapula

Humerus

Ulnar nerve

Ulna

Ulnar nerve

Median nerve

Superficial branch
of ulnar nerve

Digital branch
of ulnar nerve

Digital branch
of median nerve

(c) Right upper limb, anterior view

Pathways of Brachial Plexus Branches
Figure 16.9c

Posterior cord

Teres minor

Axillary nerve

Deltoid

Branches of the Brachial Plexus—Axillary Nerve
Table 16.3

239

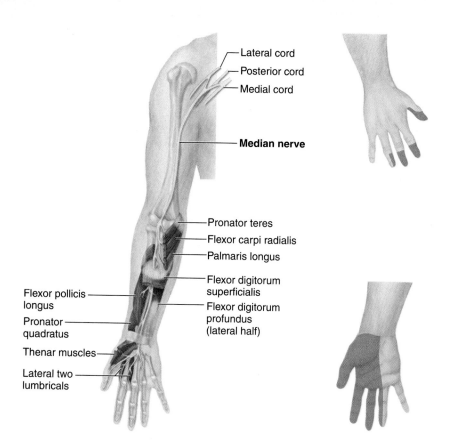

Lateral cord
Posterior cord
Medial cord

Median nerve

Pronator teres
Flexor carpi radialis
Palmaris longus
Flexor digitorum superficialis
Flexor digitorum profundus (lateral half)

Flexor pollicis longus
Pronator quadratus
Thenar muscles
Lateral two lumbricals

Branches of the Brachial Plexus—Median Nerve
Table 16.3

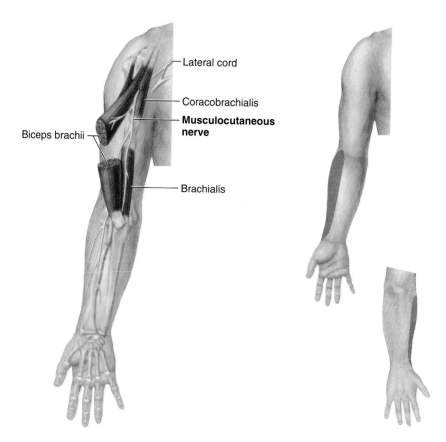

Lateral cord
Coracobrachialis
Musculocutaneous nerve
Biceps brachii
Brachialis

Branches of the Brachial Plexus—Musculocutaneous Nerve
Table 16.3

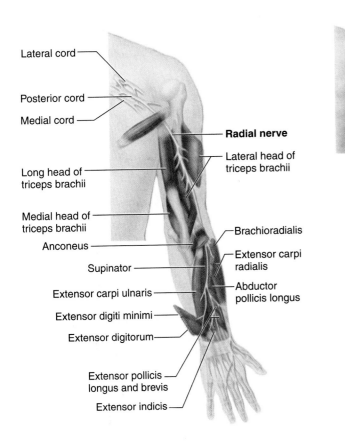

Lateral cord

Posterior cord

Medial cord

Radial nerve

Lateral head of
triceps brachii

Long head of
triceps brachii

Medial head of
triceps brachii

Anconeus

Supinator

Extensor carpi ulnaris

Extensor digiti minimi

Extensor digitorum

Extensor pollicis
longus and brevis

Extensor indicis

Brachioradialis

Extensor carpi
radialis

Abductor
pollicis longus

Branches of the Brachial Plexus—Radial Nerve
Table 16.3

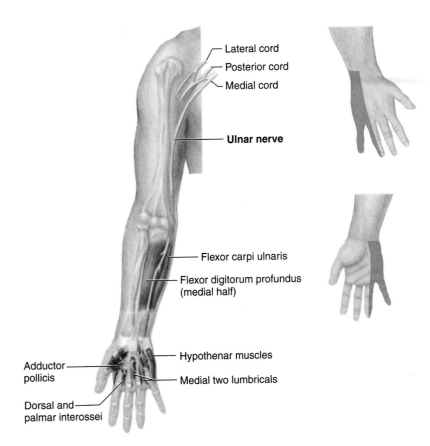

Lateral cord

Posterior cord

Medial cord

Ulnar nerve

Flexor carpi ulnaris

Flexor digitorum profundus
(medial half)

Hypothenar muscles

Adductor
pollicis

Medial two lumbricals

Dorsal and
palmar interossei

Branches of the Brachial Plexus—Ulnar Nerve
Table 16.3

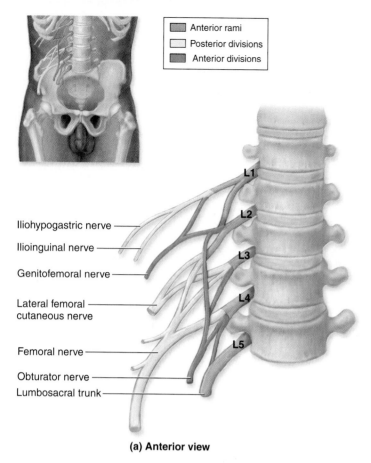

Anterior rami
Posterior divisions
Anterior divisions

L1

Iliohypogastric nerve

Ilioinguinal nerve

Genitofemoral nerve

L2

L3

Lateral femoral
cutaneous nerve

L4

Femoral nerve

L5

Obturator nerve

Lumbosacral trunk

(a) Anterior view

Lumbar Plexus
Figure 16.10a

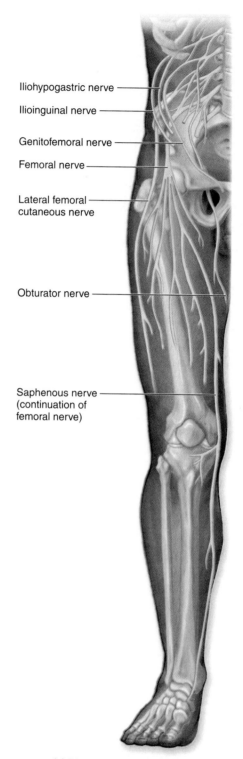

Iliohypogastric nerve

Ilioinguinal nerve

Genitofemoral nerve

Femoral nerve

Lateral femoral
cutaneous nerve

Obturator nerve

Saphenous nerve
(continuation of
femoral nerve)

(c) Right lower limb, anterior view

Pathways of Lumbar Plexus Nerves
Figure 16.10c

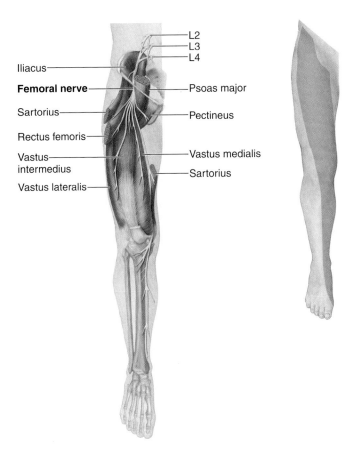

Iliacus

Femoral nerve

Sartorius

Rectus femoris

Vastus intermedius

Vastus lateralis

L2
L3
L4

Psoas major

Pectineus

Vastus medialis

Sartorius

Branches of the Lumbar Plexus—Femoral Nerve
Table 16.4

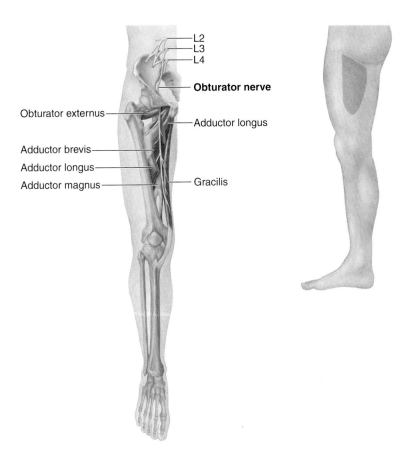

L2
L3
L4

Obturator nerve

Obturator externus

Adductor brevis

Adductor longus

Adductor magnus

Adductor longus

Gracilis

Branches of the Lumbar Plexus—Obturator Nerve
Table 16.4

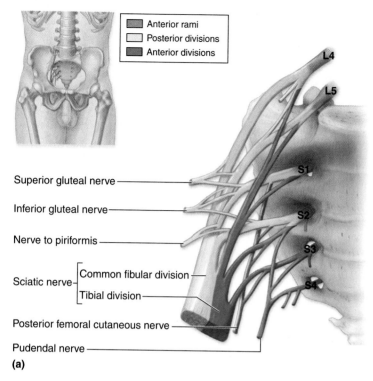

Anterior rami
Posterior divisions
Anterior divisions

L4

L5

S1

S2

S3

S4

Superior gluteal nerve

Inferior gluteal nerve

Nerve to piriformis

Sciatic nerve — Common fibular division
Tibial division

Posterior femoral cutaneous nerve

Pudendal nerve

(a)

Sacral Plexus
Figure 16.11a

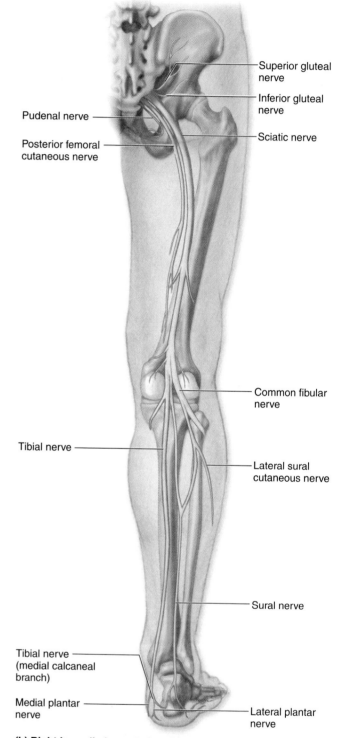

Pudenal nerve

Posterior femoral
cutaneous nerve

Superior gluteal
nerve

Inferior gluteal
nerve

Sciatic nerve

Common fibular
nerve

Tibial nerve

Lateral sural
cutaneous nerve

Sural nerve

Tibial nerve
(medial calcaneal
branch)

Medial plantar
nerve

Lateral plantar
nerve

(b) Right lower limb, posterior view

Distribution of Sacral Plexus Nerves
Figure 16.11b

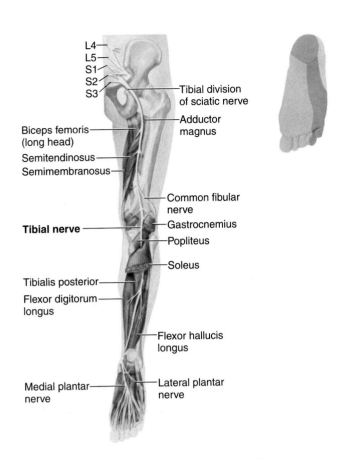

Branches of the Sacral Plexus—Tibial Nerve
Table 16.5

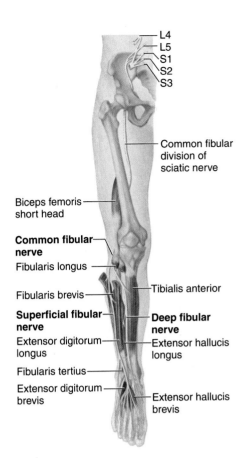

Branches of the Sacral Plexus—Common Fibular Nerve
Table 16.5

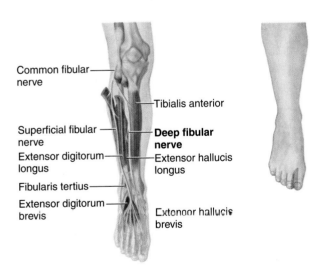

Branches of the Sacral Plexus—Deep Fibular Nerve
Table 16.5

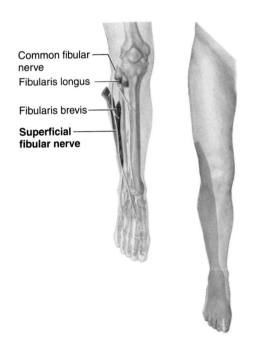

Branches of the Sacral Plexus—Superficial Fibular Nerve
Table 16.5

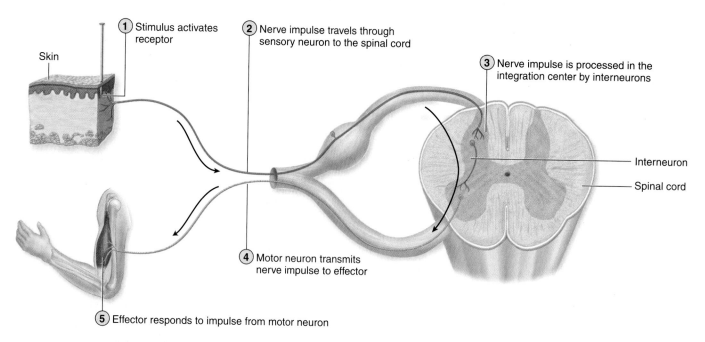

① Stimulus activates receptor

② Nerve impulse travels through sensory neuron to the spinal cord

③ Nerve impulse is processed in the integration center by interneurons

Skin

Interneuron

Spinal cord

④ Motor neuron transmits nerve impulse to effector

⑤ Effector responds to impulse from motor neuron

Simple Reflex Arcs
Figure 16.12

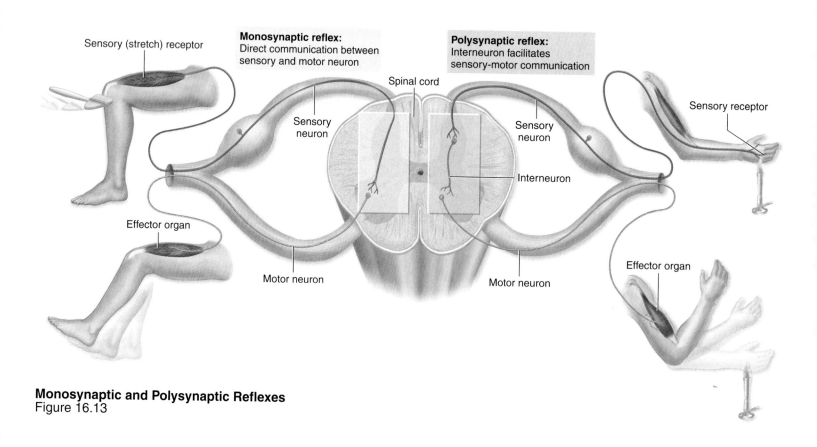

Sensory (stretch) receptor

Monosynaptic reflex:
Direct communication between sensory and motor neuron

Polysynaptic reflex:
Interneuron facilitates sensory-motor communication

Spinal cord

Sensory receptor

Sensory neuron

Sensory neuron

Interneuron

Effector organ

Motor neuron

Motor neuron

Effector organ

Monosynaptic and Polysynaptic Reflexes
Figure 16.13

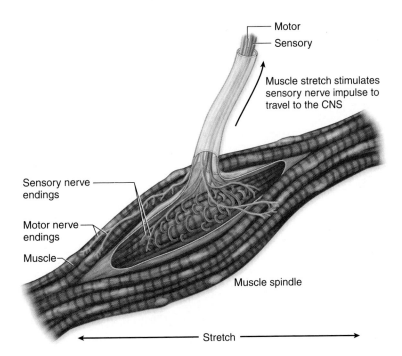

Motor

Sensory

Muscle stretch stimulates
sensory nerve impulse to
travel to the CNS

Sensory nerve
endings

Motor nerve
endings

Muscle

Muscle spindle

Stretch

Stretch Reflexes
Figure 16.14

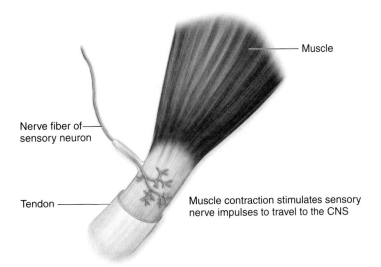

Muscle

Nerve fiber of
sensory neuron

Tendon

Muscle contraction stimulates sensory
nerve impulses to travel to the CNS

Golgi Tendon Reflex
Figure 16.15

247

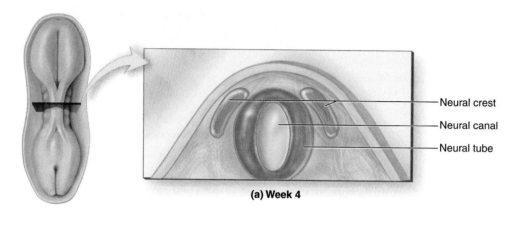

Neural crest
Neural canal
Neural tube

(a) Week 4

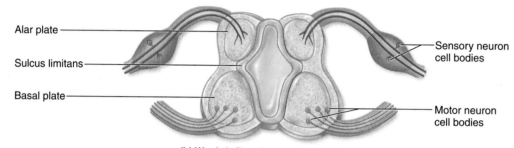

Alar plate

Sulcus limitans

Basal plate

Sensory neuron
cell bodies

Motor neuron
cell bodies

(b) Week 6: Basal and alar plates form

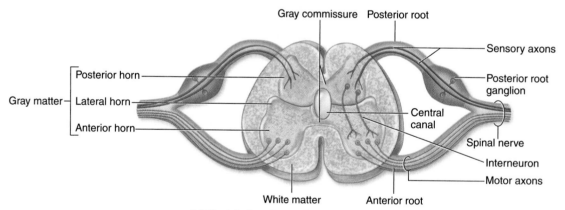

Gray commissure   Posterior root

Sensory axons

Posterior root
ganglion

Posterior horn

Gray matter — Lateral horn

Anterior horn

Central
canal

Spinal nerve

Interneuron

Motor axons

White matter   Anterior root

(c) Week 9: Gray horns form from basal and alar plates

Spinal Cord Development
Figure 16.16

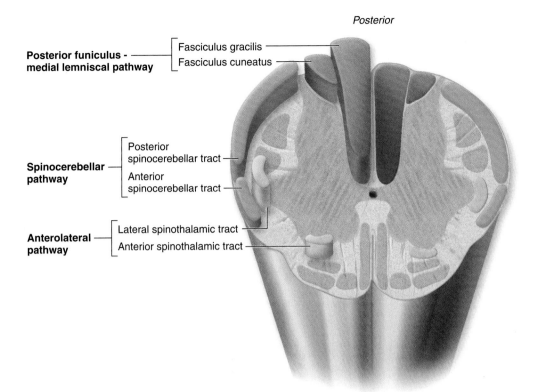

Posterior

Posterior funiculus -
medial lemniscal pathway
├─ Fasciculus gracilis
└─ Fasciculus cuneatus

Spinocerebellar
pathway
├─ Posterior spinocerebellar tract
└─ Anterior spinocerebellar tract

Anterolateral
pathway
├─ Lateral spinothalamic tract
└─ Anterior spinothalamic tract

Anterior

Sensory Pathways in the Spinal Cord
Figure 17.1

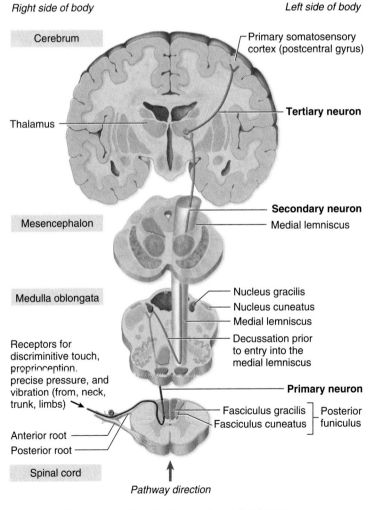

Right side of body   Left side of body

Cerebrum

Primary somatosensory cortex (postcentral gyrus)

Tertiary neuron

Thalamus

Mesencephalon

Secondary neuron
Medial lemniscus

Medulla oblongata

Nucleus gracilis
Nucleus cuneatus
Medial lemniscus
Decussation prior to entry into the medial lemniscus

Receptors for discriminitive touch, proprioception, precise pressure, and vibration (from, neck, trunk, limbs)

Primary neuron

Fasciculus gracilis ┐ Posterior
Fasciculus cuneatus ┘ funiculus

Anterior root
Posterior root

Spinal cord

Pathway direction

Posterior Funiculus-Medial Lemniscal Pathway
Figure 17.2

249

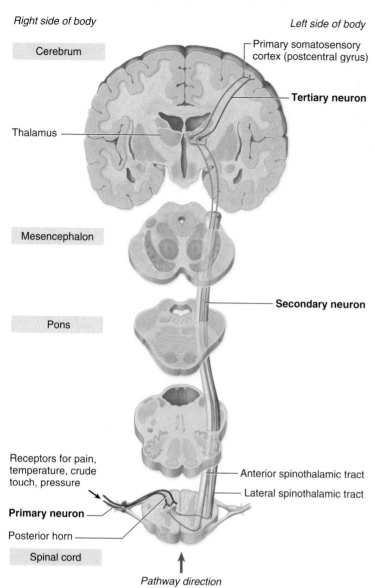

Right side of body   *Left side of body*

Cerebrum

Primary somatosensory
cortex (postcentral gyrus)

Tertiary neuron

Thalamus

Mesencephalon

Secondary neuron

Pons

Receptors for pain,
temperature, crude
touch, pressure

Anterior spinothalamic tract

Lateral spinothalamic tract

Primary neuron

Posterior horn

Spinal cord

Pathway direction

Anterolateral Pathway
Figure 17.3

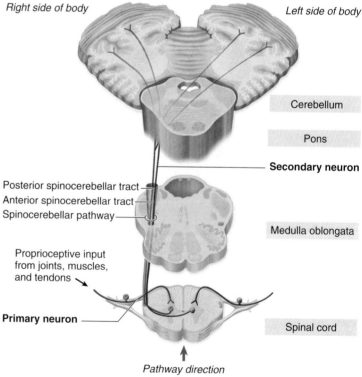

Right side of body   *Left side of body*

Cerebellum

Pons

Secondary neuron

Posterior spinocerebellar tract

Anterior spinocerebellar tract

Spinocerebellar pathway

Medulla oblongata

Proprioceptive input
from joints, muscles,
and tendons

Primary neuron

Spinal cord

Pathway direction

Spinocerebellar Pathway
Figure 17.4

250

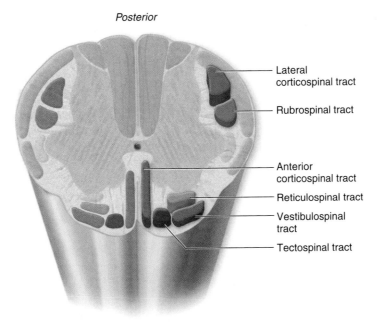

Posterior

— Lateral corticospinal tract

— Rubrospinal tract

— Anterior corticospinal tract

— Reticulospinal tract

— Vestibulospinal tract

— Tectospinal tract

Anterior

Motor Pathways in the Spinal Cord
Figure 17.5

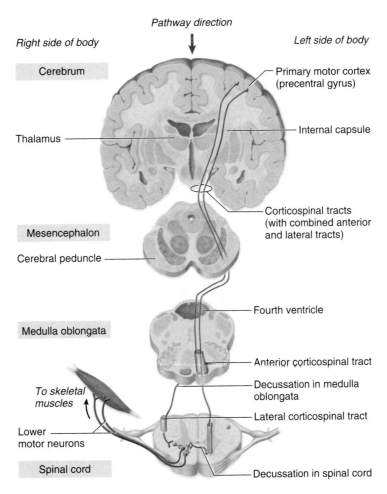

Pathway direction

Right side of body   *Left side of body*

Cerebrum

— Primary motor cortex (precentral gyrus)

Thalamus —

— Internal capsule

— Corticospinal tracts (with combined anterior and lateral tracts)

Mesencephalon

Cerebral peduncle —

— Fourth ventricle

Medulla oblongata

To skeletal muscles

— Anterior corticospinal tract

— Decussation in medulla oblongata

— Lateral corticospinal tract

Lower motor neurons

Spinal cord

— Decussation in spinal cord

Corticospinal Tracts
Figure 17.6

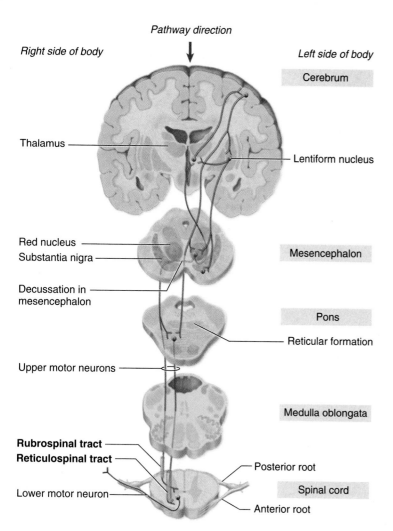

Pathway direction

Right side of body

Left side of body

Cerebrum

Thalamus

Lentiform nucleus

Red nucleus

Mesencephalon

Substantia nigra

Decussation in
mesencephalon

Pons

Reticular formation

Upper motor neurons

Medulla oblongata

Rubrospinal tract

Reticulospinal tract

Posterior root

Spinal cord

Lower motor neuron

Anterior root

Indirect Motor Pathways in the Spinal Cord
Figure 17.7

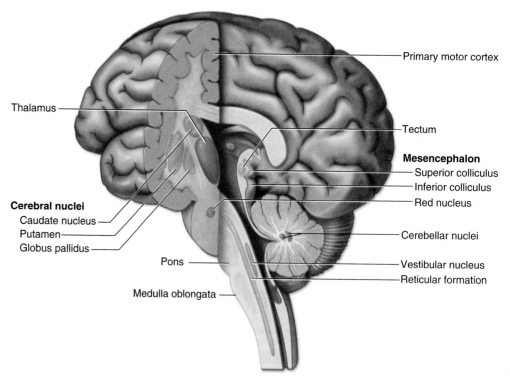

Primary motor cortex

Thalamus

Tectum

Mesencephalon

Superior colliculus

Inferior colliculus

Red nucleus

Cerebral nuclei

Caudate nucleus

Putamen

Globus pallidus

Cerebellar nuclei

Pons

Vestibular nucleus

Reticular formation

Medulla oblongata

Cerebral Nuclei and Selected Indirect Motor System Components
Figure 17.8

Voluntary movements
The primary motor cortex and the basal nuclei in the forebrain send impulses through the nuclei of the pons to the cerebellum.

Assessment of voluntary movements
Proprioceptors in skeletal muscles and joints report degree of movement to the cerebellum.

Integration and analysis
The cerebellum compares the planned movements (motor signals) against the results of the actual movements (sensory signals).

Error-correcting signals
The cerebellum sends impulses through the thalamus to the primary motor cortex and to motor nuclei in the brainstem.

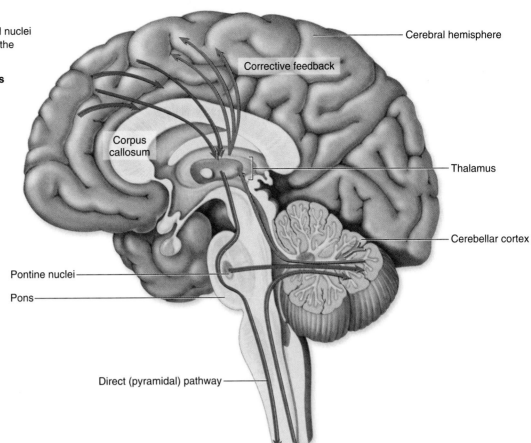

Cerebral hemisphere
Corrective feedback
Corpus callosum
Thalamus
Cerebellar cortex
Pontine nuclei
Pons
Direct (pyramidal) pathway

Cerebellar Pathways
Figure 17.9

Sagittal section

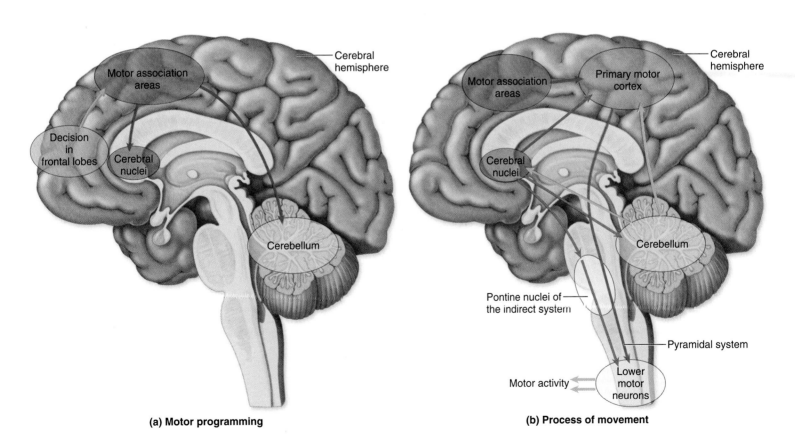

Cerebral hemisphere
Motor association areas
Decision in frontal lobes
Cerebral nuclei
Cerebellum

Cerebral hemisphere
Motor association areas
Primary motor cortex
Cerebral nuclei
Cerebellum
Pontine nuclei of the indirect system
Pyramidal system
Motor activity
Lower motor neurons

(a) Motor programming

(b) Process of movement

Somatic Motor Control
Figure 17.10

253

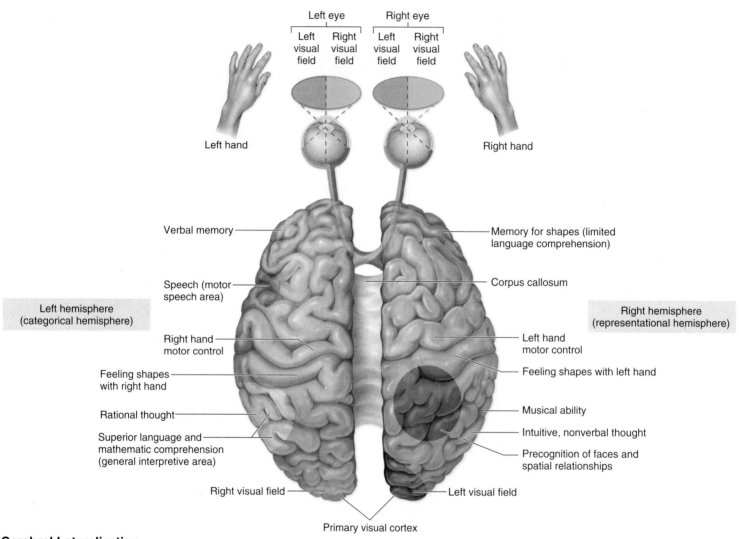

Left eye   Right eye

Left | Right   Left | Right
visual | visual   visual | visual
field | field   field | field

Left hand   Right hand

Verbal memory —————   ————— Memory for shapes (limited
                                             language comprehension)

Speech (motor —————   ————— Corpus callosum
speech area)

Left hemisphere   Right hemisphere
(categorical hemisphere)   (representational hemisphere)

Right hand —————   ————— Left hand
motor control   motor control

Feeling shapes —————   ————— Feeling shapes with left hand
with right hand

Rational thought —————   ————— Musical ability

Superior language and —————   ————— Intuitive, nonverbal thought
mathematic comprehension
(general interpretive area)   ————— Precognition of faces and
                                             spatial relationships

Right visual field —————   ————— Left visual field

Primary visual cortex

Cerebral Lateralization
Figure 17.12

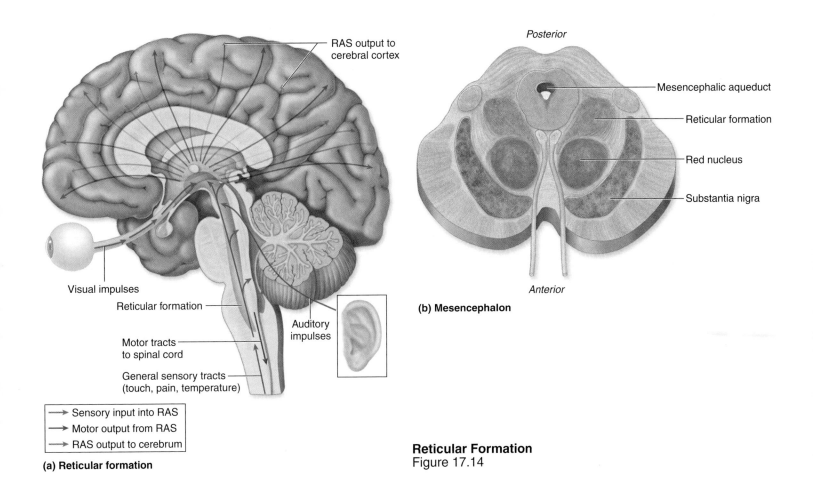

RAS output to
cerebral cortex

Visual impulses

Reticular formation

Motor tracts
to spinal cord

General sensory tracts
(touch, pain, temperature)

Auditory
impulses

→ Sensory input into RAS
→ Motor output from RAS
→ RAS output to cerebrum

(a) Reticular formation

Posterior

Mesencephalic aqueduct

Reticular formation

Red nucleus

Substantia nigra

Anterior

(b) Mesencephalon

Reticular Formation
Figure 17.14

255

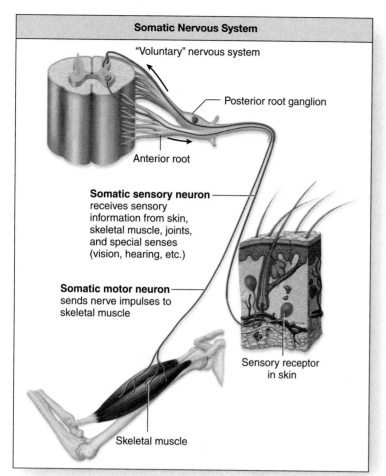

Somatic Nervous System

"Voluntary" nervous system

Posterior root ganglion

Anterior root

Somatic sensory neuron
receives sensory
information from skin,
skeletal muscle, joints,
and special senses
(vision, hearing, etc.)

Somatic motor neuron
sends nerve impulses to
skeletal muscle

Sensory receptor
in skin

Skeletal muscle

Autonomic (Visceral) Nervous System

"Involuntary" nervous system

Autonomic
ganglion

Preganglionic autonomic motor neuron

Motor information is passed through
preganglionic and ganglionic neurons

Ganglionic autonomic motor neuron
sends nerve impulses to smooth muscle,
cardiac muscle, and glands

Visceral sensory neuron
receives sensory information from
blood vessels and visceral walls

Smooth muscle in trachea

Sensory
receptor
in viscera

Comparison of Somatic and Autonomic Nervous Systems
Figure 18.1

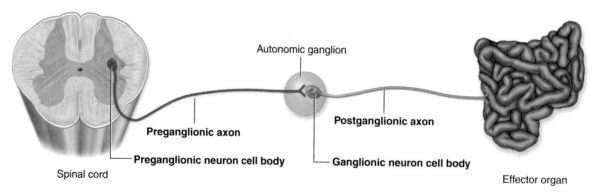

Autonomic ganglion

Preganglionic axon

Preganglionic neuron cell body

Spinal cord

Postganglionic axon

Ganglionic neuron cell body

Effector organ

Components of the Autonomic Nervous System
Figure 18.2

① "Preganglionic" flight from Indianapolis to Chicago (autonomic ganglion).

Chicago

Indianapolis

Miami

② "Postganglionic" flight from Chicago to Miami (effector organ).

ANS-Airline Analogy
Study Tip p. 548

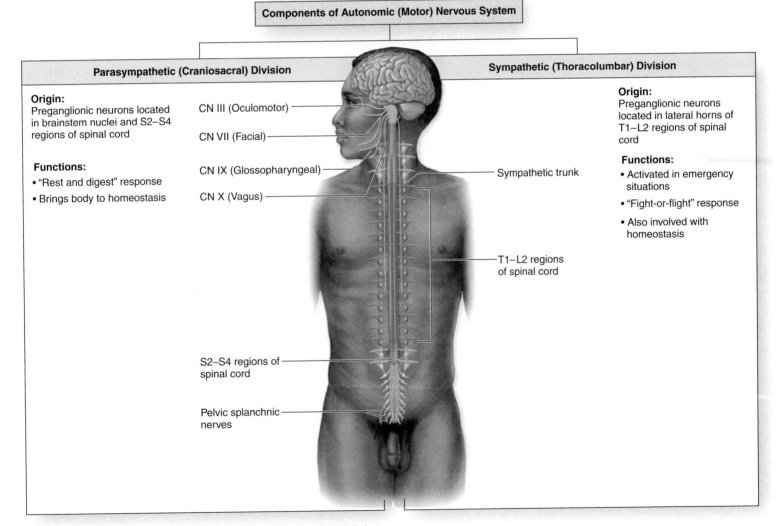

Components of Autonomic (Motor) Nervous System

Parasympathetic (Craniosacral) Division

Origin:
Preganglionic neurons located in brainstem nuclei and S2–S4 regions of spinal cord

Functions:
• "Rest and digest" response
• Brings body to homeostasis

CN III (Oculomotor)

CN VII (Facial)

CN IX (Glossopharyngeal)

CN X (Vagus)

S2–S4 regions of spinal cord

Pelvic splanchnic nerves

Sympathetic (Thoracolumbar) Division

Origin:
Preganglionic neurons located in lateral horns of T1–L2 regions of spinal cord

Functions:
• Activated in emergency situations
• "Fight-or-flight" response
• Also involved with homeostasis

Sympathetic trunk

T1–L2 regions of spinal cord

Comparison of Parasympathetic and Sympathetic Divisions
Figure 18.3

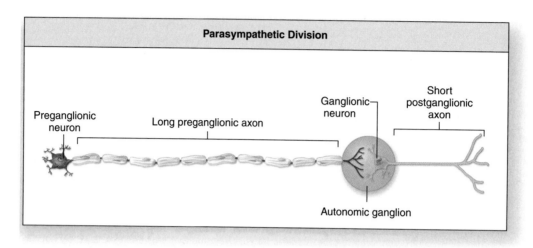

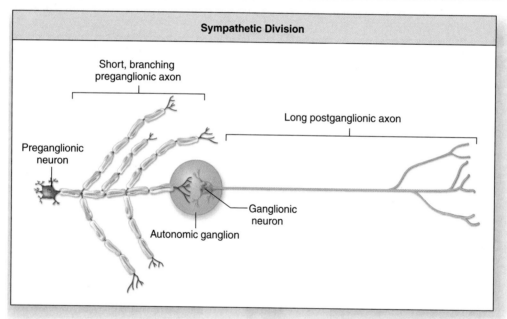

Anatomic Differences Between Parasympathetic and Sympathetic Neurons
Figure 18.4

258

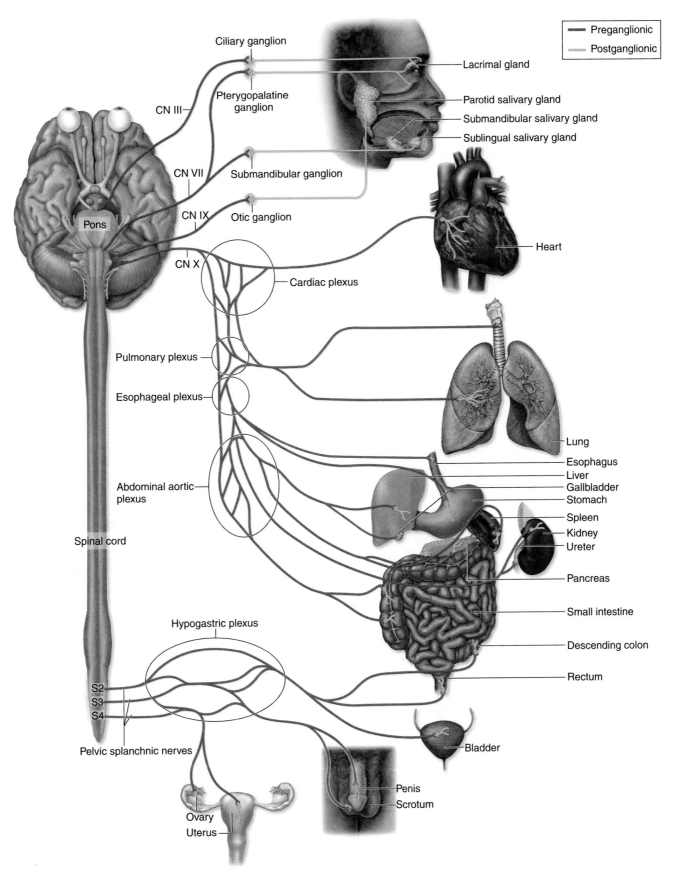

Overview of Parasympathetic Pathways
Figure 18.5

259

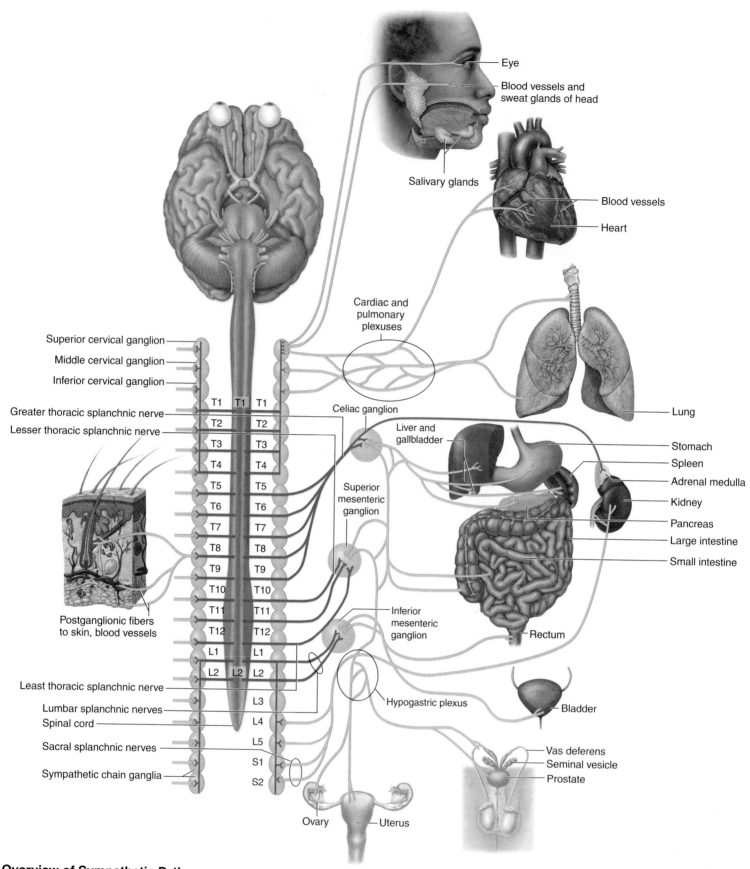

Eye

Blood vessels and
sweat glands of head

Salivary glands

Blood vessels

Heart

Cardiac and
pulmonary
plexuses

Lung

Superior cervical ganglion

Middle cervical ganglion

Inferior cervical ganglion

Celiac ganglion

Greater thoracic splanchnic nerve

Lesser thoracic splanchnic nerve

Liver and
gallbladder

Stomach

Spleen

Adrenal medulla

Kidney

Pancreas

Large intestine

Small intestine

Superior
mesenteric
ganglion

T1 T1 T1
T2 T2
T3 T3
T4 T4
T5 T5
T6 T6
T7 T7
T8 T8
T9 T9
T10 T10
T11 T11
T12 T12
L1 L1
L2 L2 L2
L3
L4
L5
S1
S2

Postganglionic fibers
to skin, blood vessels

Inferior
mesenteric
ganglion

Rectum

Least thoracic splanchnic nerve

Lumbar splanchnic nerves

Spinal cord

Sacral splanchnic nerves

Sympathetic chain ganglia

Hypogastric plexus

Bladder

Vas deferens

Seminal vesicle

Prostate

Ovary

Uterus

Overview of Sympathetic Pathways
Figure 18.6

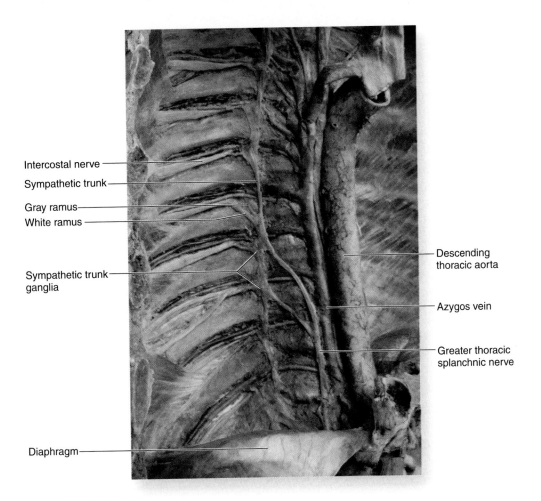

Intercostal nerve
Sympathetic trunk
Gray ramus
White ramus

Sympathetic trunk
ganglia

Diaphragm

Descending
thoracic aorta

Azygos vein

Greater thoracic
splanchnic nerve

Sympathetic Trunk
Figure 18.7
Figure 18.7: © The McGraw-Hill Companies, Inc./Photo and Dissection by Christine Eckel

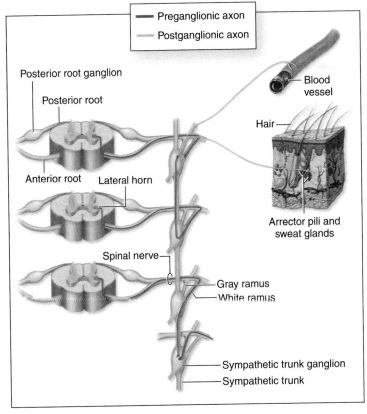

Preganglionic axon
Postganglionic axon

Posterior root ganglion
Posterior root

Blood
vessel

Hair

Anterior root   Lateral horn

Arrector pili and
sweat glands

Spinal nerve

Gray ramus
White ramus

Sympathetic trunk ganglion
Sympathetic trunk

(a) Spinal nerve pathway

Spinal Nerve Pathway
Figure 18.8a

261

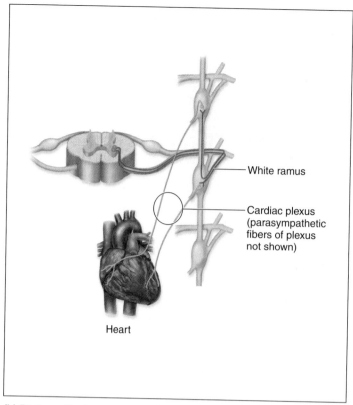

(b) Postganglionic sympathetic nerve pathway

Postganglionic Sympathetic Nerve Pathway
Figure 18.8b

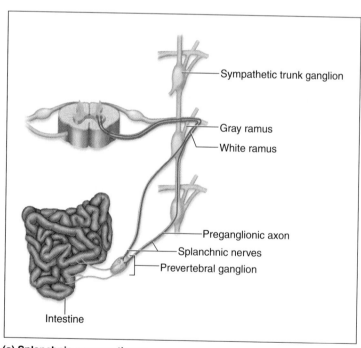

(c) Splanchnic nerve pathway

Splanchnic Nerve Pathway
Figure 18.8c

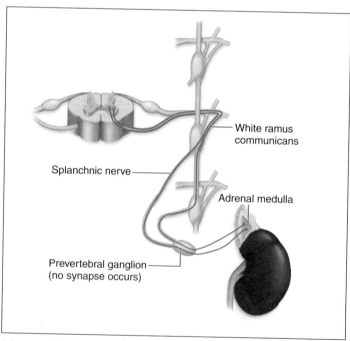

(d) Adrenal medulla pathway

Adrenal Medulla Pathway
Figure 18.8d

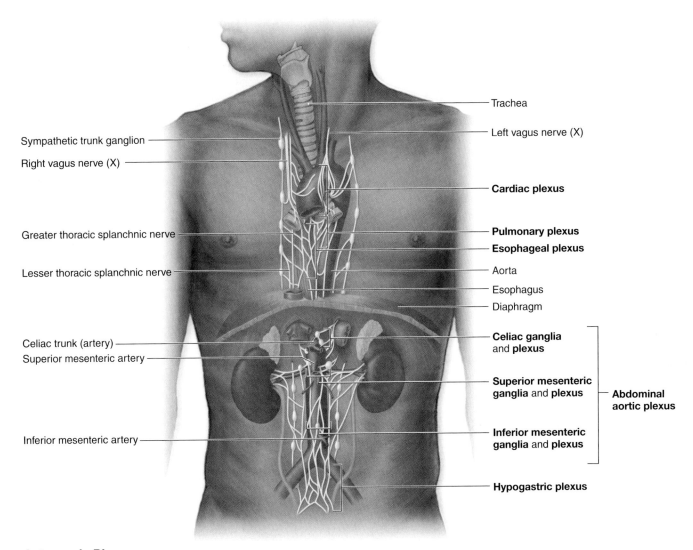

Trachea

Sympathetic trunk ganglion

Left vagus nerve (X)

Right vagus nerve (X)

Cardiac plexus

Greater thoracic splanchnic nerve

Pulmonary plexus

Esophageal plexus

Lesser thoracic splanchnic nerve

Aorta

Esophagus

Diaphragm

Celiac trunk (artery)

Celiac ganglia
and **plexus**

Superior mesenteric artery

Superior mesenteric
ganglia and **plexus**

Abdominal
aortic plexus

Inferior mesenteric artery

Inferior mesenteric
ganglia and **plexus**

Hypogastric plexus

Autonomic Plexuses
Figure 18.9

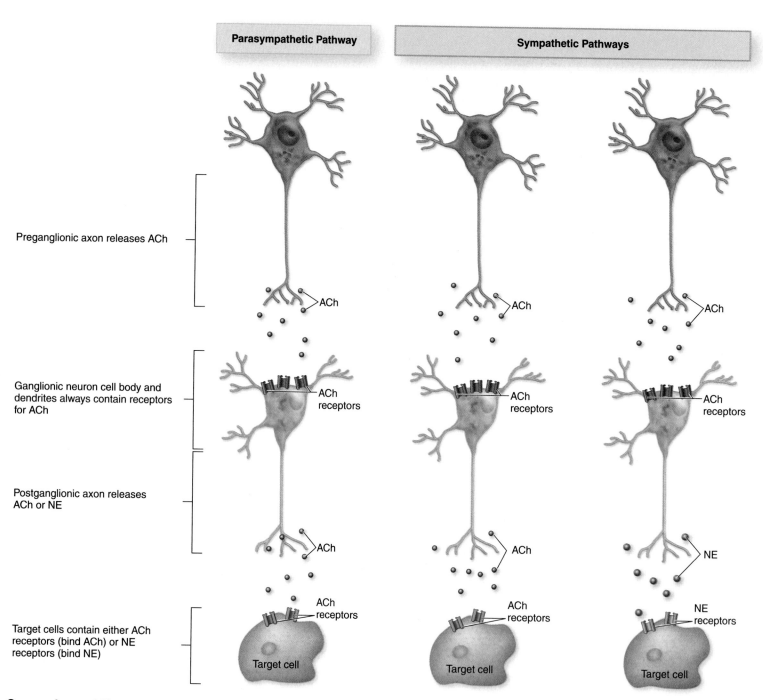

Parasympathetic Pathway

Sympathetic Pathways

Preganglionic axon releases ACh

ACh

ACh

ACh

Ganglionic neuron cell body and dendrites always contain receptors for ACh

ACh receptors

ACh receptors

ACh receptors

Postganglionic axon releases ACh or NE

ACh

ACh

NE

Target cells contain either ACh receptors (bind ACh) or NE receptors (bind NE)

ACh receptors

ACh receptors

NE receptors

Target cell

Target cell

Target cell

Comparison of Neurotransmitters in the Autonomic Nervous System
Figure 18.10

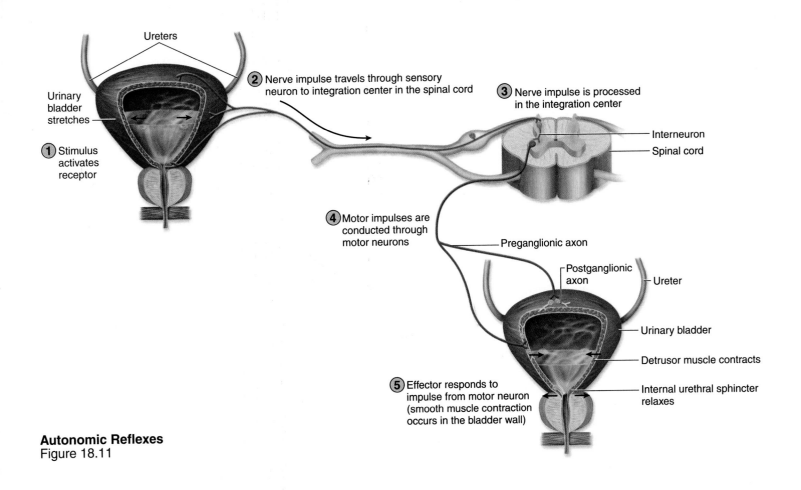

① Stimulus activates receptor

Ureters

Urinary bladder stretches

② Nerve impulse travels through sensory neuron to integration center in the spinal cord

③ Nerve impulse is processed in the integration center

Interneuron

Spinal cord

④ Motor impulses are conducted through motor neurons

Preganglionic axon

Postganglionic axon

Ureter

Urinary bladder

Detrusor muscle contracts

⑤ Effector responds to impulse from motor neuron (smooth muscle contraction occurs in the bladder wall)

Internal urethral sphincter relaxes

Autonomic Reflexes
Figure 18.11

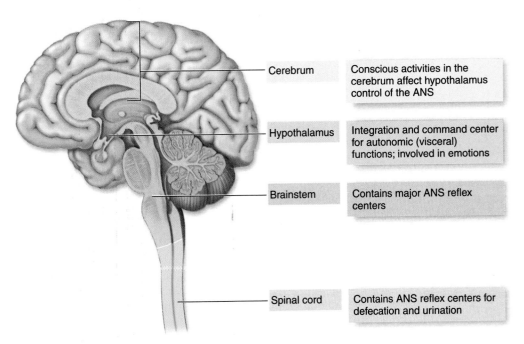

Cerebrum — Conscious activities in the cerebrum affect hypothalamus control of the ANS

Hypothalamus — Integration and command center for autonomic (visceral) functions; involved in emotions

Brainstem — Contains major ANS reflex centers

Spinal cord — Contains ANS reflex centers for defecation and urination

Control of Autonomic Functions by Higher Brain Centers
Figure 18.12

265

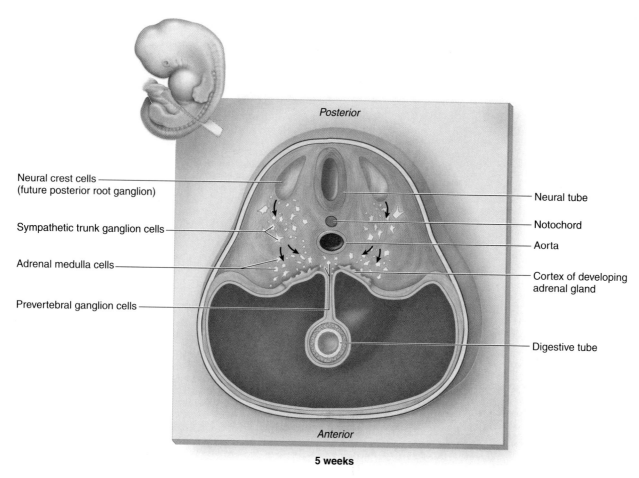

Posterior

Neural crest cells
(future posterior root ganglion)

Sympathetic trunk ganglion cells

Adrenal medulla cells

Prevertebral ganglion cells

Neural tube

Notochord

Aorta

Cortex of developing
adrenal gland

Digestive tube

Anterior

5 weeks

Neural Crest Cell Derivatives
Figure 18.13

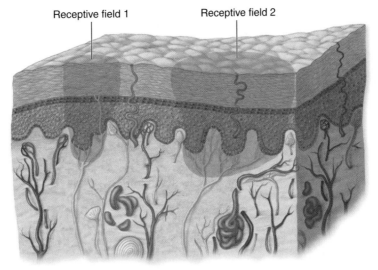

Receptive Fields
Figure 19.1

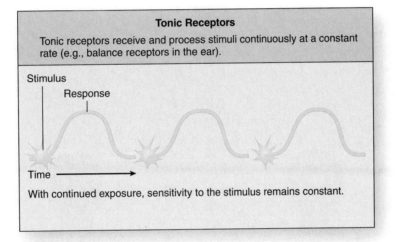

Tonic Receptors

Tonic receptors receive and process stimuli continuously at a constant rate (e.g., balance receptors in the ear).

Stimulus

Response

Time ⟶

With continued exposure, sensitivity to the stimulus remains constant.

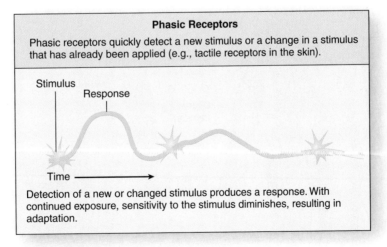

Phasic Receptors

Phasic receptors quickly detect a new stimulus or a change in a stimulus that has already been applied (e.g., tactile receptors in the skin).

Stimulus

Response

Time ⟶

Detection of a new or changed stimulus produces a response. With continued exposure, sensitivity to the stimulus diminishes, resulting in adaptation.

Receptor Activities
Figure 19.2

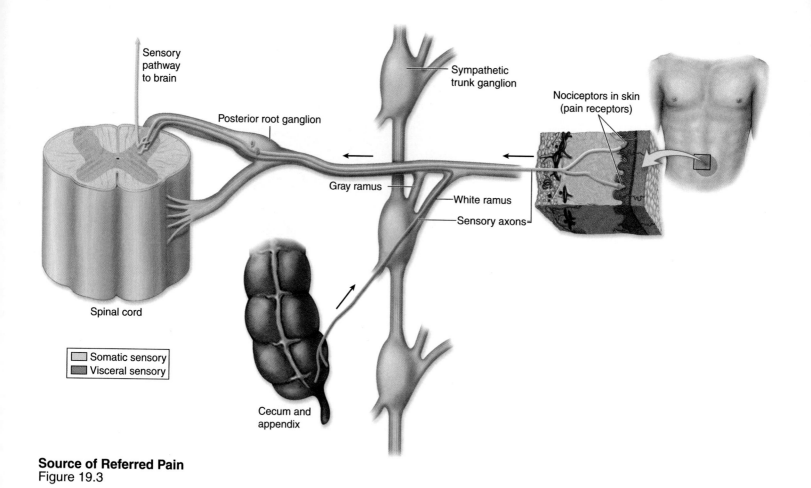

Source of Referred Pain
Figure 19.3

Sensory
pathway
to brain

Posterior root ganglion

Sympathetic
trunk ganglion

Nociceptors in skin
(pain receptors)

Gray ramus

White ramus

Sensory axons

Spinal cord

Somatic sensory
Visceral sensory

Cecum and
appendix

Common Sites of Referred Pain
Figure 19.4

Liver and
gallbladder

Heart

Stomach

Pancreas

Umbilicus

Appendix

Ovary
(female)

Kidney

Urinary
bladder

Ureter

Liver and
gallbladder

268

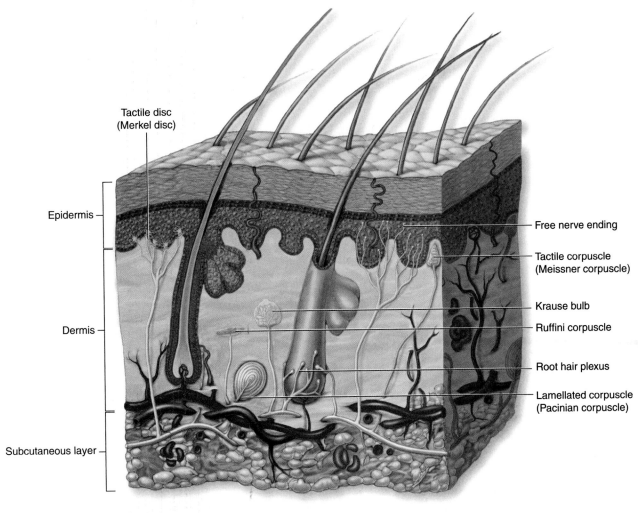

Tactile disc
(Merkel disc)

Epidermis

Dermis

Subcutaneous layer

Free nerve ending

Tactile corpuscle
(Meissner corpuscle)

Krause bulb

Ruffini corpuscle

Root hair plexus

Lamellated corpuscle
(Pacinian corpuscle)

General Cutaneous Receptors
Figure 19.5

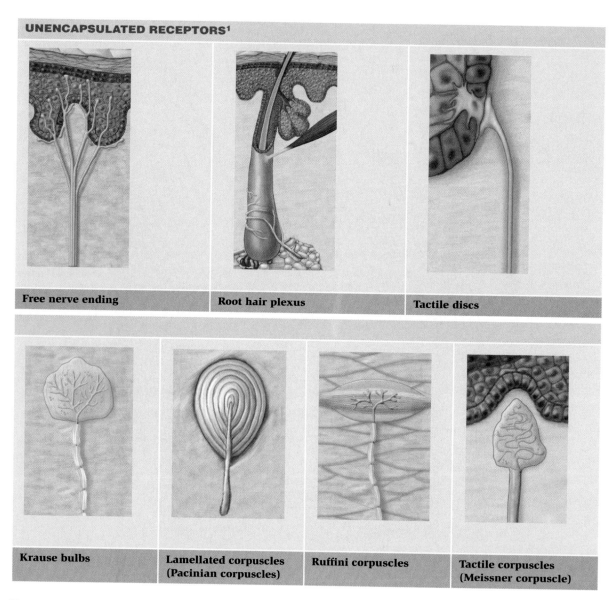

UNENCAPSULATED RECEPTORS[1]

| Free nerve ending | Root hair plexus | Tactile discs |

| Krause bulbs | Lamellated corpuscles (Pacinian corpuscles) | Ruffini corpuscles | Tactile corpuscles (Meissner corpuscle) |

Types of Tactile Receptors
Table 19.2

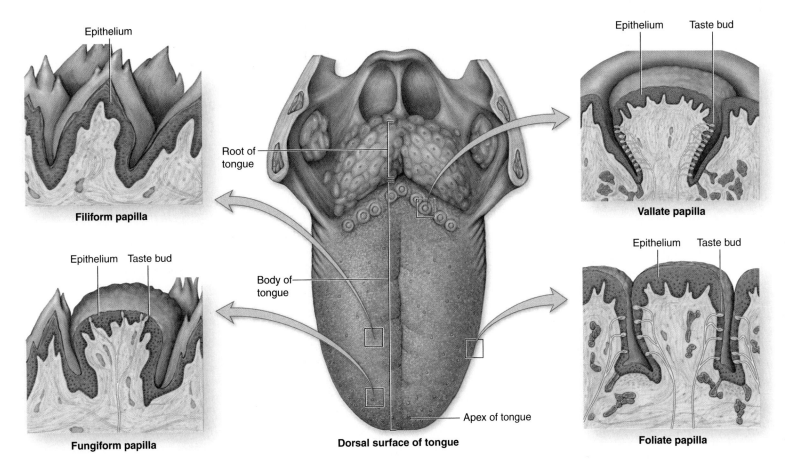

Epithelium

Filiform papilla

Epithelium   Taste bud

Fungiform papilla

Root of tongue

Body of tongue

Apex of tongue

Dorsal surface of tongue

Epithelium   Taste bud

Vallate papilla

Epithelium   Taste bud

Foliate papilla

Tongue Papillae
Figure 19.6

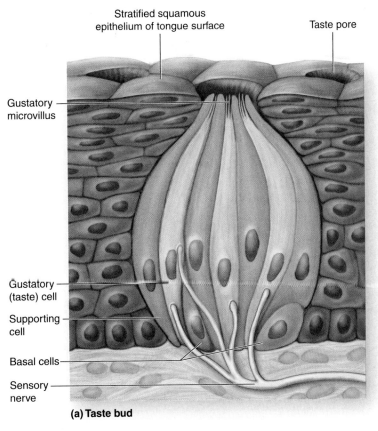

Stratified squamous epithelium of tongue surface

Taste pore

Gustatory microvillus

Gustatory (taste) cell

Supporting cell

Basal cells

Sensory nerve

(a) Taste bud

Taste Bud
Figure 19.7a

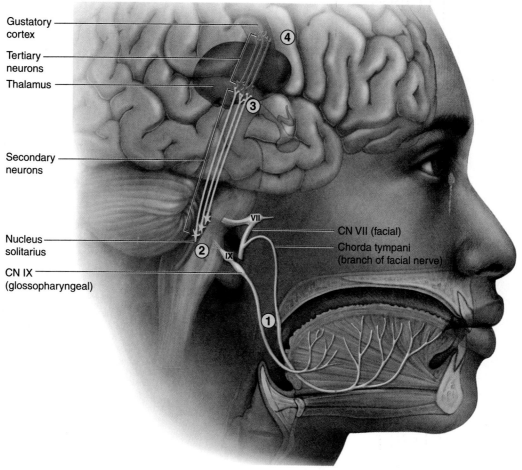

Gustatory Pathway
Figure 19.8

Labels on figure:
- Gustatory cortex
- Tertiary neurons
- Thalamus
- Secondary neurons
- Nucleus solitarius
- CN IX (glossopharyngeal)
- CN VII (facial)
- Chorda tympani (branch of facial nerve)

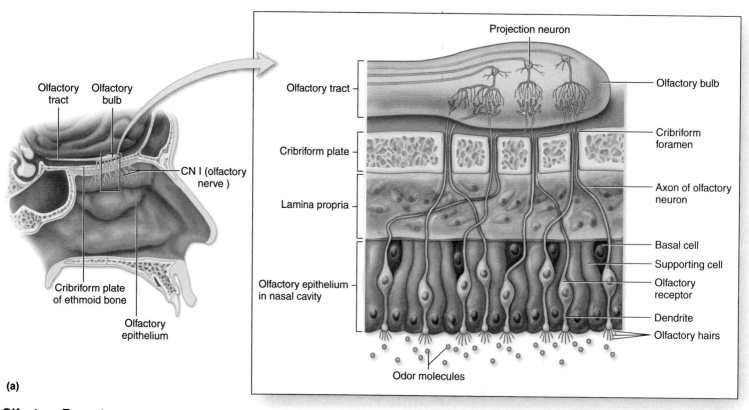

Labels on figure:
- Projection neuron
- Olfactory tract
- Olfactory bulb
- Cribriform plate
- Cribriform foramen
- Lamina propria
- Axon of olfactory neuron
- Olfactory epithelium in nasal cavity
- Basal cell
- Supporting cell
- Olfactory receptor
- Dendrite
- Olfactory hairs
- Odor molecules

- Olfactory tract
- Olfactory bulb
- CN I (olfactory nerve)
- Cribriform plate of ethmoid bone
- Olfactory epithelium

(a)

Olfactory Receptors
Figure 19.9a

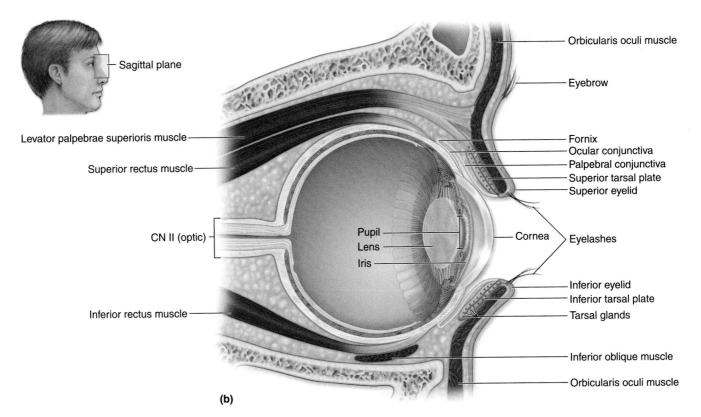

Sagittal plane

Orbicularis oculi muscle

Eyebrow

Levator palpebrae superioris muscle

Superior rectus muscle

Fornix
Ocular conjunctiva
Palpebral conjunctiva
Superior tarsal plate
Superior eyelid

CN II (optic)

Pupil
Lens
Iris

Cornea

Eyelashes

Inferior eyelid
Inferior tarsal plate
Tarsal glands

Inferior rectus muscle

Inferior oblique muscle

Orbicularis oculi muscle

(b)

Accessory Structures of the Eye
Figure 19.10b

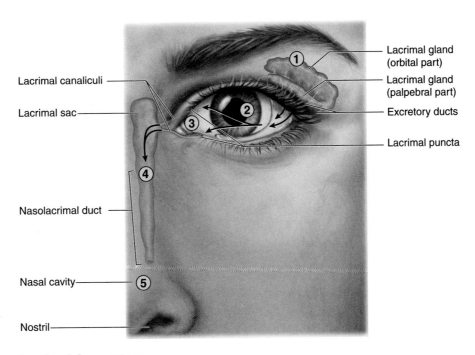

Lacrimal canaliculi

Lacrimal sac

Lacrimal gland (orbital part)

Lacrimal gland (palpebral part)

Excretory ducts

Lacrimal puncta

Nasolacrimal duct

Nasal cavity

Nostril

Lacrimal Apparatus
Figure 19.11

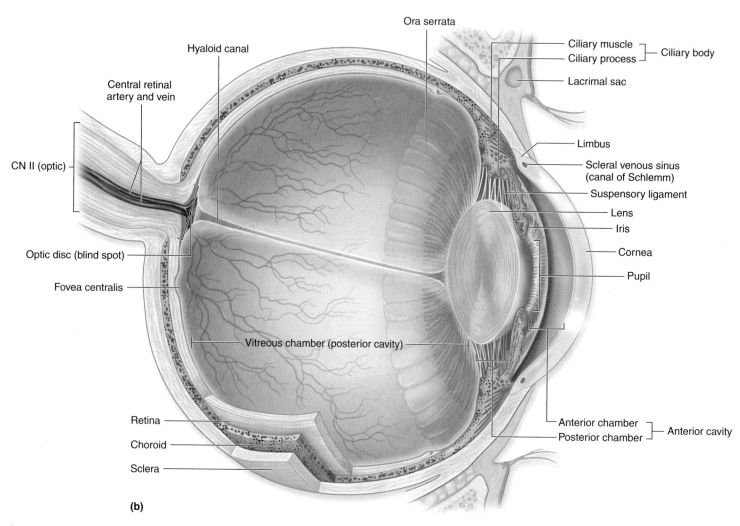

Ora serrata

Hyaloid canal

Ciliary muscle ⎤
Ciliary process ⎦ Ciliary body

Central retinal
artery and vein

Lacrimal sac

Limbus

CN II (optic)

Scleral venous sinus
(canal of Schlemm)

Suspensory ligament

Lens

Iris

Optic disc (blind spot)

Cornea

Fovea centralis

Pupil

Vitreous chamber (posterior cavity)

Retina

Choroid

Sclera

Anterior chamber ⎤
Posterior chamber ⎦ Anterior cavity

(b)

Anatomy of the Internal Eye
Figure 19.12b

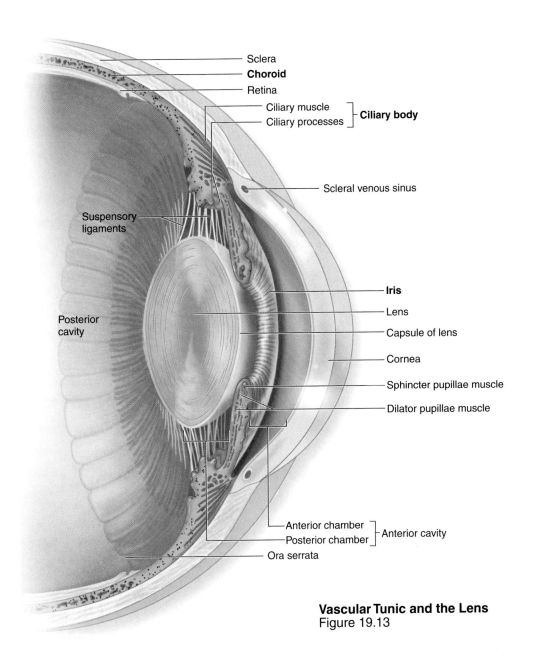

Sclera

Choroid

Retina

Ciliary muscle ⎫
Ciliary processes ⎭ **Ciliary body**

Scleral venous sinus

Suspensory
ligaments

Iris

Lens

Posterior
cavity

Capsule of lens

Cornea

Sphincter pupillae muscle

Dilator pupillae muscle

Anterior chamber ⎫
Posterior chamber ⎭ Anterior cavity

Ora serrata

Vascular Tunic and the Lens
Figure 19.13

Pupillary constriction—Contraction of sphincter pupillae muscle (parasympathetic innervation)

Sphincter pupillae muscle contracts

Dilator pupillae muscle relaxes

Pupillary dilation—Contraction of dilator pupillae muscle (sympathetic innervation)

Dilator pupillae muscle contracts

Sphincter pupillae muscle relaxes

Pupil Diameter
Figure 19.14

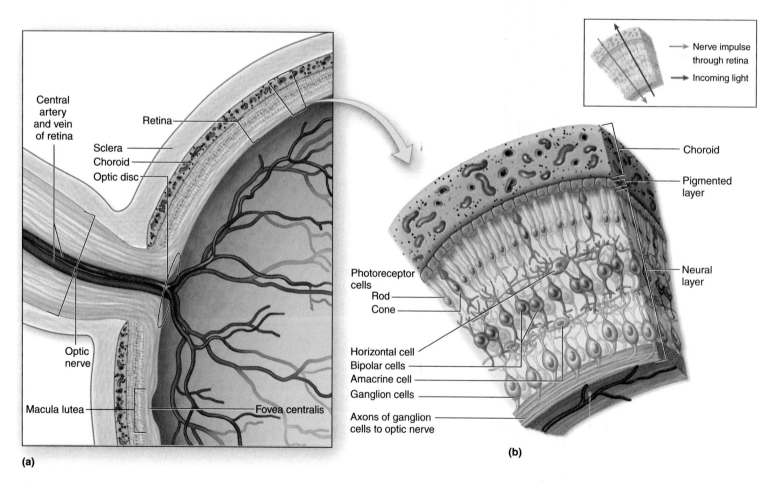

Central artery and vein of retina

Retina
Sclera
Choroid
Optic disc

Optic nerve

Macula lutea

Fovea centralis

(a)

Nerve impulse through retina
Incoming light

Choroid
Pigmented layer
Neural layer

Photoreceptor cells
Rod
Cone

Horizontal cell
Bipolar cells
Amacrine cell
Ganglion cells

Axons of ganglion cells to optic nerve

(b)

Structure and Organization of the Retina
Figure 19.15a,b

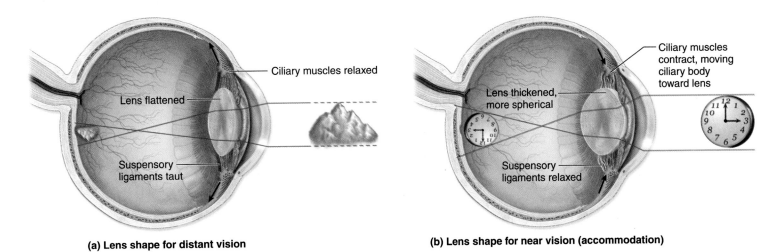

(a) Lens shape for distant vision

(b) Lens shape for near vision (accommodation)

Lens Shape in Far Vision and Near Vision
Figure 19.17

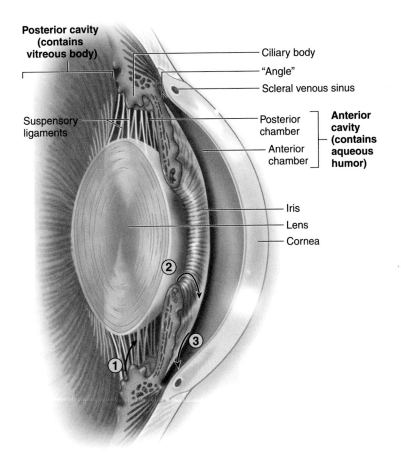

Aqueous Humor: Secretion and Reabsorption
Figure 19.18

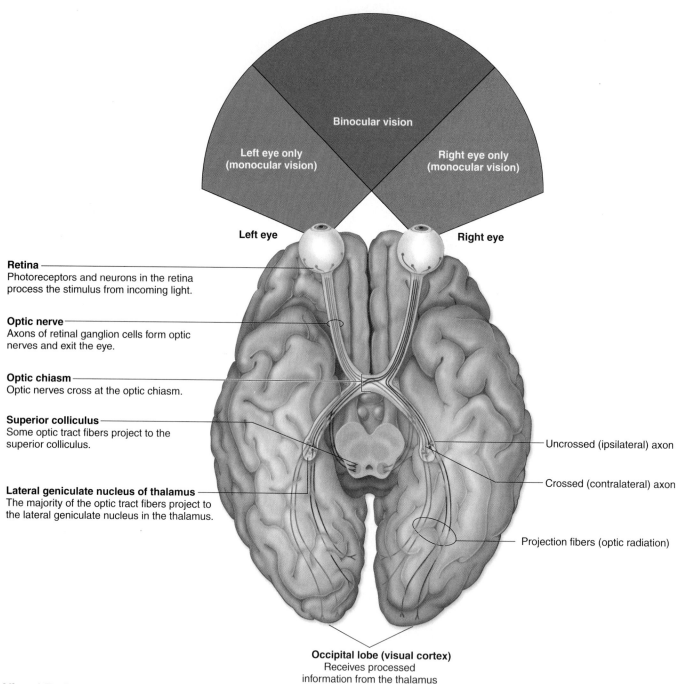

Binocular vision

Left eye only
(monocular vision)

Right eye only
(monocular vision)

Left eye

Right eye

Retina
Photoreceptors and neurons in the retina
process the stimulus from incoming light.

Optic nerve
Axons of retinal ganglion cells form optic
nerves and exit the eye.

Optic chiasm
Optic nerves cross at the optic chiasm.

Superior colliculus
Some optic tract fibers project to the
superior colliculus.

Lateral geniculate nucleus of thalamus
The majority of the optic tract fibers project to
the lateral geniculate nucleus in the thalamus.

Uncrossed (ipsilateral) axon

Crossed (contralateral) axon

Projection fibers (optic radiation)

Occipital lobe (visual cortex)
Receives processed
information from the thalamus

Visual Pathways
Figure 19.19

Emmetropia (normal vision)	Hyperopia (farsightedness): Eyeball is too short so near objects are blurry.	Myopia (nearsightedness): Eyeball is too long so far objects are blurry.

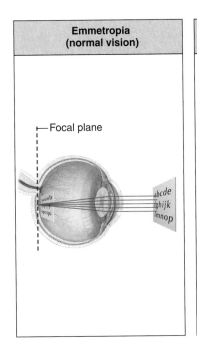

Focal plane

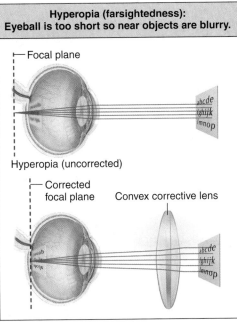

Focal plane

Hyperopia (uncorrected)

Corrected focal plane   Convex corrective lens

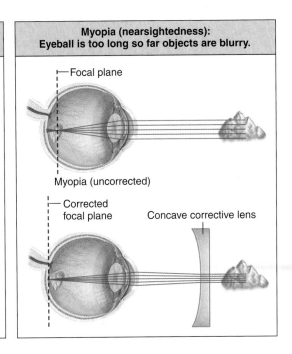

Focal plane

Myopia (uncorrected)

Corrected focal plane   Concave corrective lens

Emmetropia, Hyperopia and Myopia
Clinical View p. 592

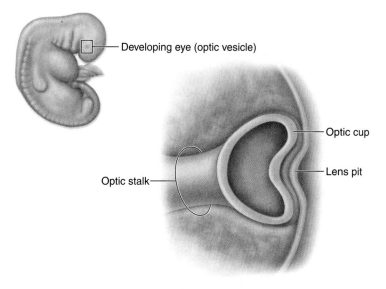

(a) **Early week 4: Optic vesicle forms on an optic cup; ectoderm forms a lens pit.**

Developing eye (optic vesicle)
Optic cup
Lens pit
Optic stalk

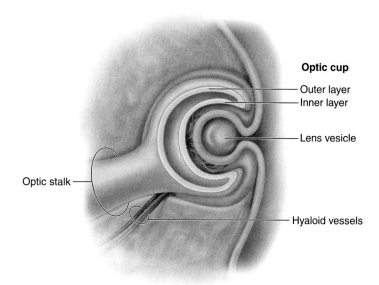

(b) **Late week 4: Optic cup deepens and forms inner and outer layers; lens pit forms lens vesicle.**

Optic cup
Outer layer
Inner layer
Lens vesicle
Optic stalk
Hyaloid vessels

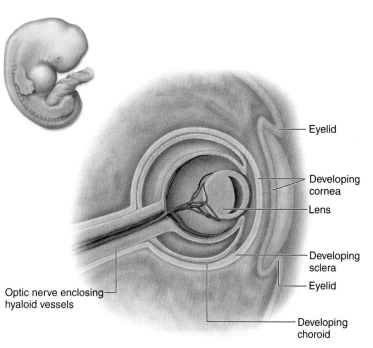

(c) **Week 6: Lens becomes an internal structure: corneas, sclera, and choroid start to form.**

Eyelid
Developing cornea
Lens
Developing sclera
Eyelid
Optic nerve enclosing hyaloid vessels
Developing choroid

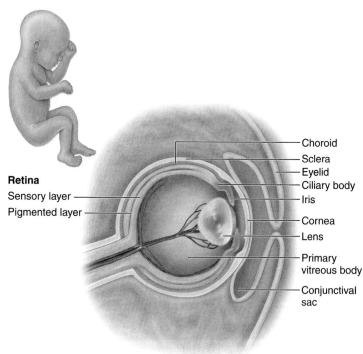

(d) **Week 20: Three tunics of the eye have formed.**

Retina
Sensory layer
Pigmented layer
Choroid
Sclera
Eyelid
Ciliary body
Iris
Cornea
Lens
Primary vitreous body
Conjunctival sac

Eye Development
Figure 19.20

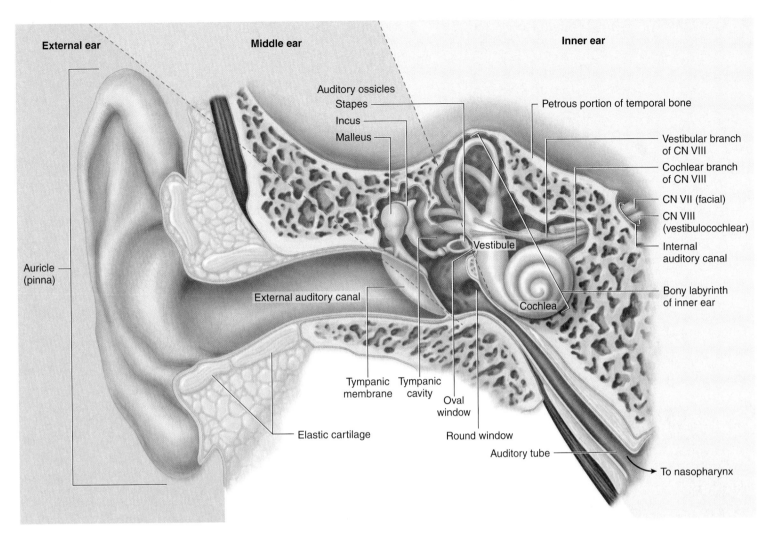

Anatomic Regions of the Right Ear
Figure 19.21

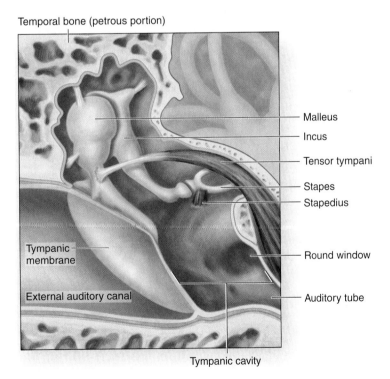

Right Middle Ear
Figure 19.22

281

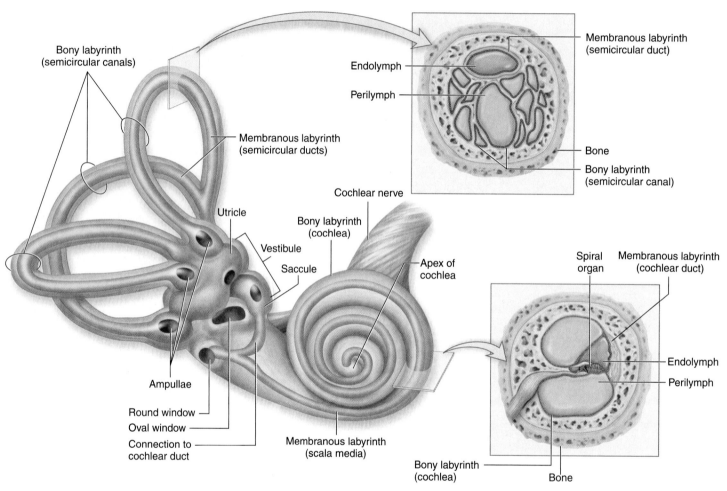

Bony labyrinth
(semicircular canals)

Membranous labyrinth
(semicircular ducts)

Membranous labyrinth
(semicircular duct)

Endolymph

Perilymph

Bone

Bony labyrinth
(semicircular canal)

Cochlear nerve

Utricle

Bony labyrinth
(cochlea)

Vestibule

Saccule

Apex of
cochlea

Spiral
organ

Membranous labyrinth
(cochlear duct)

Endolymph

Perilymph

Ampullae

Round window

Oval window

Connection to
cochlear duct

Membranous labyrinth
(scala media)

Bony labyrinth
(cochlea)

Bone

Right Inner Ear
Figure 19.23

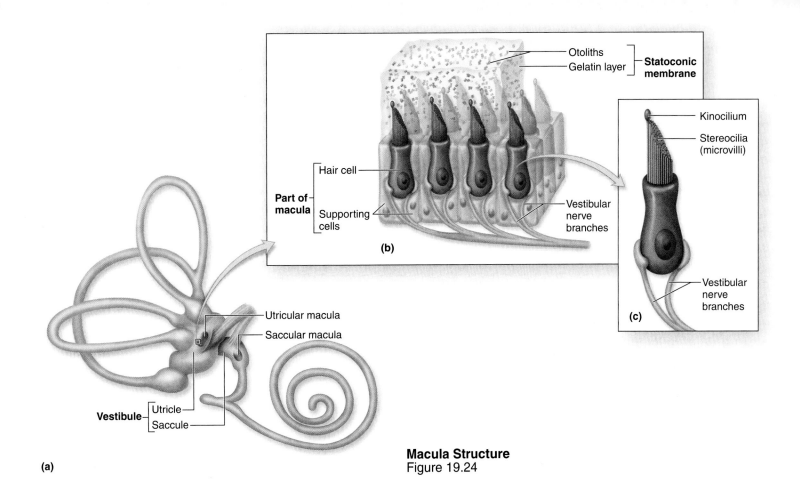

Otoliths
Gelatin layer
Statoconic membrane

Hair cell

Part of macula

Supporting cells

Vestibular nerve branches

(b)

Kinocilium

Stereocilia (microvilli)

Vestibular nerve branches

(c)

Utricular macula

Saccular macula

Vestibule — Utricle
Saccule

(a)

Macula Structure
Figure 19.24

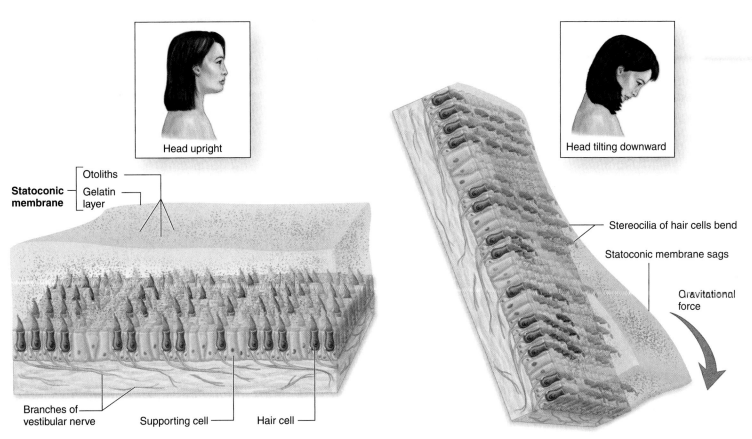

Head upright

Statoconic membrane — Otoliths
Gelatin layer

Branches of vestibular nerve

Supporting cell

Hair cell

Head tilting downward

Stereocilia of hair cells bend

Statoconic membrane sags

Gravitational force

Detecting Head Position
Figure 19.25

283

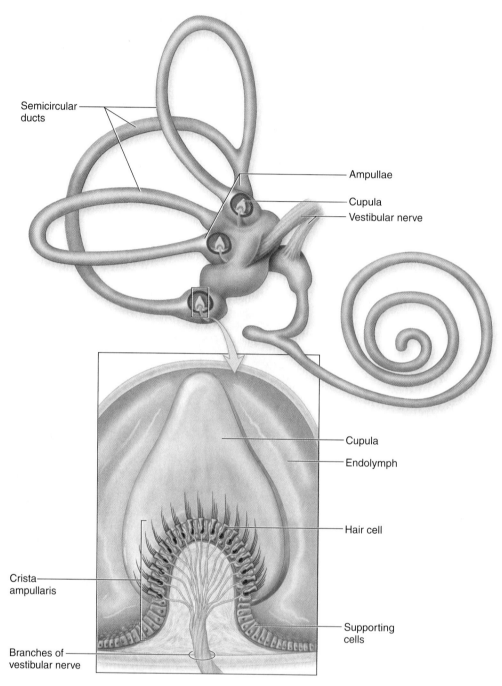

Semicircular ducts

Ampullae

Cupula

Vestibular nerve

Cupula

Endolymph

Hair cell

Crista ampullaris

Supporting cells

Branches of vestibular nerve

Ampulla Structure
Figure 19.26

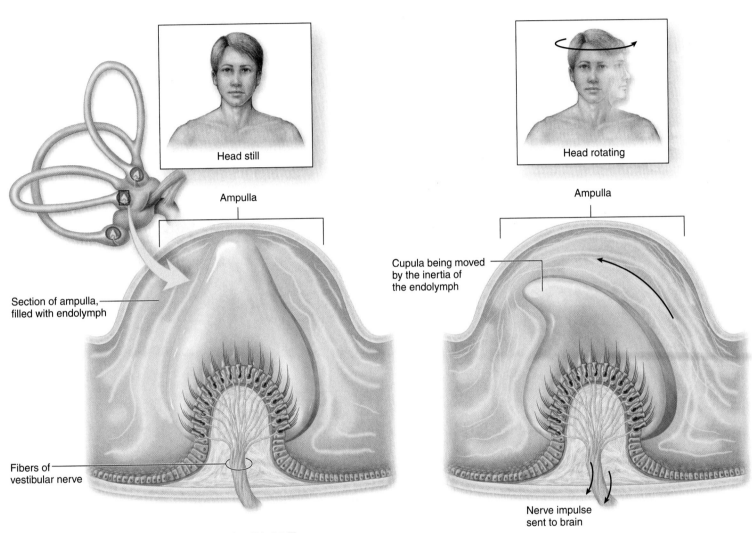

Head still

Head rotating

Ampulla

Ampulla

Cupula being moved
by the inertia of
the endolymph

Section of ampulla,
filled with endolymph

Fibers of
vestibular nerve

Nerve impulse
sent to brain

Function of the Crista Ampullaris in the Right Ear
Figure 19.27

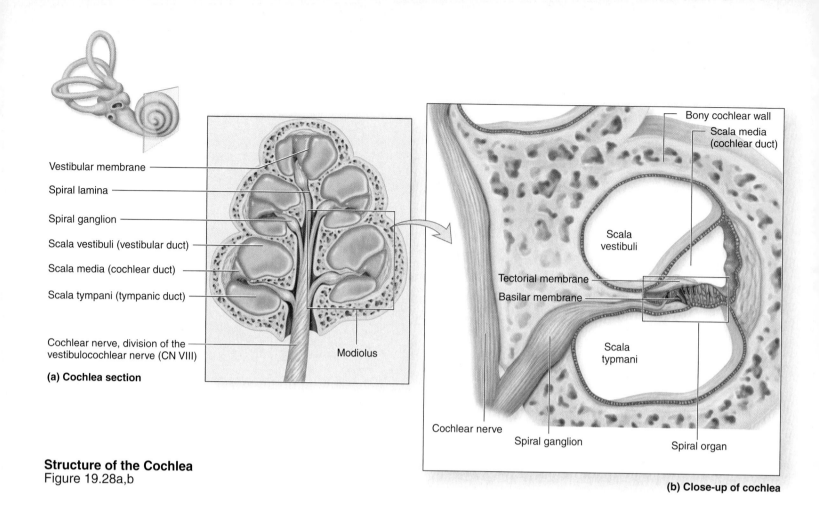

Vestibular membrane

Spiral lamina

Spiral ganglion

Scala vestibuli (vestibular duct)

Scala media (cochlear duct)

Scala tympani (tympanic duct)

Cochlear nerve, division of the vestibulocochlear nerve (CN VIII)

Modiolus

(a) Cochlea section

Bony cochlear wall

Scala media (cochlear duct)

Scala vestibuli

Tectorial membrane

Basilar membrane

Scala typmani

Cochlear nerve

Spiral ganglion

Spiral organ

(b) Close-up of cochlea

Structure of the Cochlea
Figure 19.28a,b

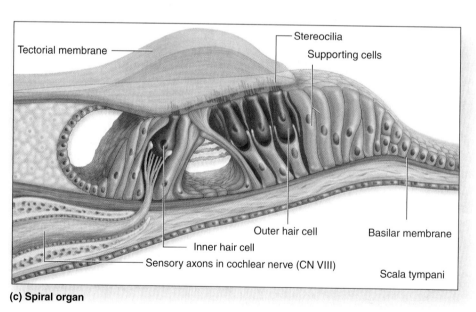

Tectorial membrane

Stereocilia

Supporting cells

Outer hair cell

Basilar membrane

Inner hair cell

Sensory axons in cochlear nerve (CN VIII)

Scala tympani

(c) Spiral organ

Spiral Organ
Figure 19.28c

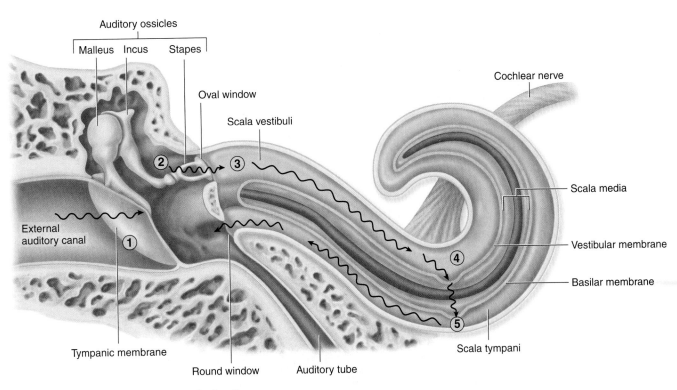

Auditory ossicles

Malleus Incus Stapes

Oval window

Scala vestibuli

Cochlear nerve

External
auditory canal

Scala media

Vestibular membrane

Basilar membrane

Tympanic membrane

Round window

Auditory tube

Scala tympani

Sound Wave Pathways Through the Ear
Figure 19.29

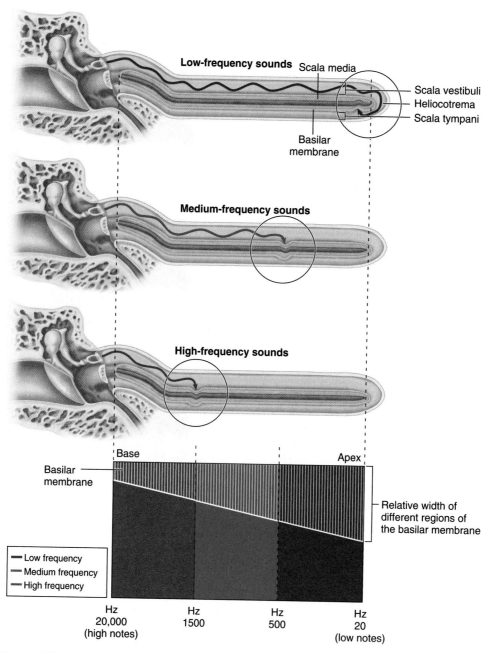

Low-frequency sounds Scala media

Scala vestibuli
Heliocotrema
Scala tympani

Basilar
membrane

Medium-frequency sounds

High-frequency sounds

Base

Apex

Basilar
membrane

Relative width of
different regions of
the basilar membrane

- Low frequency
- Medium frequency
- High frequency

Hz
20,000
(high notes)

Hz
1500

Hz
500

Hz
20
(low notes)

Sound Wave Interpretations at the Basilar Membrane
Figure 19.30

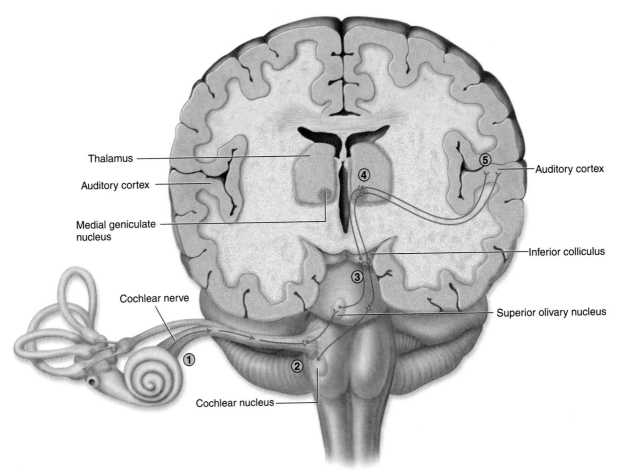

Thalamus

Auditory cortex

Medial geniculate nucleus

Cochlear nerve

④

⑤

Auditory cortex

Inferior colliculus

③

Superior olivary nucleus

①

②

Cochlear nucleus

Central Nervous System Pathways for Hearing
Figure 19.31

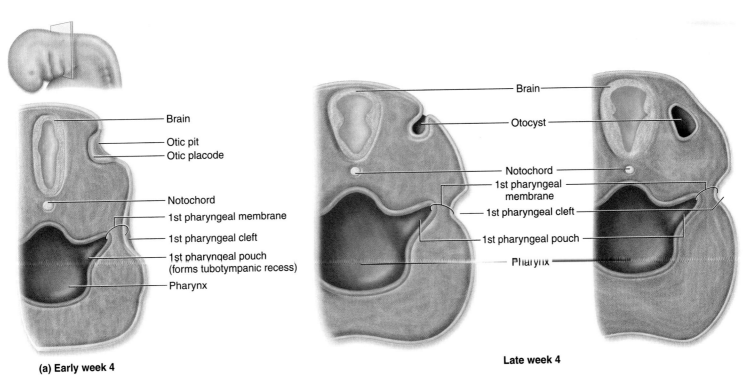

Brain

Otic pit
Otic placode

Notochord

1st pharyngeal membrane

1st pharyngeal cleft

1st pharyngeal pouch
(forms tubotympanic recess)

Pharynx

(a) Early week 4

Brain

Otocyst

Notochord

1st pharyngeal membrane

1st pharyngeal cleft

1st pharyngeal pouch

Pharynx

Late week 4

Development of the Ear—Week 4
Figure 19.32a

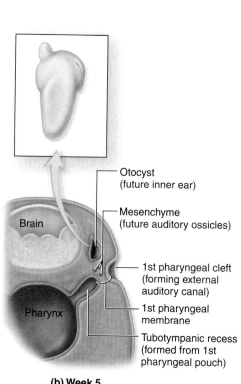

(b) Week 5

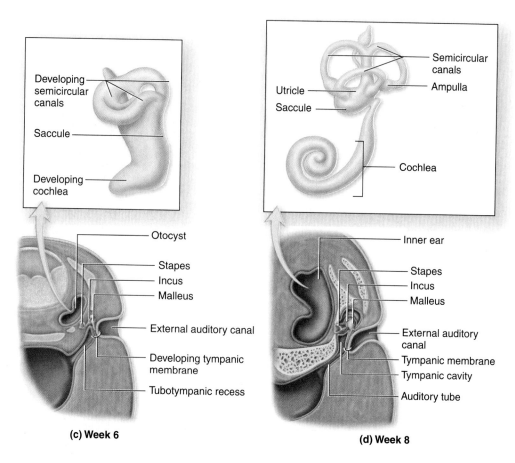

(c) Week 6

(d) Week 8

Development of the Ear -Weeks 6-8
Figure 19.32b,c,d

290

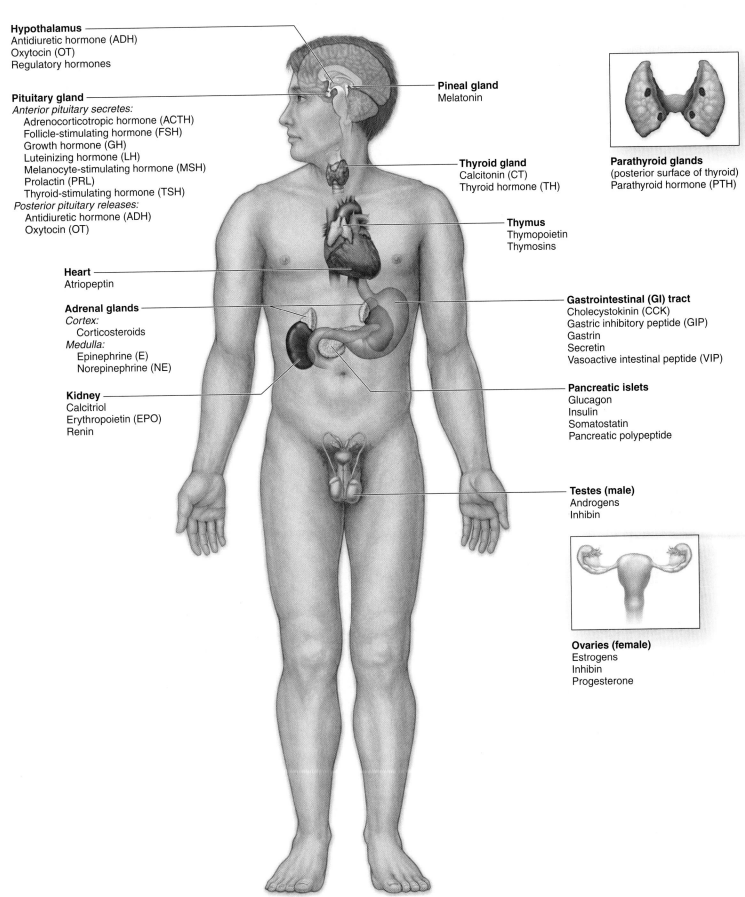

Hypothalamus
Antidiuretic hormone (ADH)
Oxytocin (OT)
Regulatory hormones

Pituitary gland
Anterior pituitary secretes:
  Adrenocorticotropic hormone (ACTH)
  Follicle-stimulating hormone (FSH)
  Growth hormone (GH)
  Luteinizing hormone (LH)
  Melanocyte-stimulating hormone (MSH)
  Prolactin (PRL)
  Thyroid-stimulating hormone (TSH)
Posterior pituitary releases:
  Antidiuretic hormone (ADH)
  Oxytocin (OT)

Heart
Atriopeptin

Adrenal glands
Cortex:
  Corticosteroids
Medulla:
  Epinephrine (E)
  Norepinephrine (NE)

Kidney
Calcitriol
Erythropoietin (EPO)
Renin

Pineal gland
Melatonin

Thyroid gland
Calcitonin (CT)
Thyroid hormone (TH)

Thymus
Thymopoietin
Thymosins

Parathyroid glands
(posterior surface of thyroid)
Parathyroid hormone (PTH)

Gastrointestinal (GI) tract
Cholecystokinin (CCK)
Gastric inhibitory peptide (GIP)
Gastrin
Secretin
Vasoactive intestinal peptide (VIP)

Pancreatic islets
Glucagon
Insulin
Somatostatin
Pancreatic polypeptide

Testes (male)
Androgens
Inhibin

Ovaries (female)
Estrogens
Inhibin
Progesterone

Endocrine System
Figure 20.1

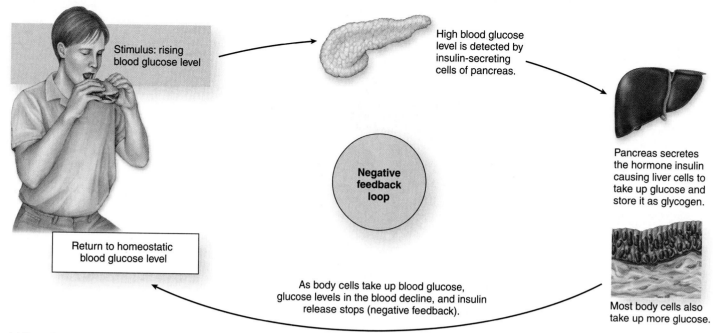

(a) Negative feedback

Negative Feedback Loop
Figure 20.2a

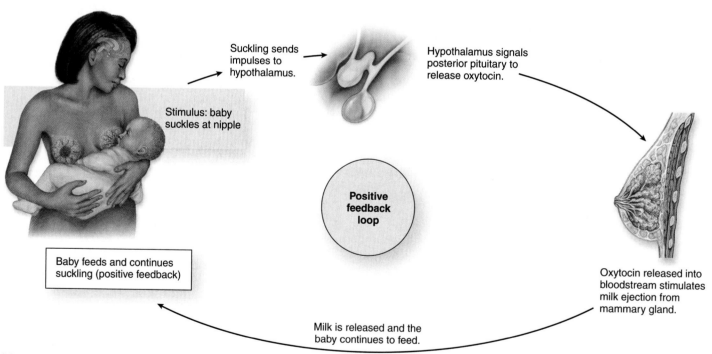

(b) Positive feedback

Positive Feedback Loop
Figure 20.2b

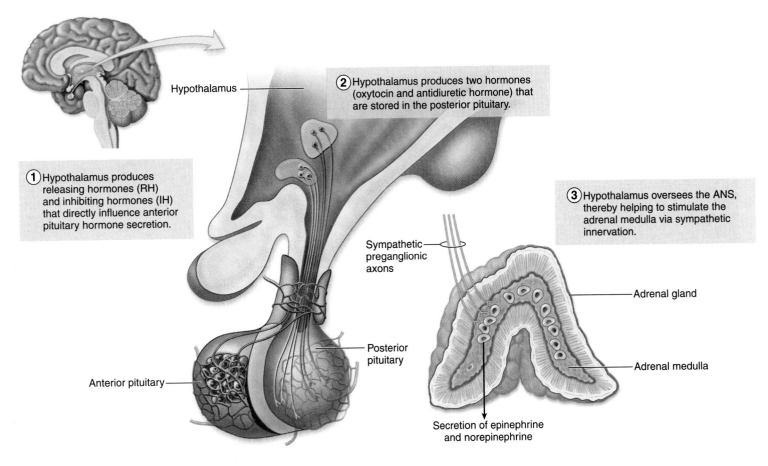

① Hypothalamus produces releasing hormones (RH) and inhibiting hormones (IH) that directly influence anterior pituitary hormone secretion.

Hypothalamus

② Hypothalamus produces two hormones (oxytocin and antidiuretic hormone) that are stored in the posterior pituitary.

③ Hypothalamus oversees the ANS, thereby helping to stimulate the adrenal medulla via sympathetic innervation.

Sympathetic preganglionic axons

Adrenal gland

Posterior pituitary

Adrenal medulla

Anterior pituitary

Secretion of epinephrine and norepinephrine

How the Hypothalamus Controls Endocrine Function
Figure 20.3

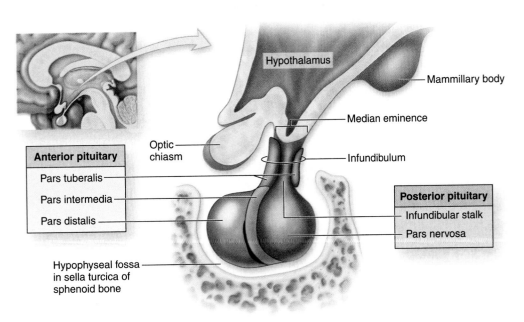

Hypothalamus

Mammillary body

Median eminence

Optic chiasm

Infundibulum

Anterior pituitary

Pars tuberalis

Pars intermedia

Pars distalis

Posterior pituitary

Infundibular stalk

Pars nervosa

Hypophyseal fossa in sella turcica of sphenoid bone

Pituitary Gland
Figure 20.4

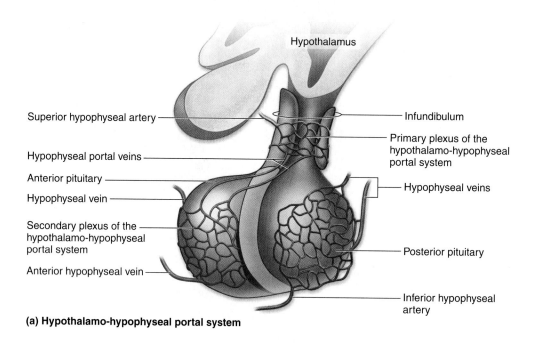

Hypothalamus

Superior hypophyseal artery

Hypophyseal portal veins

Anterior pituitary

Hypophyseal vein

Secondary plexus of the hypothalamo-hypophyseal portal system

Anterior hypophyseal vein

Infundibulum

Primary plexus of the hypothalamo-hypophyseal portal system

Hypophyseal veins

Posterior pituitary

Inferior hypophyseal artery

(a) Hypothalamo-hypophyseal portal system

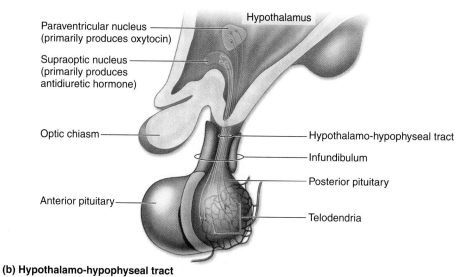

Hypothalamus

Paraventricular nucleus (primarily produces oxytocin)

Supraoptic nucleus (primarily produces antidiuretic hormone)

Optic chiasm

Anterior pituitary

Hypothalamo-hypophyseal tract

Infundibulum

Posterior pituitary

Telodendria

(b) Hypothalamo-hypophyseal tract

Hypothalamic Control of the Pituitary Gland
Figure 20.6

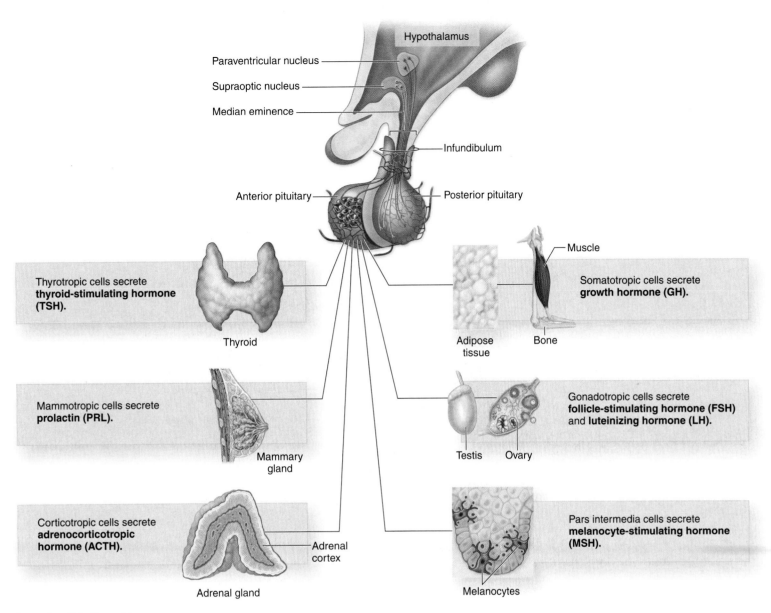

Hypothalamus

Paraventricular nucleus
Supraoptic nucleus
Median eminence

Infundibulum

Anterior pituitary

Posterior pituitary

Thyrotropic cells secrete **thyroid-stimulating hormone (TSH).**

Thyroid

Muscle

Somatotropic cells secrete **growth hormone (GH).**

Adipose tissue

Bone

Mammotropic cells secrete **prolactin (PRL).**

Mammary gland

Gonadotropic cells secrete **follicle-stimulating hormone (FSH)** and **luteinizing hormone (LH).**

Testis   Ovary

Corticotropic cells secrete **adrenocorticotropic hormone (ACTH).**

Adrenal cortex

Adrenal gland

Pars intermedia cells secrete **melanocyte-stimulating hormone (MSH).**

Melanocytes

Anterior Pituitary Hormones
Figure 20.7

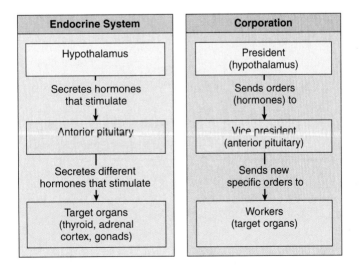

Endocrine System	Corporation
Hypothalamus	President (hypothalamus)
Secretes hormones that stimulate ↓	Sends orders (hormones) to ↓
Anterior pituitary	Vice president (anterior pituitary)
Secretes different hormones that stimulate ↓	Sends new specific orders to ↓
Target organs (thyroid, adrenal cortex, gonads)	Workers (target organs)

Endocrine System vs. Corporation
Study Tip p. 625

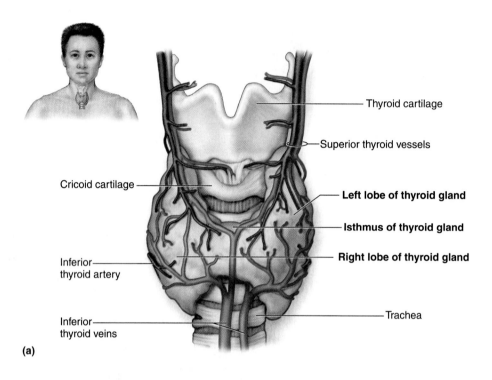

Thyroid cartilage

Superior thyroid vessels

Cricoid cartilage

Left lobe of thyroid gland

Isthmus of thyroid gland

Right lobe of thyroid gland

Inferior thyroid artery

Trachea

Inferior thyroid veins

(a)

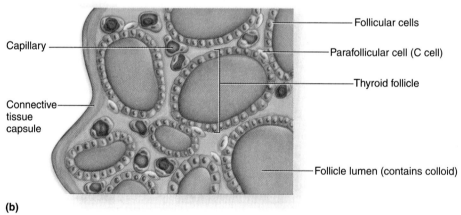

Capillary

Follicular cells

Parafollicular cell (C cell)

Thyroid follicle

Connective tissue capsule

Follicle lumen (contains colloid)

(b)

Thyroid Gland
Figure 20.8

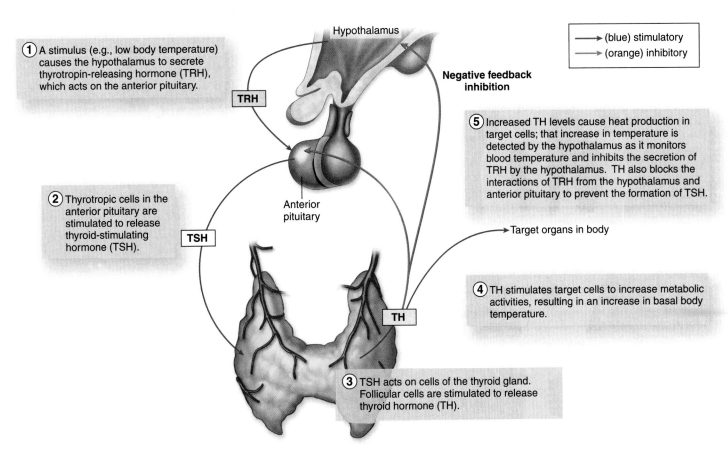

① A stimulus (e.g., low body temperature) causes the hypothalamus to secrete thyrotropin-releasing hormone (TRH), which acts on the anterior pituitary.

Hypothalamus

Negative feedback inhibition

→ (blue) stimulatory
→ (orange) inhibitory

TRH

② Thyrotropic cells in the anterior pituitary are stimulated to release thyroid-stimulating hormone (TSH).

TSH

Anterior pituitary

⑤ Increased TH levels cause heat production in target cells; that increase in temperature is detected by the hypothalamus as it monitors blood temperature and inhibits the secretion of TRH by the hypothalamus. TH also blocks the interactions of TRH from the hypothalamus and anterior pituitary to prevent the formation of TSH.

Target organs in body

④ TH stimulates target cells to increase metabolic activities, resulting in an increase in basal body temperature.

TH

③ TSH acts on cells of the thyroid gland. Follicular cells are stimulated to release thyroid hormone (TH).

Thyroid Gland-Pituitary Gland Negative Feedback Loop
Figure 20.9

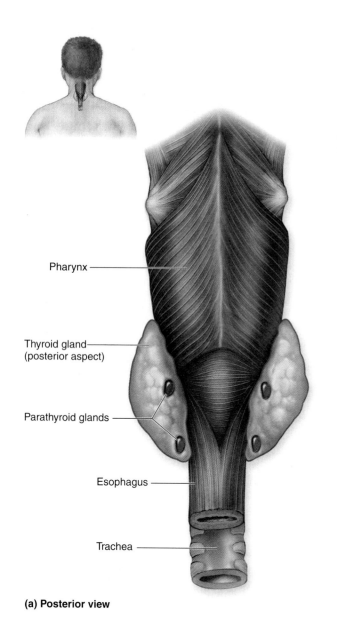

Pharynx

Thyroid gland
(posterior aspect)

Parathyroid glands

Esophagus

Trachea

(a) Posterior view

Connective tissue capsule
of parathyroid gland

Oxyphil cell

Chief cells

Capillary

(b) Histological view

Parathyroid Glands
Figure 20.10a, b

298

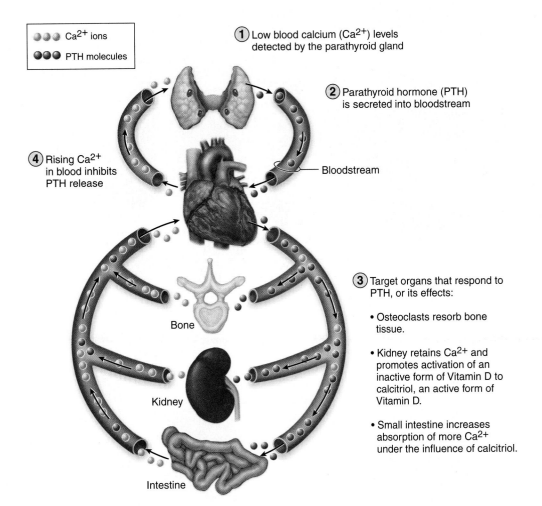

① Low blood calcium (Ca^{2+}) levels detected by the parathyroid gland

② Parathyroid hormone (PTH) is secreted into bloodstream

Bloodstream

④ Rising Ca^{2+} in blood inhibits PTH release

○○○ Ca^{2+} ions

●●● PTH molecules

Bone

Kidney

Intestine

③ Target organs that respond to PTH, or its effects:

- Osteoclasts resorb bone tissue.

- Kidney retains Ca^{2+} and promotes activation of an inactive form of Vitamin D to calcitriol, an active form of Vitamin D.

- Small intestine increases absorption of more Ca^{2+} under the influence of calcitriol.

Parathyroid Hormone
Figure 20.11

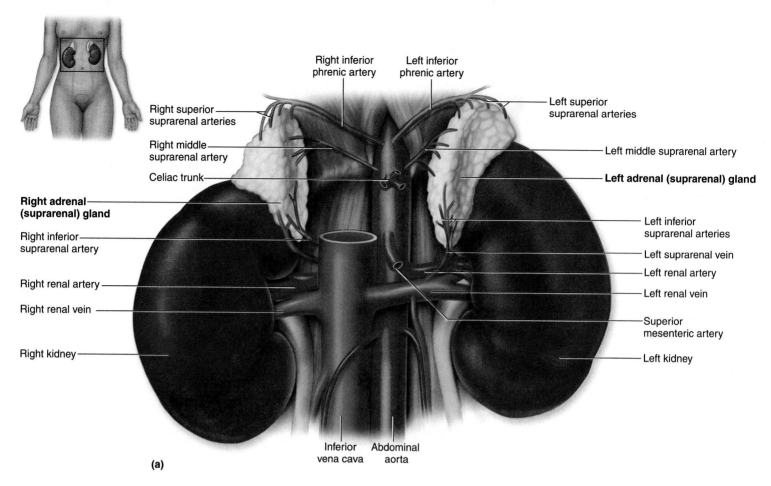

Right inferior phrenic artery

Left inferior phrenic artery

Right superior suprarenal arteries

Left superior suprarenal arteries

Right middle suprarenal artery

Left middle suprarenal artery

Celiac trunk

Left adrenal (suprarenal) gland

Right adrenal (suprarenal) gland

Left inferior suprarenal arteries

Right inferior suprarenal artery

Left suprarenal vein

Left renal artery

Right renal artery

Left renal vein

Right renal vein

Superior mesenteric artery

Right kidney

Left kidney

Inferior vena cava

Abdominal aorta

(a)

Adrenal Gland—Anterior View
Figure 20.12a

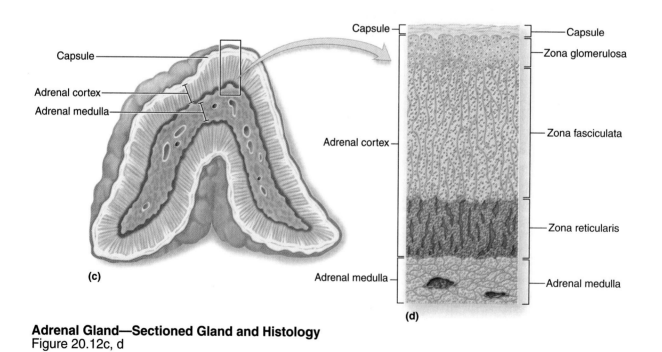

Capsule

Capsule

Capsule

Zona glomerulosa

Adrenal cortex

Adrenal medulla

Adrenal cortex

Zona fasciculata

Zona reticularis

Adrenal medulla

Adrenal medulla

(c)

(d)

Adrenal Gland—Sectioned Gland and Histology
Figure 20.12c, d

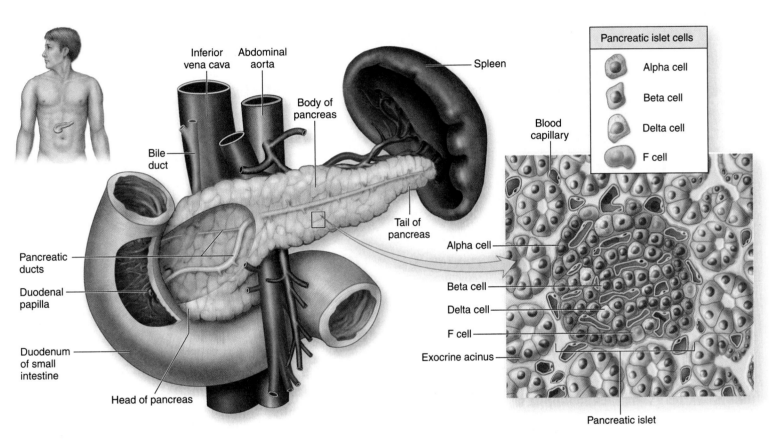

Inferior vena cava

Abdominal aorta

Body of pancreas

Spleen

Bile duct

Pancreatic ducts

Duodenal papilla

Tail of pancreas

Duodenum of small intestine

Head of pancreas

Pancreatic islet cells

Alpha cell

Beta cell

Delta cell

F cell

Blood capillary

Alpha cell

Beta cell

Delta cell

F cell

Exocrine acinus

Pancreatic islet

Pancreas
Figure 20.13

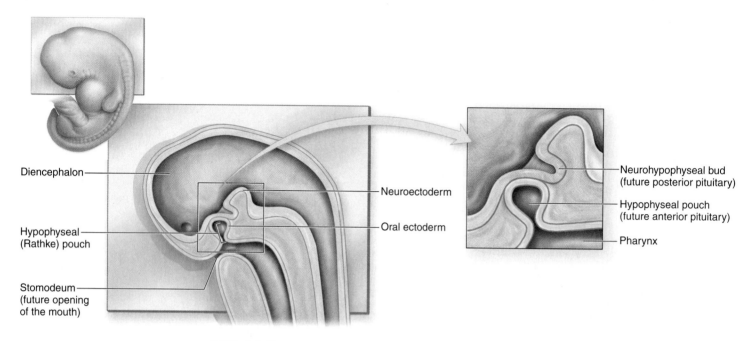

Diencephalon

Hypophyseal (Rathke) pouch

Stomodeum (future opening of the mouth)

Neuroectoderm

Oral ectoderm

Neurohypophyseal bud (future posterior pituitary)

Hypophyseal pouch (future anterior pituitary)

Pharynx

(a) Week 3: Neurohypophyseal bud and hypophyseal pouch form.

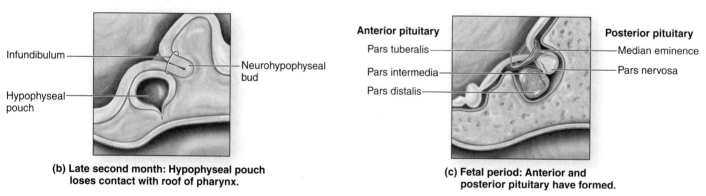

Infundibulum

Hypophyseal pouch

Neurohypophyseal bud

(b) Late second month: Hypophyseal pouch loses contact with roof of pharynx.

Anterior pituitary

Pars tuberalis

Pars intermedia

Pars distalis

Posterior pituitary

Median eminence

Pars nervosa

(c) Fetal period: Anterior and posterior pituitary have formed.

Pituitary Gland Development
Figure 20.14

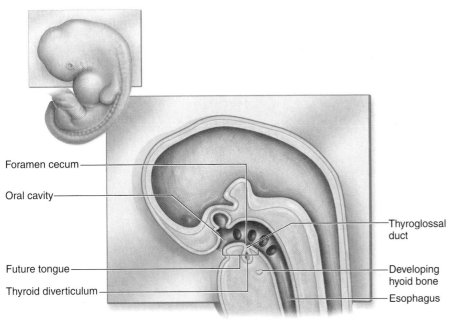

Foramen cecum

Oral cavity

Future tongue

Thyroid diverticulum

Thyroglossal duct

Developing hyoid bone

Esophagus

(a) Week 4: Thyroid diverticulum forms.

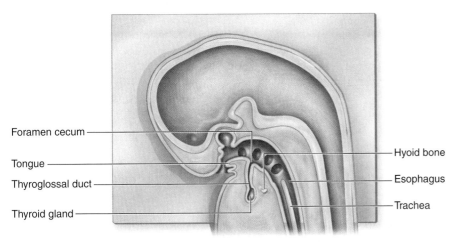

Foramen cecum

Tongue

Thyroglossal duct

Thyroid gland

Hyoid bone

Esophagus

Trachea

(b) Week 7: Thyroid gland migrates inferiorly.

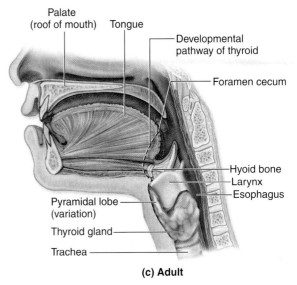

Palate (roof of mouth)

Tongue

Developmental pathway of thyroid

Foramen cecum

Hyoid bone

Larynx

Esophagus

Pyramidal lobe (variation)

Thyroid gland

Trachea

(c) Adult

Thyroid Gland Development
Figure 20.15

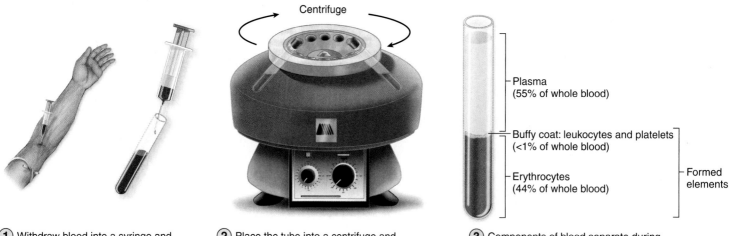

1. Withdraw blood into a syringe and place in a glass tube.

2. Place the tube into a centrifuge and spin for about 10 minutes.

3. Components of blood separate during centrifugation to reveal plasma, buffy coat, and erythrocytes.

Centrifuge

Plasma
(55% of whole blood)

Buffy coat: leukocytes and platelets
(<1% of whole blood)

Erythrocytes
(44% of whole blood)

Formed elements

Whole Blood Separation
Figure 21.1

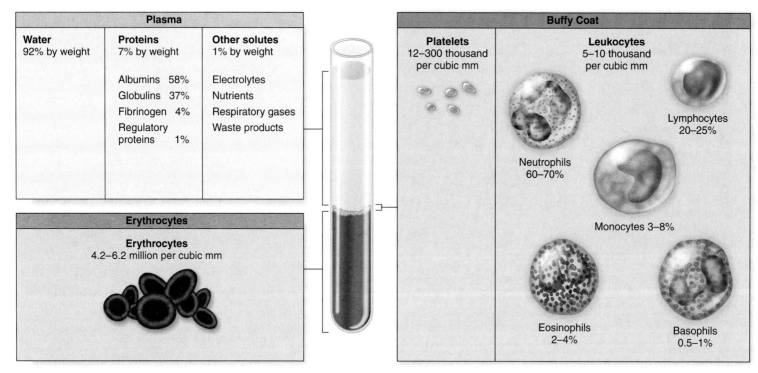

Plasma		
Water 92% by weight	**Proteins** 7% by weight	**Other solutes** 1% by weight
	Albumins 58%	Electrolytes
	Globulins 37%	Nutrients
	Fibrinogen 4%	Respiratory gases
	Regulatory proteins 1%	Waste products

Erythrocytes

Erythrocytes
4.2–6.2 million per cubic mm

Buffy Coat

Platelets
12–300 thousand per cubic mm

Leukocytes
5–10 thousand per cubic mm

Neutrophils 60–70%

Lymphocytes 20–25%

Monocytes 3–8%

Eosinophils 2–4%

Basophils 0.5–1%

Whole Blood Composition
Figure 21.2

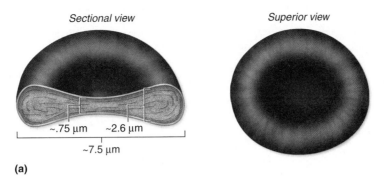

Sectional view

Superior view

~.75 μm ~2.6 μm

~7.5 μm

(a)

Erythrocyte Structure
Figure 21.4a

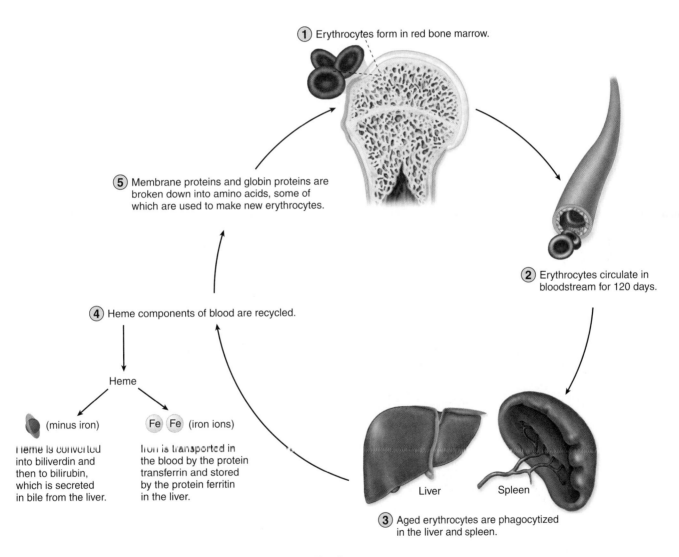

(1) Erythrocytes form in red bone marrow.

(5) Membrane proteins and globin proteins are broken down into amino acids, some of which are used to make new erythrocytes.

(2) Erythrocytes circulate in bloodstream for 120 days.

(4) Heme components of blood are recycled.

Heme

(minus iron)

Fe Fe (iron ions)

Heme is converted into biliverdin and then to bilirubin, which is secreted in bile from the liver.

Iron is transported in the blood by the protein transferrin and stored by the protein ferritin in the liver.

Liver   Spleen

(3) Aged erythrocytes are phagocytized in the liver and spleen.

Recycling the Components of Aged or Damaged Erythrocytes
Figure 21.6

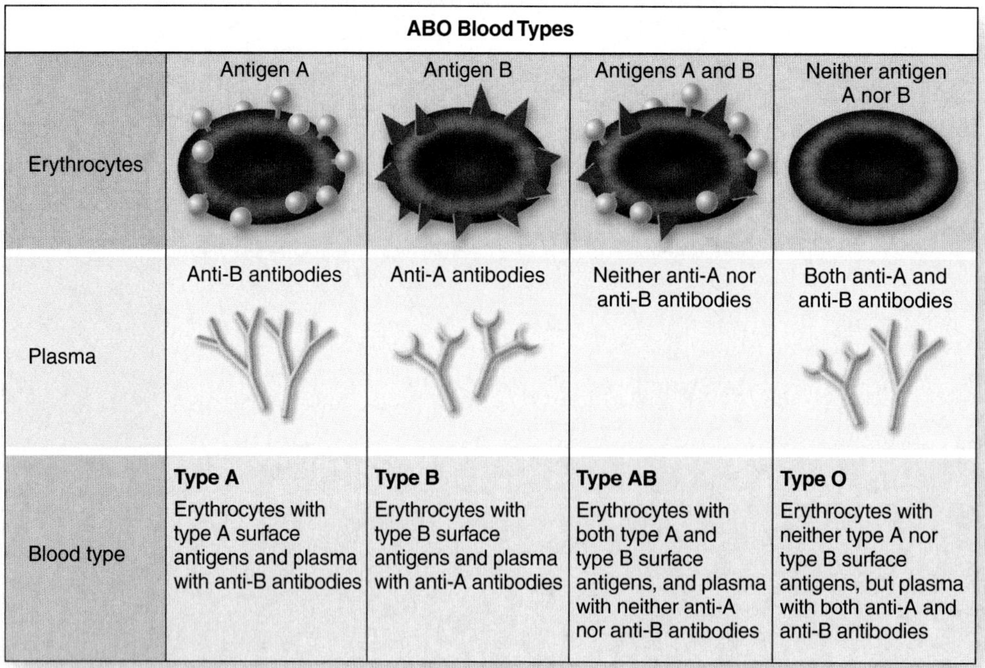

ABO Blood Types

	Antigen A	Antigen B	Antigens A and B	Neither antigen A nor B
Erythrocytes				
Plasma	Anti-B antibodies	Anti-A antibodies	Neither anti-A nor anti-B antibodies	Both anti-A and anti-B antibodies
Blood type	**Type A** Erythrocytes with type A surface antigens and plasma with anti-B antibodies	**Type B** Erythrocytes with type B surface antigens and plasma with anti-A antibodies	**Type AB** Erythrocytes with both type A and type B surface antigens, and plasma with neither anti-A nor anti-B antibodies	**Type O** Erythrocytes with neither type A nor type B surface antigens, but plasma with both anti-A and anti-B antibodies

(a)

ABO Blood Types
Figure 21.7a

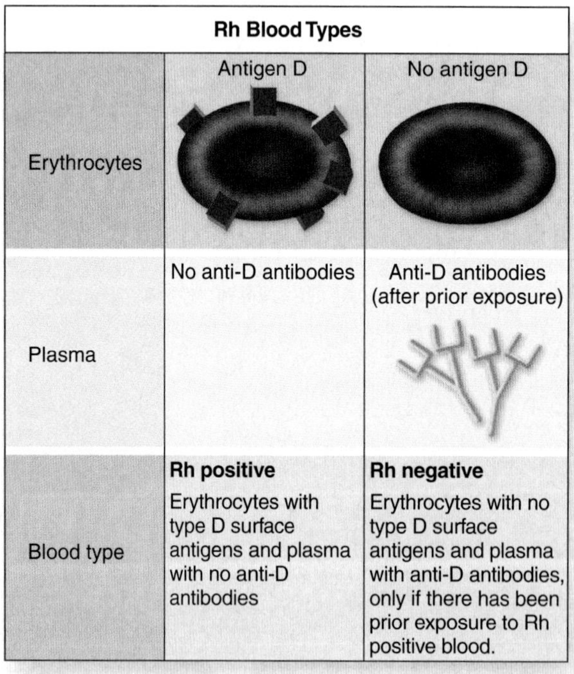

Rh Blood Types

	Antigen D	No antigen D
Erythrocytes		
Plasma	No anti-D antibodies	Anti-D antibodies (after prior exposure)
Blood type	**Rh positive** Erythrocytes with type D surface antigens and plasma with no anti-D antibodies	**Rh negative** Erythrocytes with no type D surface antigens and plasma with anti-D antibodies, only if there has been prior exposure to Rh positive blood.

(b)

Rh Blood Types
Figure 21.7b

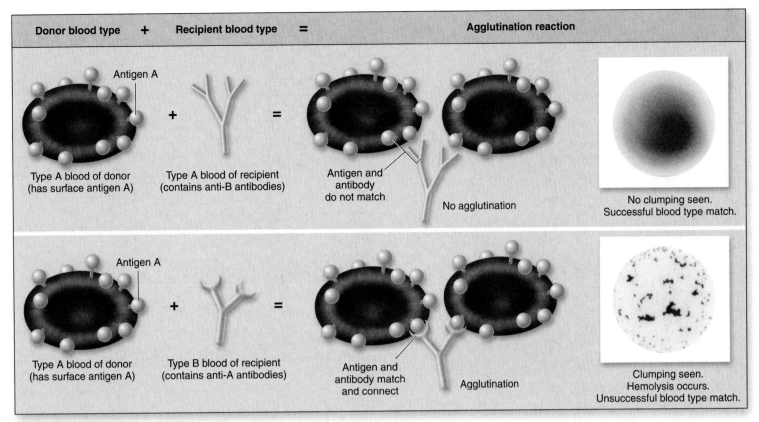

Donor blood type + Recipient blood type = Agglutination reaction

Antigen A

Type A blood of donor (has surface antigen A)

+

Type A blood of recipient (contains anti-B antibodies)

=

Antigen and antibody do not match

No agglutination

No clumping seen. Successful blood type match.

Antigen A

Type A blood of donor (has surface antigen A)

+

Type B blood of recipient (contains anti-A antibodies)

=

Antigen and antibody match and connect

Agglutination

Clumping seen. Hemolysis occurs. Unsuccessful blood type match.

(a)

Agglutination Reaction
Figure 21.8a

Blood from type A donor

Type B blood recipient with anti-A antibodies in plasma

Type A donor erythrocyte

Agglutinated erythrocytes block small vessels

(b)

Agglutination Reaction as seen in Blood Vessel
Figure 21.8b

307

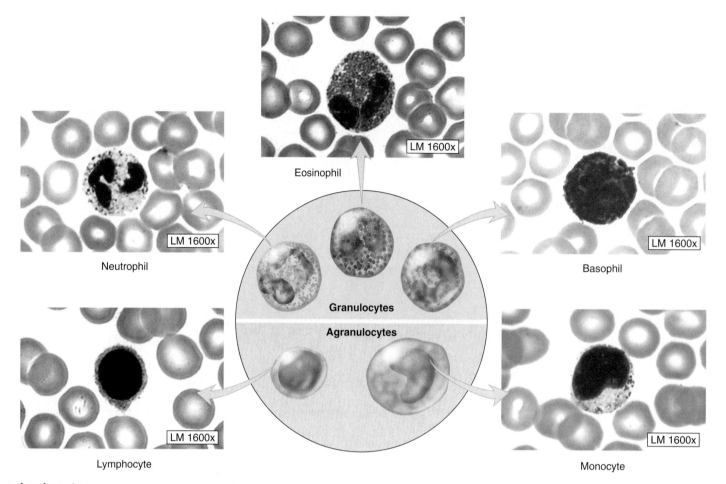

LM 1600x

Eosinophil

LM 1600x

Neutrophil

LM 1600x

Basophil

Granulocytes

Agranulocytes

LM 1600x

Lymphocyte

LM 1600x

Monocyte

Leukocytes
Table 21.3
Table 21.3: © The McGraw-Hill Companies, Inc./Photo by Dr. Alvin Telser

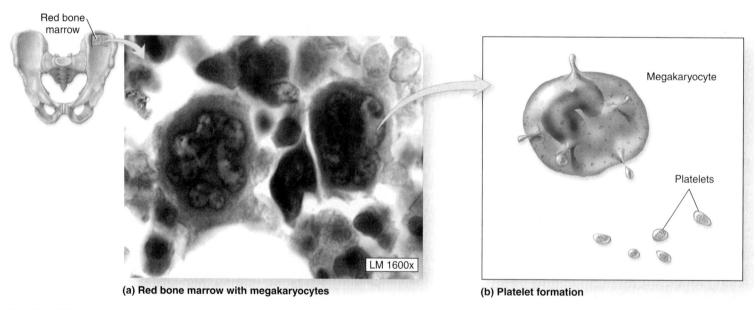

Red bone marrow

Megakaryocyte

Platelets

LM 1600x

(a) Red bone marrow with megakaryocytes

(b) Platelet formation

Origin of Platelets
Figure 21.9
Figure 21.9: © The McGraw-Hill Companies, Inc./Photo by Dr. Alvin Telser

Origin, Differentiation, and Maturation of Formed Elements
Figure 21.11

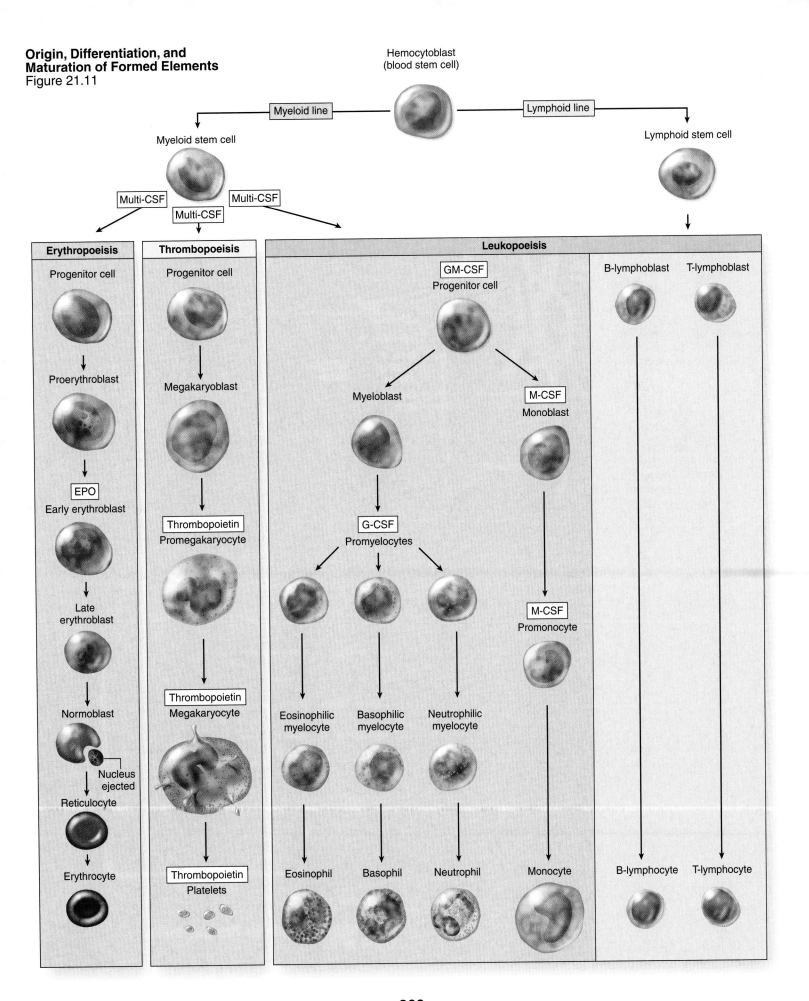

Hemocytoblast
(blood stem cell)

Myeloid line   Lymphoid line

Myeloid stem cell

Lymphoid stem cell

Multi-CSF   Multi-CSF   Multi-CSF

Erythropoeisis

Progenitor cell

Proerythroblast

EPO
Early erythroblast

Late
erythroblast

Normoblast

Nucleus
ejected

Reticulocyte

Erythrocyte

Thrombopoeisis

Progenitor cell

Megakaryoblast

Thrombopoietin
Promegakaryocyte

Thrombopoietin
Megakaryocyte

Thrombopoietin
Platelets

Leukopoeisis

GM-CSF
Progenitor cell

Myeloblast

M-CSF
Monoblast

G-CSF
Promyelocytes

M-CSF
Promonocyte

Eosinophilic
myelocyte

Basophilic
myelocyte

Neutrophilic
myelocyte

Eosinophil   Basophil   Neutrophil   Monocyte

B-lymphoblast   T-lymphoblast

B-lymphocyte   T-lymphocyte

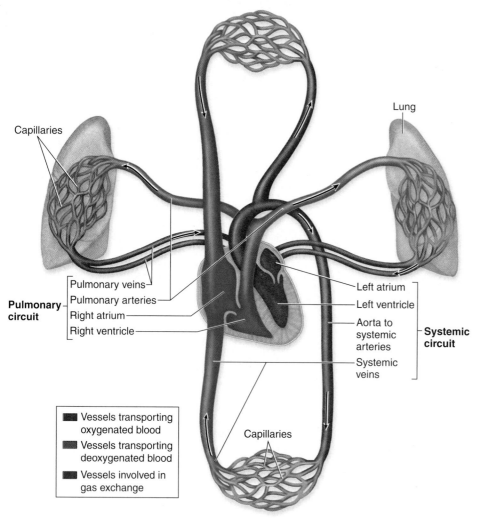

Capillaries

Lung

Pulmonary circuit
- Pulmonary veins
- Pulmonary arteries
- Right atrium
- Right ventricle

Left atrium
Left ventricle
Aorta to systemic arteries
Systemic circuit
Systemic veins

Capillaries

- ■ Vessels transporting oxygenated blood
- ■ Vessels transporting deoxygenated blood
- ■ Vessels involved in gas exchange

Cardiovascular System
Figure 22.1

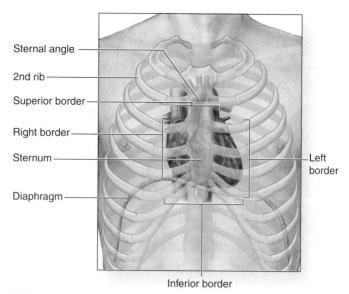

Sternal angle
2nd rib
Superior border
Right border
Sternum
Diaphragm
Left border
Inferior border

(a) Anterior view

Heart Position Within the Thoracic Cavity
Figure 22.2a

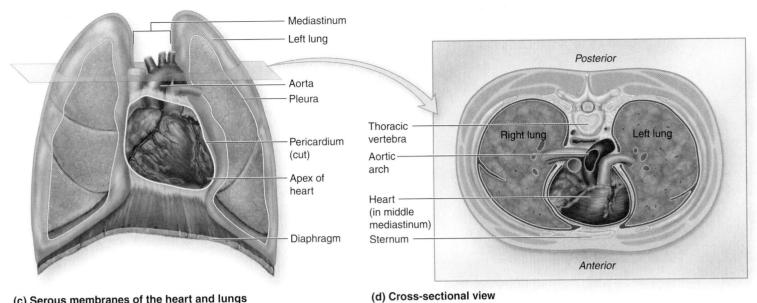

Mediastinum

Left lung

Aorta

Pleura

Pericardium (cut)

Apex of heart

Diaphragm

(c) Serous membranes of the heart and lungs

Posterior

Thoracic vertebra

Aortic arch

Right lung

Left lung

Heart (in middle mediastinum)

Sternum

Anterior

(d) Cross-sectional view

Serous Membranes of the Heart and Lungs
Figure 22.2c,d

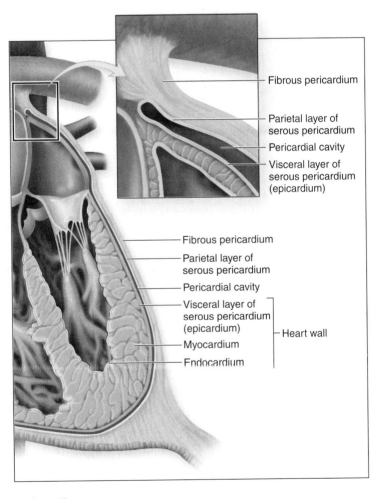

Fibrous pericardium

Parietal layer of serous pericardium

Pericardial cavity

Visceral layer of serous pericardium (epicardium)

Fibrous pericardium

Parietal layer of serous pericardium

Pericardial cavity

Visceral layer of serous pericardium (epicardium)

Myocardium

Endocardium

Heart wall

Pericardium
Figure 22.3

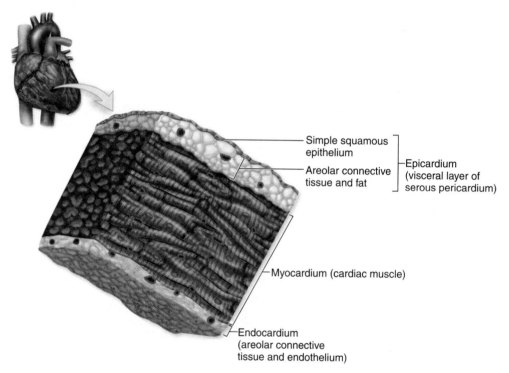

Simple squamous epithelium

Areolar connective tissue and fat

Epicardium (visceral layer of serous pericardium)

Myocardium (cardiac muscle)

Endocardium (areolar connective tissue and endothelium)

Organization of the Heart Wall
Figure 22.4

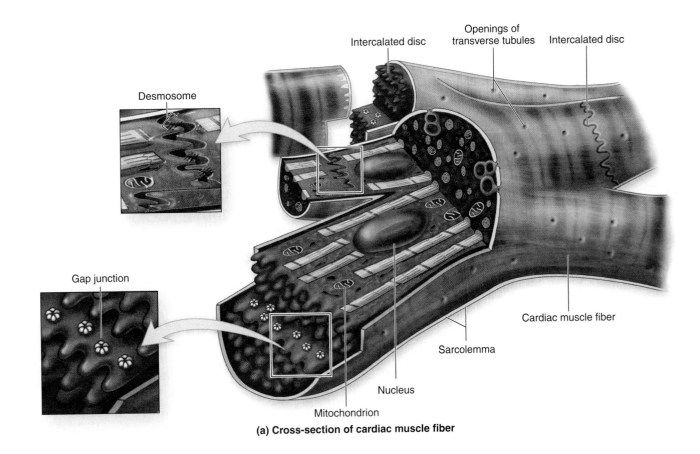

Intercalated disc

Openings of
transverse tubules

Intercalated disc

Desmosome

Cardiac muscle fiber

Gap junction

Sarcolemma

Nucleus

Mitochondrion

(a) Cross-section of cardiac muscle fiber

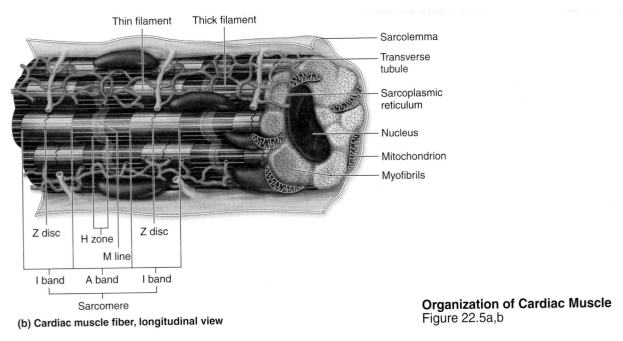

Thin filament

Thick filament

Sarcolemma

Transverse
tubule

Sarcoplasmic
reticulum

Nucleus

Mitochondrion

Myofibrils

Z disc

H zone

M line

Z disc

I band   A band   I band

Sarcomere

(b) Cardiac muscle fiber, longitudinal view

Organization of Cardiac Muscle
Figure 22.5a,b

313

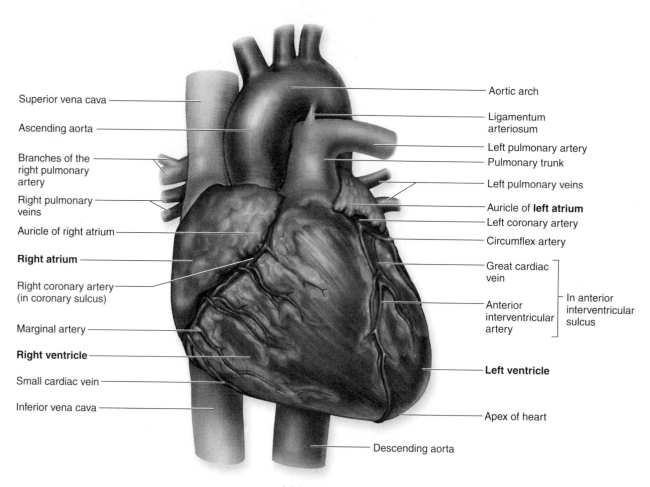

Superior vena cava

Ascending aorta

Branches of the
right pulmonary
artery

Right pulmonary
veins

Auricle of right atrium

Right atrium

Right coronary artery
(in coronary sulcus)

Marginal artery

Right ventricle

Small cardiac vein

Inferior vena cava

Aortic arch

Ligamentum
arteriosum

Left pulmonary artery

Pulmonary trunk

Left pulmonary veins

Auricle of **left atrium**

Left coronary artery

Circumflex artery

Great cardiac
vein

Anterior
interventricular
artery

In anterior
interventricular
sulcus

Left ventricle

Apex of heart

Descending aorta

(a) Anterior view

External Anatomy and Features of the Heart—Anterior View
Figure 22.6a

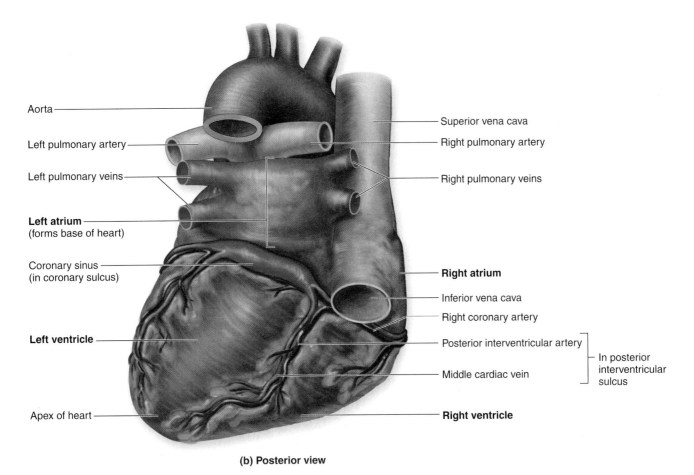

Aorta

Left pulmonary artery

Left pulmonary veins

Left atrium
(forms base of heart)

Coronary sinus
(in coronary sulcus)

Left ventricle

Apex of heart

Superior vena cava

Right pulmonary artery

Right pulmonary veins

Right atrium

Inferior vena cava

Right coronary artery

Posterior interventricular artery

Middle cardiac vein

In posterior interventricular sulcus

Right ventricle

(b) Posterior view

External Anatomy and Features of the Heart—Posterior View
Figure 22.6b

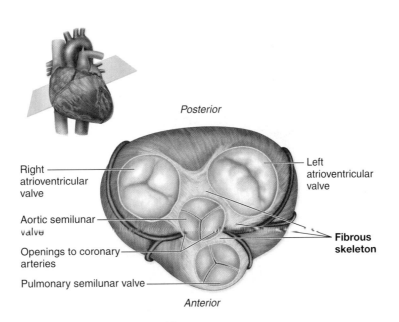

Posterior

Right atrioventricular valve

Aortic semilunar valve

Openings to coronary arteries

Pulmonary semilunar valve

Left atrioventricular valve

Fibrous skeleton

Anterior

Fibrous Skeleton of the Heart
Figure 22.7

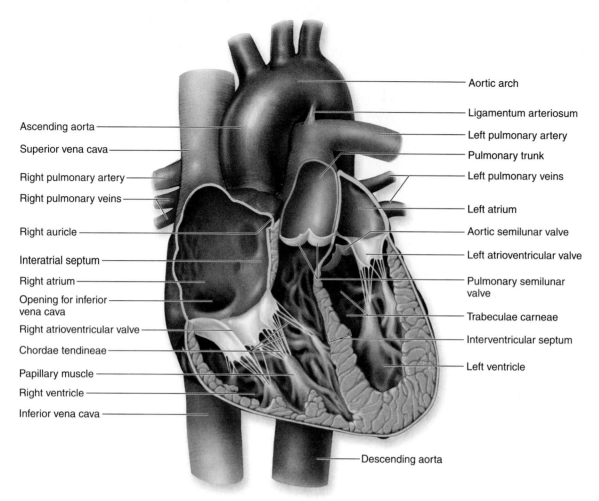

Aortic arch

Ligamentum arteriosum

Left pulmonary artery

Pulmonary trunk

Left pulmonary veins

Left atrium

Aortic semilunar valve

Left atrioventricular valve

Pulmonary semilunar valve

Trabeculae carneae

Interventricular septum

Left ventricle

Ascending aorta

Superior vena cava

Right pulmonary artery

Right pulmonary veins

Right auricle

Interatrial septum

Right atrium

Opening for inferior vena cava

Right atrioventricular valve

Chordae tendineae

Papillary muscle

Right ventricle

Inferior vena cava

Descending aorta

Coronal section, anterior view

Internal Anatomy of the Heart
Figure 22.8

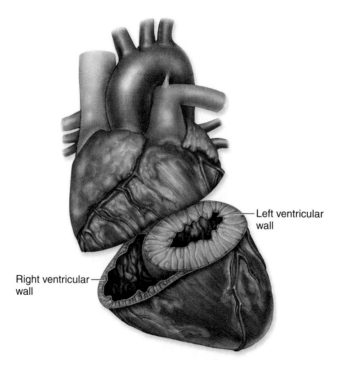

Left ventricular wall

Right ventricular wall

Comparison of Right and Left Ventricular Wall Thickness
Figure 22.9

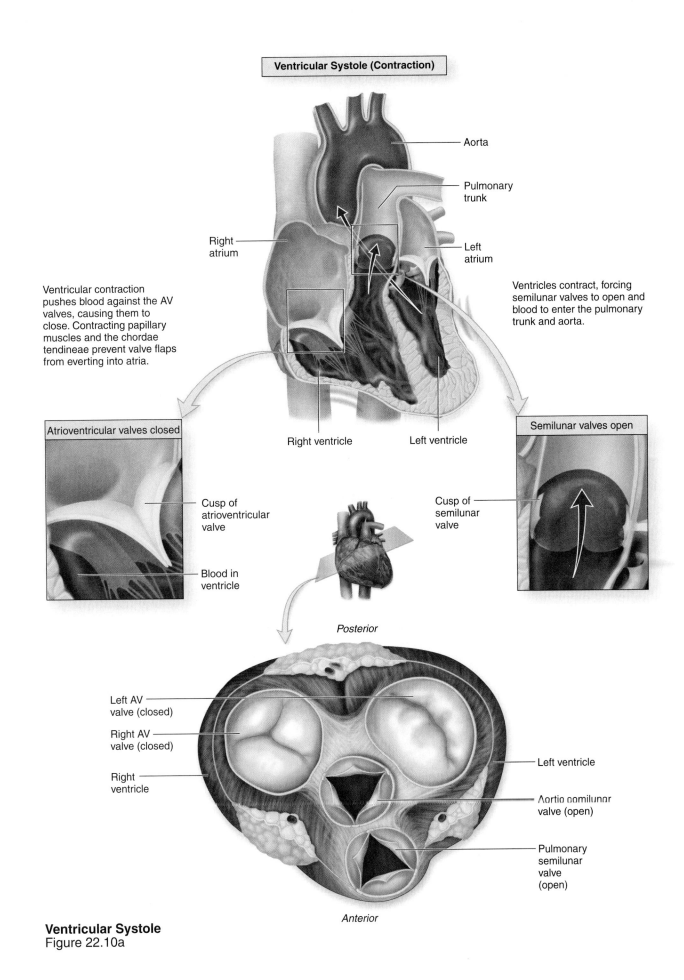

Ventricular Systole (Contraction)

Aorta

Pulmonary trunk

Right atrium

Left atrium

Ventricular contraction pushes blood against the AV valves, causing them to close. Contracting papillary muscles and the chordae tendineae prevent valve flaps from everting into atria.

Ventricles contract, forcing semilunar valves to open and blood to enter the pulmonary trunk and aorta.

Right ventricle

Left ventricle

Atrioventricular valves closed

Semilunar valves open

Cusp of atrioventricular valve

Cusp of semilunar valve

Blood in ventricle

Posterior

Left AV valve (closed)

Right AV valve (closed)

Right ventricle

Left ventricle

Aortic semilunar valve (open)

Pulmonary semilunar valve (open)

Anterior

Ventricular Systole
Figure 22.10a

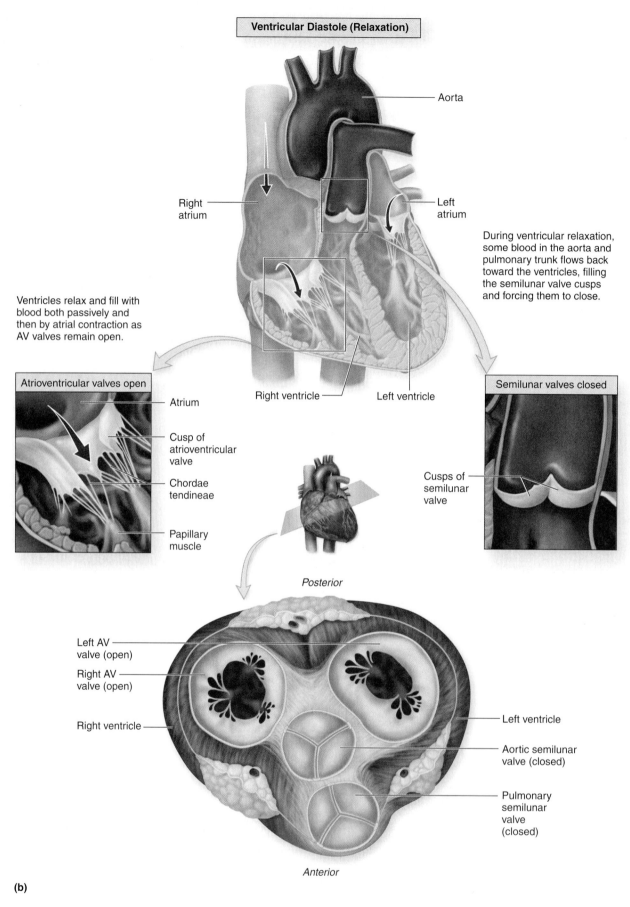

Ventricular Diastole (Relaxation)

Aorta

Right atrium

Left atrium

During ventricular relaxation, some blood in the aorta and pulmonary trunk flows back toward the ventricles, filling the semilunar valve cusps and forcing them to close.

Ventricles relax and fill with blood both passively and then by atrial contraction as AV valves remain open.

Right ventricle

Left ventricle

Atrioventricular valves open

Atrium

Cusp of atrioventricular valve

Chordae tendineae

Papillary muscle

Semilunar valves closed

Cusps of semilunar valve

Posterior

Left AV valve (open)

Right AV valve (open)

Right ventricle

Left ventricle

Aortic semilunar valve (closed)

Pulmonary semilunar valve (closed)

Anterior

(b)

Ventricular Diastole
Figure 22.10b

Structure \ Phase	Atrial systole	Early ventricular systole	← →	Late ventricular systole	Early ventricular diastole	← →	Late ventricular diastole
Atria	Contract	Relax			Relax		
Ventricles	Relax	Contract			Relax		
AV valves	Open	Closed			Open		
Semilunar valves	Closed	Open			Closed		

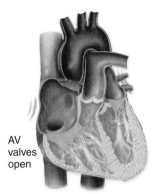 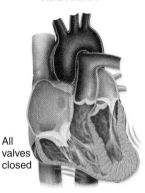 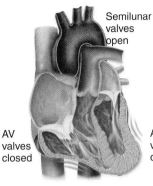 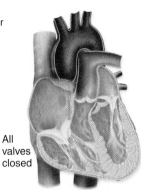 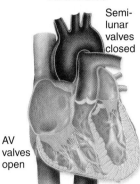

Atria contracted

Atria relaxed

Atria relaxed

Atria relaxed

Atria relaxed

Semilunar valves open

Semilunar valves closed

AV valves open

All valves closed

AV valves closed

All valves closed

AV valves open

Ventricles relaxed

Ventricles contracted

Ventricles contracted

Ventricles relaxed

Ventricles relaxed

① Atrial systole
Atria contract, AV valves open, semilunar valves closed

② Early ventricular systole
Atria relax, ventricles contract, AV valves forced closed, semilunar valves still closed

③ Late ventricular systole
Atria relax, ventricles contract, AV valves remain closed, semilunar valves forced open

④ Early ventricular diastole
Atria and ventricles relax, AV valves and semilunar valves closed, atria begin passively filling with blood

⑤ Late ventricular diastole
Atria and ventricles relax, atria passively fill with blood as AV valves open, semilunar valves closed

Cardiac Cycle
Figure 22.11

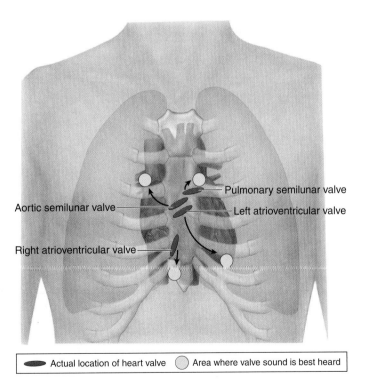

Pulmonary semilunar valve

Aortic semilunar valve

Left atrioventricular valve

Right atrioventricular valve

Actual location of heart valve   Area where valve sound is best heard

Heart Valve Locations and Auscultation Areas
Clinical View p. 681

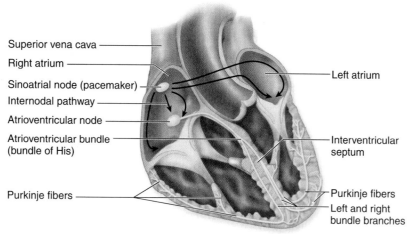

Superior vena cava
Right atrium
Sinoatrial node (pacemaker)
Internodal pathway
Atrioventricular node
Atrioventricular bundle
(bundle of His)

Left atrium

Interventricular
septum

Purkinje fibers

Purkinje fibers
Left and right
bundle branches

1. Muscle impulse is generated at the sinoatrial node. It spreads throughout the atria and to the atrioventricular node.

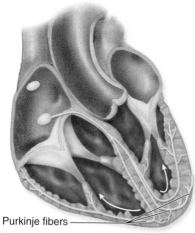

Atrioventricular
node
Atrioventricular
bundle

2. Atrioventricular node fibers delay the muscle impulse as it passes to the atrioventricular bundle.

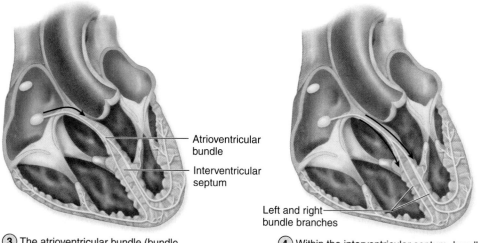

Atrioventricular
bundle
Interventricular
septum

Left and right
bundle branches

Purkinje fibers

3. The atrioventricular bundle (bundle of His) conducts the muscle impulse into the interventricular septum.

4. Within the interventricular septum, bundle branches split from the atrioventricular bundle.

5. The muscle impulse is delivered to Purkinje fibers in each ventricle and distributed throughout the ventricular myocardium.

Conduction System of the Heart
Figure 22.12

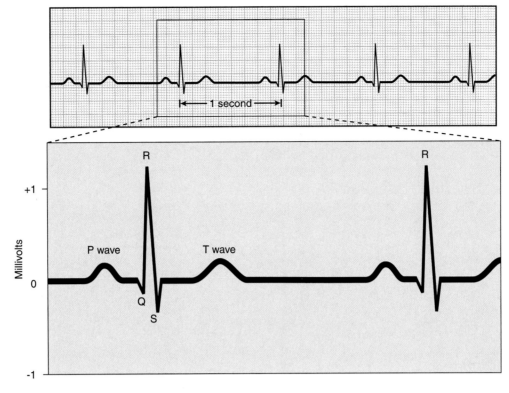

ECG
Clinical View p. 683

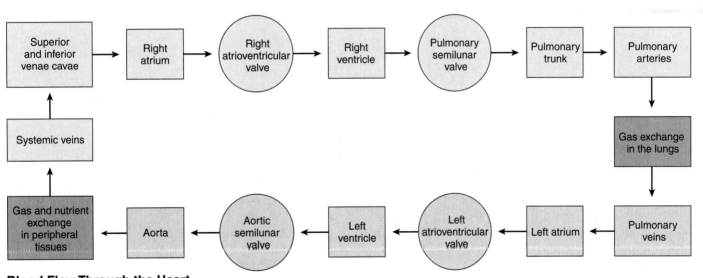

| Superior and inferior venae cavae | → | Right atrium | → | Right atrioventricular valve | → | Right ventricle | → | Pulmonary semilunar valve | → | Pulmonary trunk | → | Pulmonary arteries |

Blood Flow Through the Heart
Table 22.3

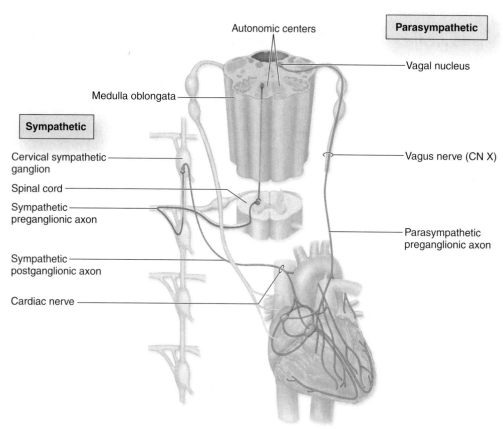

Autonomic centers

Parasympathetic

Vagal nucleus

Medulla oblongata

Sympathetic

Cervical sympathetic ganglion

Spinal cord

Sympathetic preganglionic axon

Sympathetic postganglionic axon

Cardiac nerve

Vagus nerve (CN X)

Parasympathetic preganglionic axon

Autonomic Innervation of the Heart
Figure 22.13

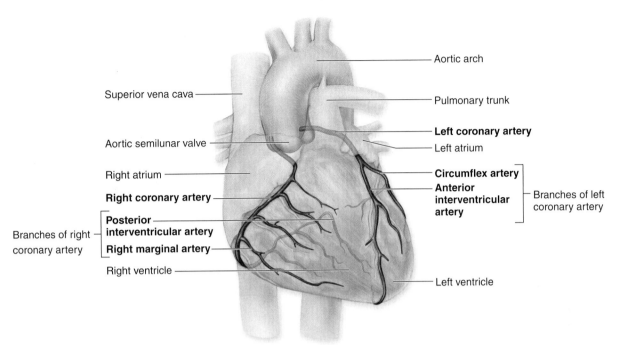

Aortic arch

Superior vena cava

Aortic semilunar valve

Right atrium

Right coronary artery

Branches of right coronary artery — **Posterior interventricular artery**

Right marginal artery

Right ventricle

Pulmonary trunk

Left coronary artery

Left atrium

Circumflex artery

Anterior interventricular artery — Branches of left coronary artery

Left ventricle

(a) Coronary arteries

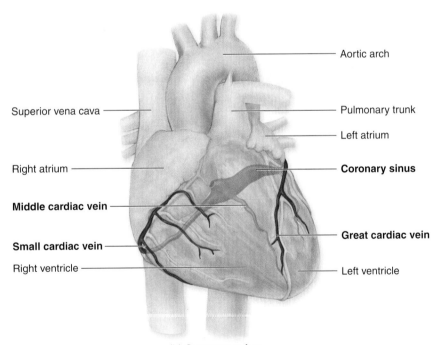

Aortic arch

Superior vena cava

Right atrium

Middle cardiac vein

Small cardiac vein

Right ventricle

Pulmonary trunk

Left atrium

Coronary sinus

Great cardiac vein

Left ventricle

(b) Coronary veins

Coronary Circulation
Figure 22.14

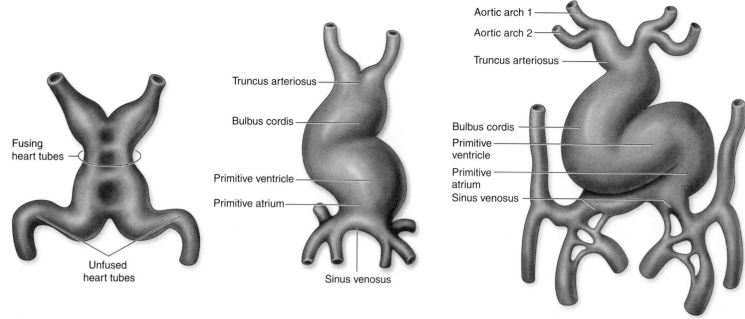

Fusing heart tubes

Unfused heart tubes

(a) 21 days: Paired heart tubes fuse.

Truncus arteriosus

Bulbus cordis

Primitive ventricle

Primitive atrium

Sinus venosus

(b) 22 days: Heart tube begins to fold.

Aortic arch 1

Aortic arch 2

Truncus arteriosus

Bulbus cordis

Primitive ventricle

Primitive atrium

Sinus venosus

(c) 28 days: S-shaped heart tube completes folding.

Development of the Heart
Figure 22.15

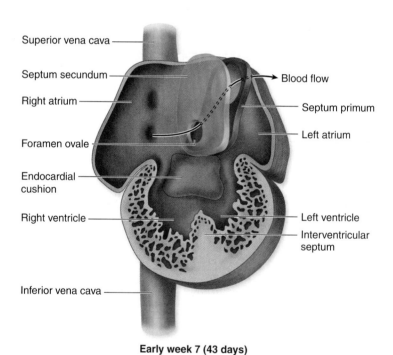

Superior vena cava

Septum secundum

Right atrium

Foramen ovale

Endocardial cushion

Right ventricle

Inferior vena cava

Blood flow

Septum primum

Left atrium

Left ventricle

Interventricular septum

Early week 7 (43 days)

Interatrial Septum
Figure 22.16

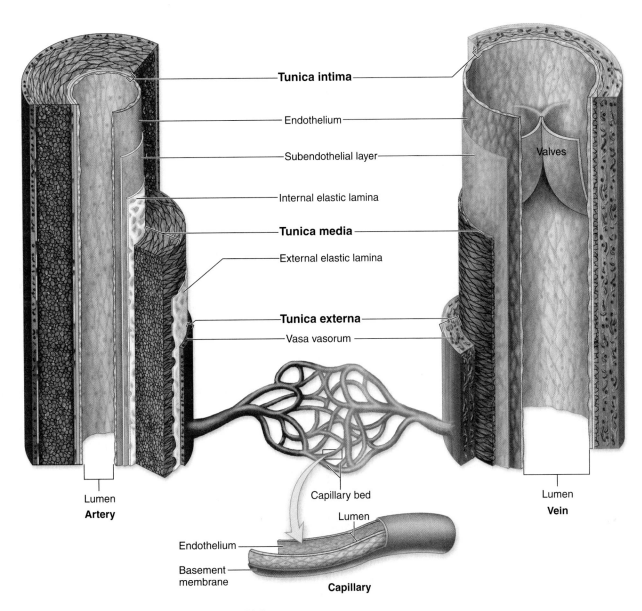

Tunica intima

Endothelium

Subendothelial layer

Internal elastic lamina

Valves

Tunica media

External elastic lamina

Tunica externa

Vasa vasorum

Capillary bed

Lumen
Artery

Lumen
Vein

Lumen

Endothelium

Basement
membrane

Capillary

Walls of an Artery, a Capillary, and a Vein
Figure 23.1

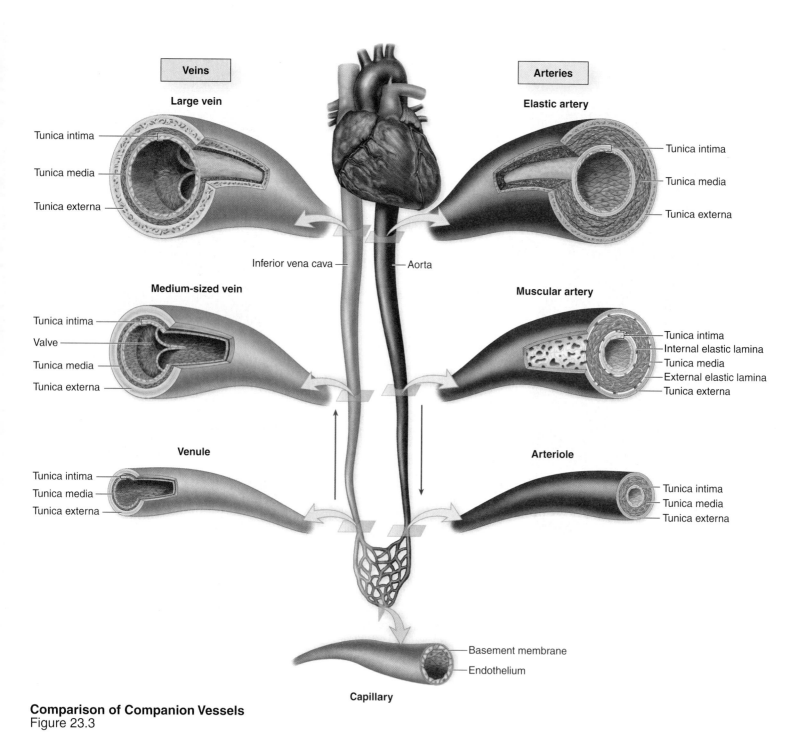

Veins

Large vein

Tunica intima

Tunica media

Tunica externa

Arteries

Elastic artery

Tunica intima

Tunica media

Tunica externa

Inferior vena cava

Aorta

Medium-sized vein

Tunica intima

Valve

Tunica media

Tunica externa

Muscular artery

Tunica intima

Internal elastic lamina

Tunica media

External elastic lamina

Tunica externa

Venule

Tunica intima

Tunica media

Tunica externa

Arteriole

Tunica intima

Tunica media

Tunica externa

Basement membrane

Endothelium

Capillary

Comparison of Companion Vessels
Figure 23.3

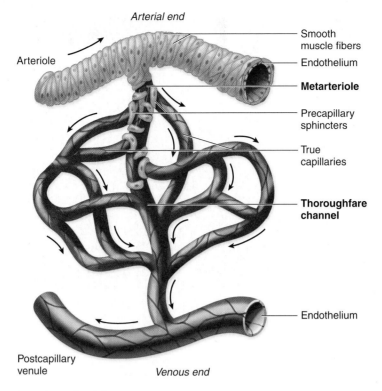

Arterial end

Arteriole

Smooth
muscle fibers

Endothelium

Metarteriole

Precapillary
sphincters

True
capillaries

**Thoroughfare
channel**

Endothelium

Postcapillary
venule

Venous end

Capillary Bed Structure
Figure 23.5

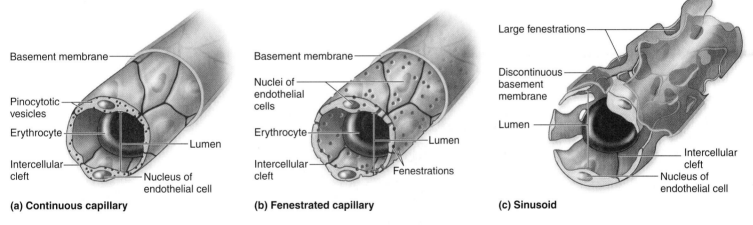

Basement membrane

Pinocytotic
vesicles

Erythrocyte

Intercellular
cleft

Lumen

Nucleus of
endothelial cell

(a) Continuous capillary

Basement membrane

Nuclei of
endothelial
cells

Erythrocyte

Intercellular
cleft

Lumen

Fenestrations

(b) Fenestrated capillary

Large fenestrations

Discontinuous
basement
membrane

Lumen

Intercellular
cleft

Nucleus of
endothelial cell

(c) Sinusoid

Types of Capillaries
Figure 23.6

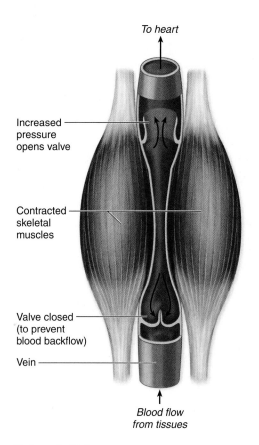

To heart

Increased pressure opens valve

Contracted skeletal muscles

Valve closed (to prevent blood backflow)

Vein

Blood flow from tissues

Valves in Veins
Figure 23.7

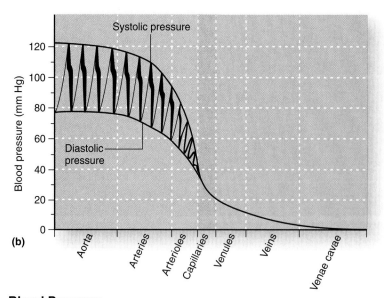

(b)

Systolic pressure

Blood pressure (mm Hg)

120

100

80

60

40

20

0

Diastolic pressure

Aorta

Arteries

Arterioles

Capillaries

Venules

Veins

Venae cavae

Blood Pressure
Figure 23.8b

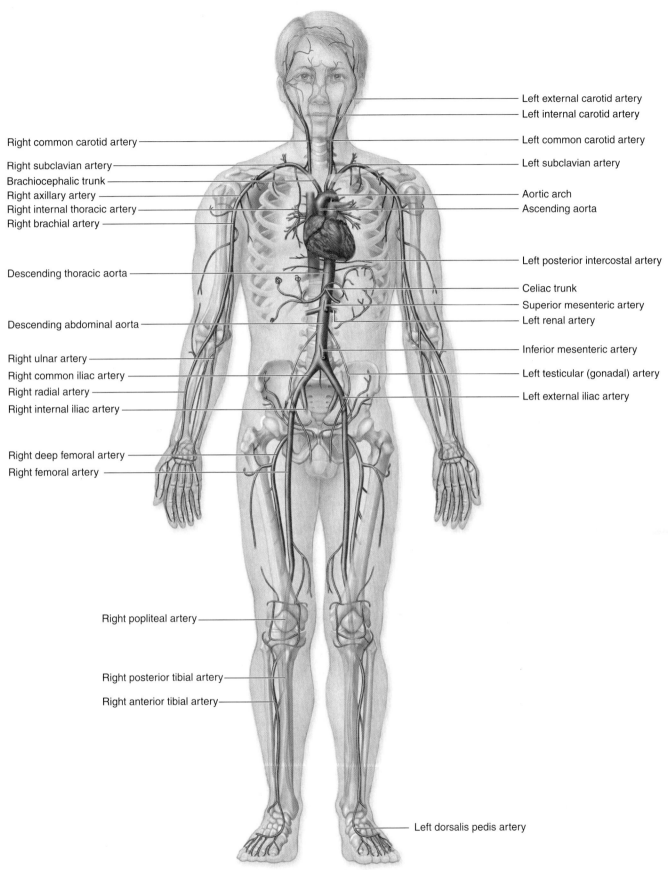

Left external carotid artery

Left internal carotid artery

Right common carotid artery

Left common carotid artery

Right subclavian artery

Left subclavian artery

Brachiocephalic trunk

Right axillary artery

Aortic arch

Right internal thoracic artery

Ascending aorta

Right brachial artery

Left posterior intercostal artery

Descending thoracic aorta

Celiac trunk

Superior mesenteric artery

Left renal artery

Descending abdominal aorta

Inferior mesenteric artery

Right ulnar artery

Right common iliac artery

Left testicular (gonadal) artery

Right radial artery

Left external iliac artery

Right internal iliac artery

Right deep femoral artery

Right femoral artery

Right popliteal artery

Right posterior tibial artery

Right anterior tibial artery

Left dorsalis pedis artery

(a) Arteries, anterior view

General Arterial Distribution
Figure 23.9a

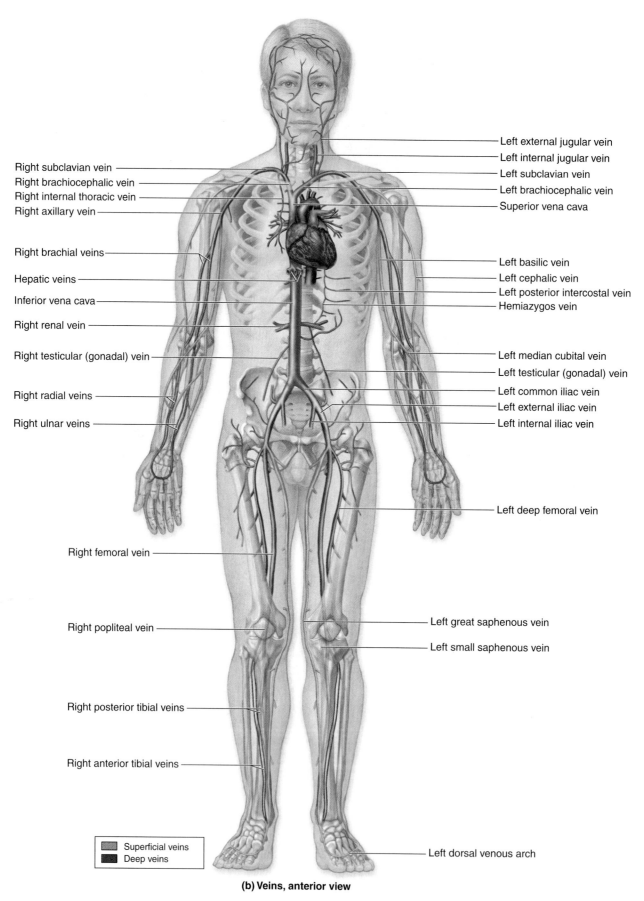

Right subclavian vein
Right brachiocephalic vein
Right internal thoracic vein
Right axillary vein

Right brachial veins

Hepatic veins

Inferior vena cava

Right renal vein

Right testicular (gonadal) vein

Right radial veins

Right ulnar veins

Left external jugular vein
Left internal jugular vein
Left subclavian vein
Left brachiocephalic vein
Superior vena cava

Left basilic vein
Left cephalic vein
Left posterior intercostal vein
Hemiazygos vein

Left median cubital vein
Left testicular (gonadal) vein
Left common iliac vein
Left external iliac vein
Left internal iliac vein

Left deep femoral vein

Right femoral vein

Right popliteal vein

Right posterior tibial veins

Right anterior tibial veins

Left great saphenous vein
Left small saphenous vein

Superficial veins
Deep veins

Left dorsal venous arch

(b) Veins, anterior view

General Venous Distribution
Figure 23.9b

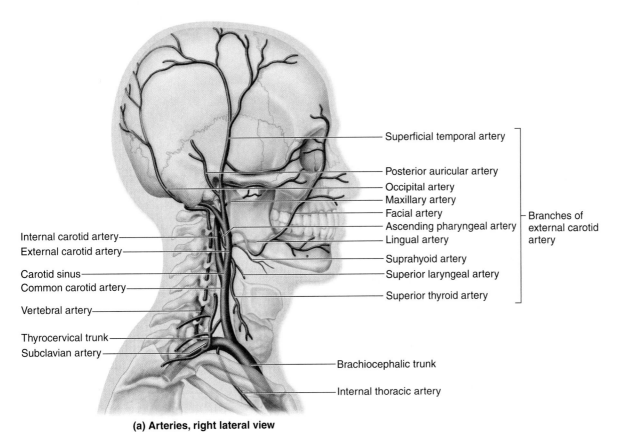

Superficial temporal artery
Posterior auricular artery
Occipital artery
Maxillary artery
Facial artery
Ascending pharyngeal artery
Lingual artery
Suprahyoid artery
Superior laryngeal artery
Superior thyroid artery

Branches of external carotid artery

Internal carotid artery
External carotid artery
Carotid sinus
Common carotid artery
Vertebral artery
Thyrocervical trunk
Subclavian artery

Brachiocephalic trunk
Internal thoracic artery

(a) Arteries, right lateral view

Blood Flow to the External Head and Neck—Major Arteries
Figure 23.10a

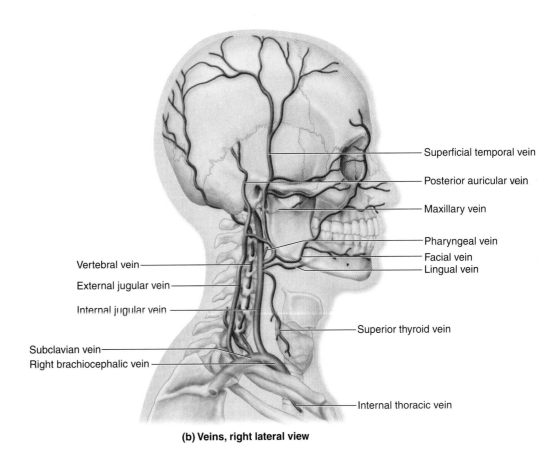

Superficial temporal vein
Posterior auricular vein
Maxillary vein
Pharyngeal vein
Facial vein
Lingual vein

Vertebral vein
External jugular vein
Internal jugular vein

Superior thyroid vein

Subclavian vein
Right brachiocephalic vein

Internal thoracic vein

(b) Veins, right lateral view

Blood Flow to the External Head and Neck—Major Veins
Figure 23.10b

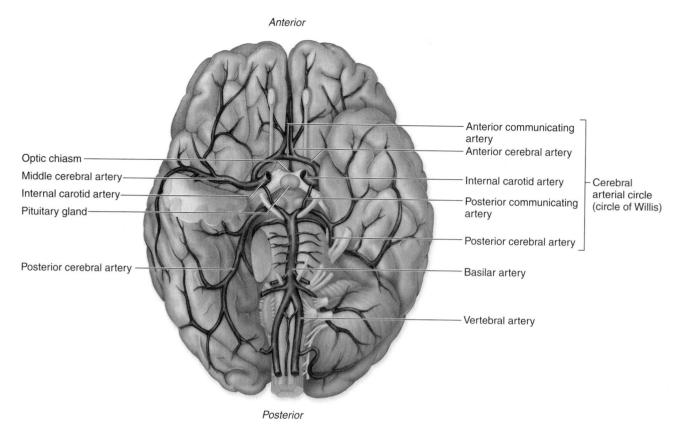

Optic chiasm

Middle cerebral artery

Internal carotid artery

Pituitary gland

Posterior cerebral artery

Anterior communicating artery

Anterior cerebral artery

Internal carotid artery

Posterior communicating artery

Posterior cerebral artery

Cerebral arterial circle (circle of Willis)

Basilar artery

Vertebral artery

Anterior

Posterior

(a) Arteries of the brain, inferior view

Blood Flow to the Cranium—Arterial Circulation
Figure 23.11a

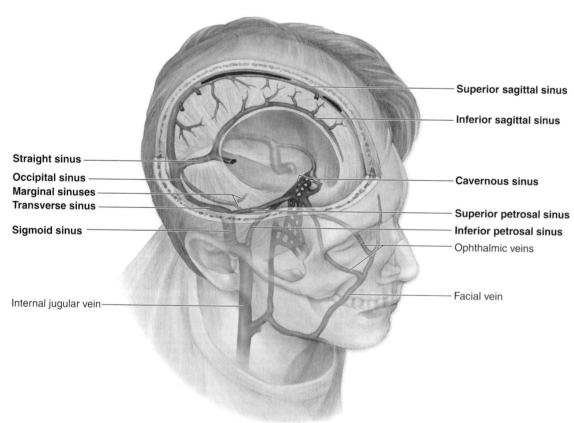

Straight sinus

Occipital sinus

Marginal sinuses

Transverse sinus

Sigmoid sinus

Internal jugular vein

Superior sagittal sinus

Inferior sagittal sinus

Cavernous sinus

Superior petrosal sinus

Inferior petrosal sinus

Ophthalmic veins

Facial vein

(b) Cranial and facial veins, right superior anterolateral view

Blood Flow to the Cranium—Venous Drainage
Figure 23.11b

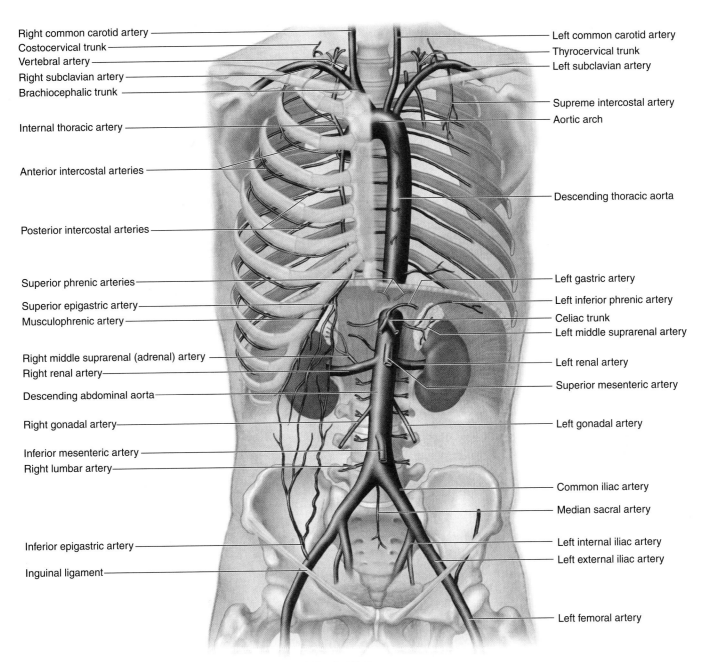

Right common carotid artery

Costocervical trunk

Vertebral artery

Right subclavian artery

Brachiocephalic trunk

Internal thoracic artery

Anterior intercostal arteries

Posterior intercostal arteries

Superior phrenic arteries

Superior epigastric artery

Musculophrenic artery

Right middle suprarenal (adrenal) artery

Right renal artery

Descending abdominal aorta

Right gonadal artery

Inferior mesenteric artery

Right lumbar artery

Inferior epigastric artery

Inguinal ligament

Left common carotid artery

Thyrocervical trunk

Left subclavian artery

Supreme intercostal artery

Aortic arch

Descending thoracic aorta

Left gastric artery

Left inferior phrenic artery

Celiac trunk

Left middle suprarenal artery

Left renal artery

Superior mesenteric artery

Left gonadal artery

Common iliac artery

Median sacral artery

Left internal iliac artery

Left external iliac artery

Left femoral artery

Arterial Circulation to Thoracic and Abdominal Body Walls
Figure 23.12

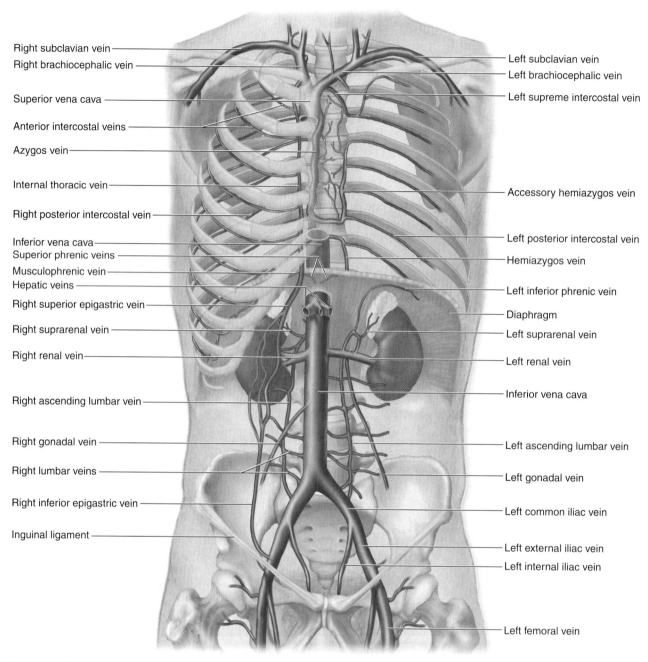

Right subclavian vein
Right brachiocephalic vein

Superior vena cava

Anterior intercostal veins

Azygos vein

Internal thoracic vein

Right posterior intercostal vein

Inferior vena cava
Superior phrenic veins
Musculophrenic vein
Hepatic veins
Right superior epigastric vein

Right suprarenal vein

Right renal vein

Right ascending lumbar vein

Right gonadal vein

Right lumbar veins

Right inferior epigastric vein

Inguinal ligament

Left subclavian vein
Left brachiocephalic vein

Left supreme intercostal vein

Accessory hemiazygos vein

Left posterior intercostal vein

Hemiazygos vein

Left inferior phrenic vein

Diaphragm

Left suprarenal vein

Left renal vein

Inferior vena cava

Left ascending lumbar vein

Left gonadal vein

Left common iliac vein

Left external iliac vein
Left internal iliac vein

Left femoral vein

Venous Circulation to Thoracic and Abdominal Body Walls
Figure 23.13

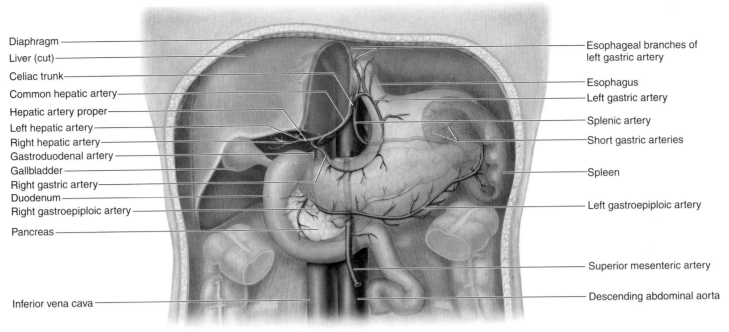

Diaphragm
Liver (cut)
Celiac trunk
Common hepatic artery
Hepatic artery proper
Left hepatic artery
Right hepatic artery
Gastroduodenal artery
Gallbladder
Right gastric artery
Duodenum
Right gastroepiploic artery
Pancreas

Inferior vena cava

Esophageal branches of left gastric artery
Esophagus
Left gastric artery
Splenic artery
Short gastric arteries
Spleen
Left gastroepiploic artery
Superior mesenteric artery
Descending abdominal aorta

(a) Celiac trunk branches

Arterial Supply to Gastrointestinal Tract and Abdominal Organs—Celiac Trunk Branches
Figure 23.15a

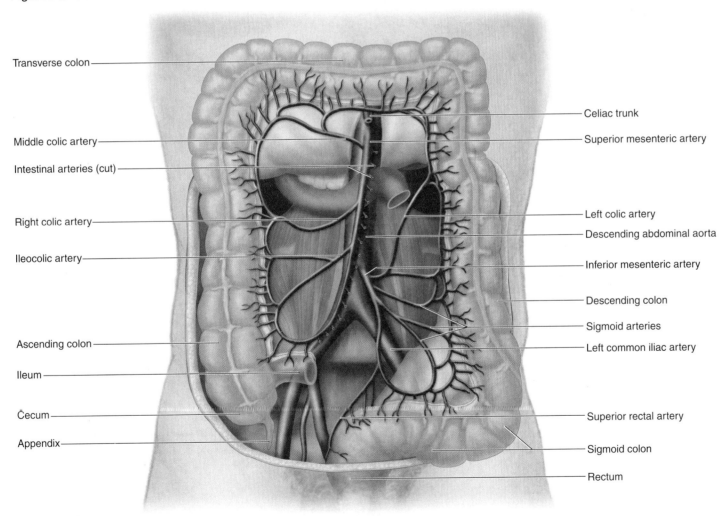

Transverse colon
Middle colic artery
Intestinal arteries (cut)
Right colic artery
Ileocolic artery
Ascending colon
Ileum
Cecum
Appendix

Celiac trunk
Superior mesenteric artery
Left colic artery
Descending abdominal aorta
Inferior mesenteric artery
Descending colon
Sigmoid arteries
Left common iliac artery
Superior rectal artery
Sigmoid colon
Rectum

(b) Superior and inferior mesenteric arteries

Arterial Supply to Gastrointestinal Tract and Abdominal Organs—Superior and Inferior Mesenteric Arteries
Figure 23.15b

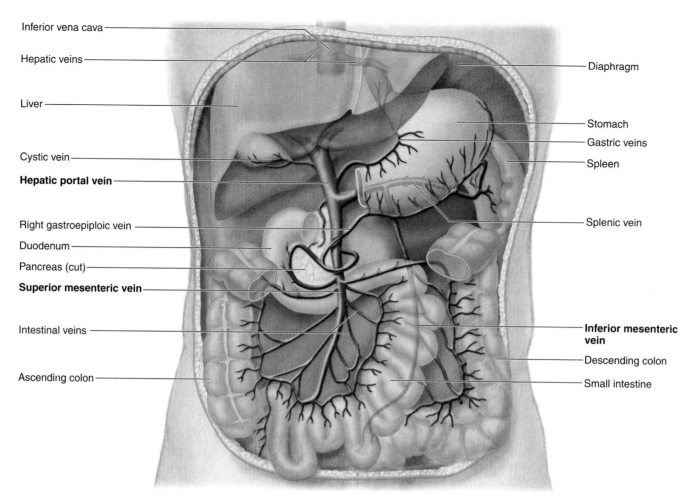

Inferior vena cava

Hepatic veins

Liver

Cystic vein

Hepatic portal vein

Right gastroepiploic vein

Duodenum

Pancreas (cut)

Superior mesenteric vein

Intestinal veins

Ascending colon

Diaphragm

Stomach

Gastric veins

Spleen

Splenic vein

Inferior mesenteric vein

Descending colon

Small intestine

Hepatic Portal System
Figure 23.16

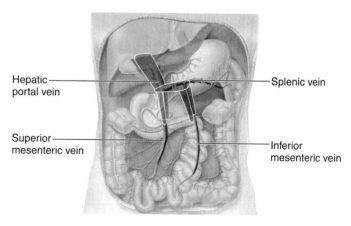

Hepatic portal vein

Superior mesenteric vein

Splenic vein

Inferior mesenteric vein

Hepatic Portal System Resembles a Chair
Study Tip p. 714

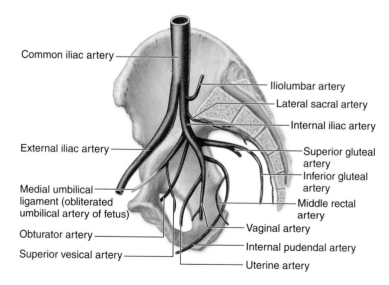

Common iliac artery

External iliac artery

Medial umbilical ligament (obliterated umbilical artery of fetus)

Obturator artery

Superior vesical artery

Iliolumbar artery

Lateral sacral artery

Internal iliac artery

Superior gluteal artery

Inferior gluteal artery

Middle rectal artery

Vaginal artery

Internal pudendal artery

Uterine artery

Arterial Supply to the Pelvis
Figure 23.18

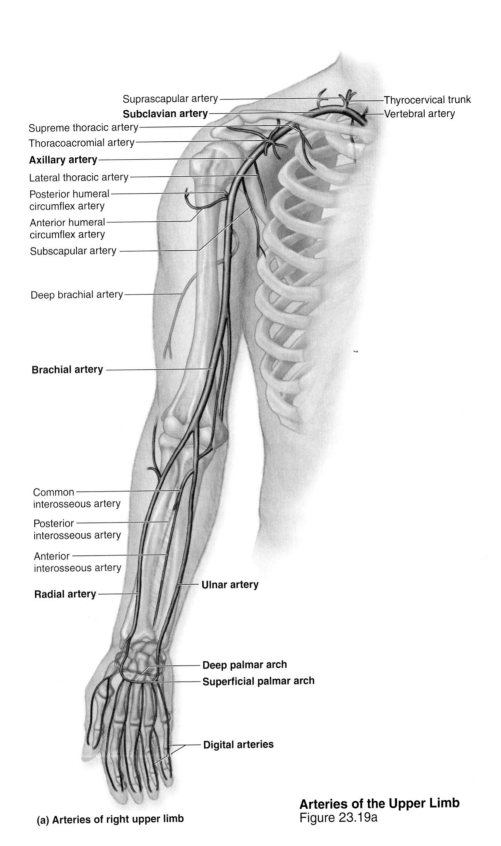

Suprascapular artery

Subclavian artery

Supreme thoracic artery

Thoracoacromial artery

Axillary artery

Lateral thoracic artery

Posterior humeral circumflex artery

Anterior humeral circumflex artery

Subscapular artery

Deep brachial artery

Brachial artery

Common interosseous artery

Posterior interosseous artery

Anterior interosseous artery

Radial artery

Thyrocervical trunk

Vertebral artery

Ulnar artery

Deep palmar arch

Superficial palmar arch

Digital arteries

(a) Arteries of right upper limb

Arteries of the Upper Limb
Figure 23.19a

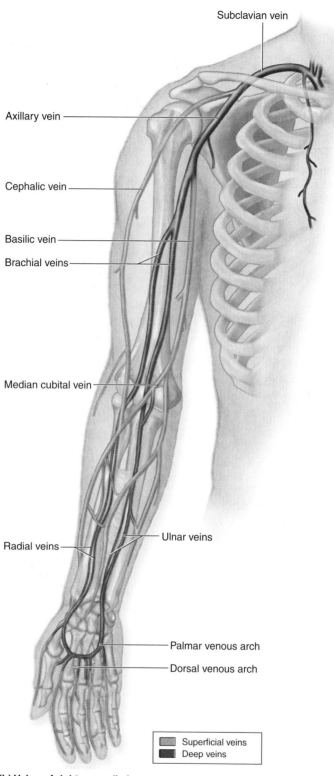

Subclavian vein

Axillary vein

Cephalic vein

Basilic vein

Brachial veins

Median cubital vein

Radial veins

Ulnar veins

Palmar venous arch

Dorsal venous arch

	Superficial veins
	Deep veins

(b) Veins of right upper limb

Veins of the Upper Limb
Figure 23.19b

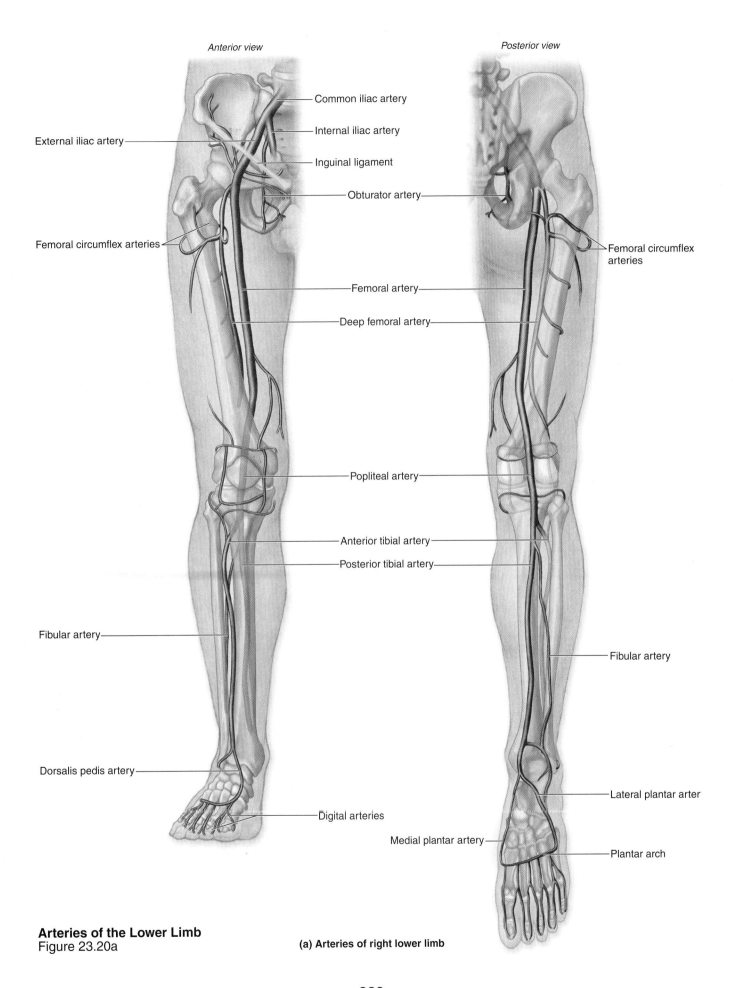

Anterior view

Posterior view

Common iliac artery

Internal iliac artery

External iliac artery

Inguinal ligament

Obturator artery

Femoral circumflex arteries

Femoral circumflex arteries

Femoral artery

Deep femoral artery

Popliteal artery

Anterior tibial artery

Posterior tibial artery

Fibular artery

Fibular artery

Dorsalis pedis artery

Lateral plantar arter

Digital arteries

Medial plantar artery

Plantar arch

Arteries of the Lower Limb
Figure 23.20a

(a) Arteries of right lower limb

339

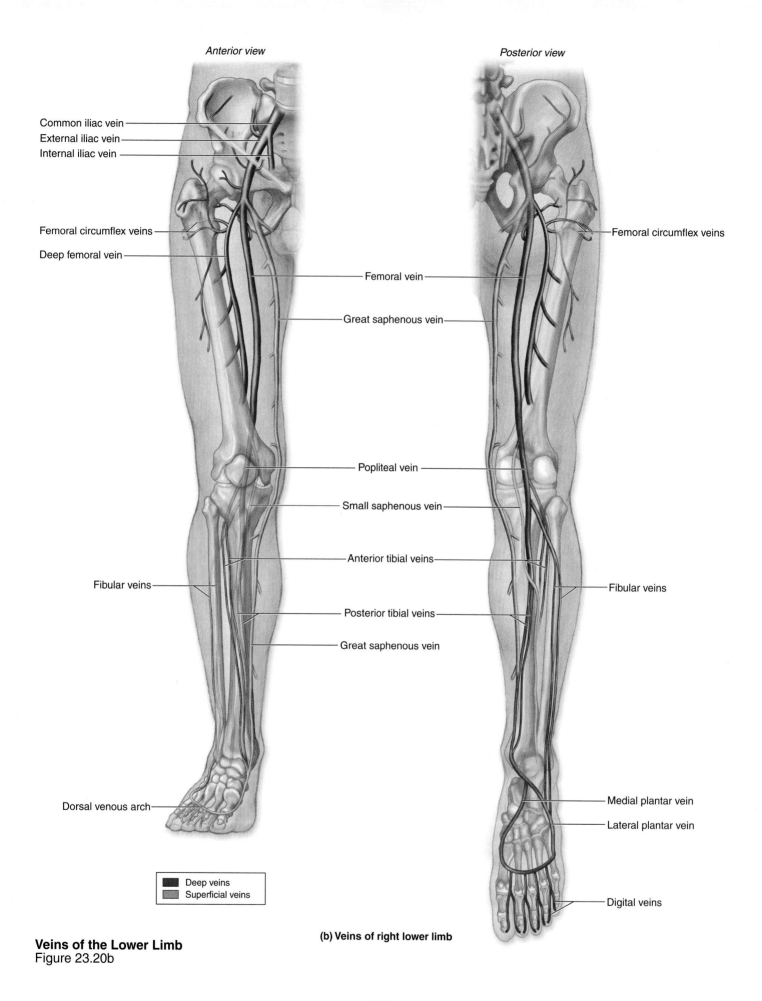

Anterior view

Posterior view

Common iliac vein

External iliac vein

Internal iliac vein

Femoral circumflex veins

Femoral circumflex veins

Deep femoral vein

Femoral vein

Great saphenous vein

Popliteal vein

Small saphenous vein

Anterior tibial veins

Fibular veins

Fibular veins

Posterior tibial veins

Great saphenous vein

Dorsal venous arch

Medial plantar vein

Lateral plantar vein

Deep veins
Superficial veins

Digital veins

(b) Veins of right lower limb

Veins of the Lower Limb
Figure 23.20b

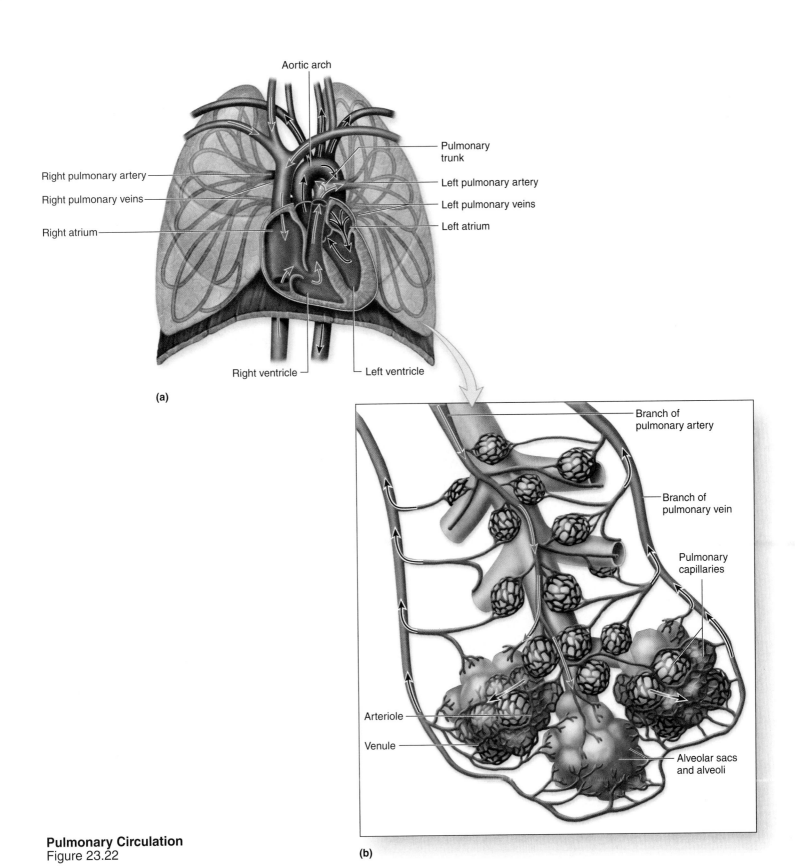

Aortic arch

Right pulmonary artery

Right pulmonary veins

Right atrium

Pulmonary trunk

Left pulmonary artery

Left pulmonary veins

Left atrium

Right ventricle

Left ventricle

(a)

Branch of pulmonary artery

Branch of pulmonary vein

Pulmonary capillaries

Arteriole

Venule

Alveolar sacs and alveoli

(b)

Pulmonary Circulation
Figure 23.22

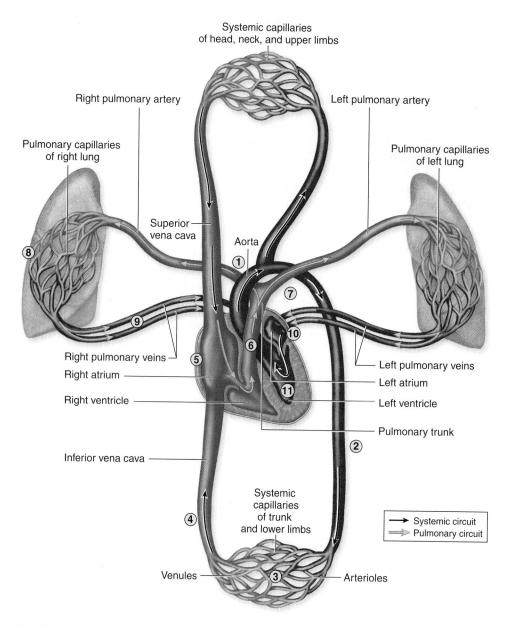

Systemic capillaries
of head, neck, and upper limbs

Right pulmonary artery

Left pulmonary artery

Pulmonary capillaries
of right lung

Pulmonary capillaries
of left lung

Superior
vena cava

Aorta

⑧

①

⑦

⑨

⑩

⑥

Right pulmonary veins

⑤

Left pulmonary veins

Right atrium

Left atrium

⑪

Right ventricle

Left ventricle

Pulmonary trunk

Inferior vena cava

②

Systemic
capillaries
of trunk
and lower limbs

④

→ Systemic circuit
⇒ Pulmonary circuit

Venules

③

Arterioles

Cardiovascular System Circulatory Routes
Figure 23.23

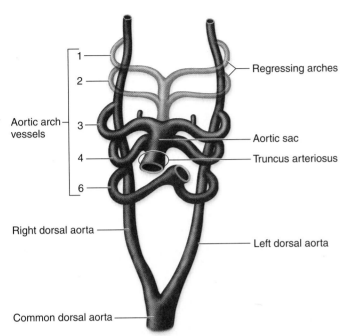

(a) Week 6: Paired aortic arch vessels connect to left and right dorsal aortae

Aortic arch vessels
1
2
3
4
6

Regressing arches
Aortic sac
Truncus arteriosus
Right dorsal aorta
Left dorsal aorta
Common dorsal aorta

Aortic Arch Vessel	Postnatal Structure Formed by Vessels
1	Small part of maxillary arteries
2	Small part of stapedial arteries
3	Left and right common carotid arteries
4	Right vessel: proximal part of right subclavian artery Left vessel: aortic arch (connects to the left dorsal aorta)
6	Right vessel: right pulmonary artery Left vessel: left pulmonary artery and ductus arteriosus

Subclavian artery
Pulmonary arteries
Degenerating right dorsal aorta
Common dorsal aorta
Ductus arteriosus
Left dorsal aorta (becomes descending thoracic aorta)

(b) Week 7: Right dorsal aorta degenerates; left dorsal aorta becomes descending thoracic aorta

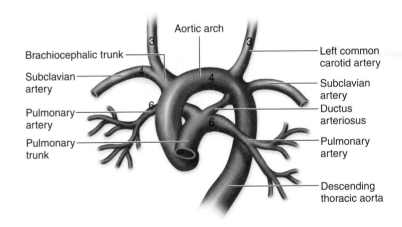

Aortic arch
Brachiocephalic trunk
Subclavian artery
Pulmonary artery
Pulmonary trunk
Left common carotid artery
Subclavian artery
Ductus arteriosus
Pulmonary artery
Descending thoracic aorta

(c) Week 8: Aortic arch and branches formed

Thoracic Artery Development
Figure 23.24

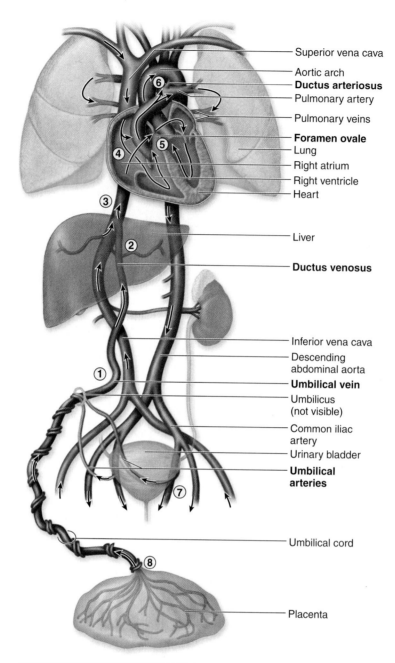

Superior vena cava

Aortic arch

Ductus arteriosus

Pulmonary artery

Pulmonary veins

Foramen ovale

Lung

Right atrium

Right ventricle

Heart

Liver

Ductus venosus

Inferior vena cava

Descending abdominal aorta

Umbilical vein

Umbilicus (not visible)

Common iliac artery

Urinary bladder

Umbilical arteries

Umbilical cord

Placenta

Fetal Cardiovascular Structure	Postnatal Structure
Ductus arteriosus	Ligamentum arteriosum
Ductus venosus	Ligamentum venosum
Foramen ovale	Fossa ovalis
Umbilical arteries	Medial umbilical ligaments
Umbilical vein	Round ligament of liver (ligamentum teres)

Fetal Circulation
Figure 23.27

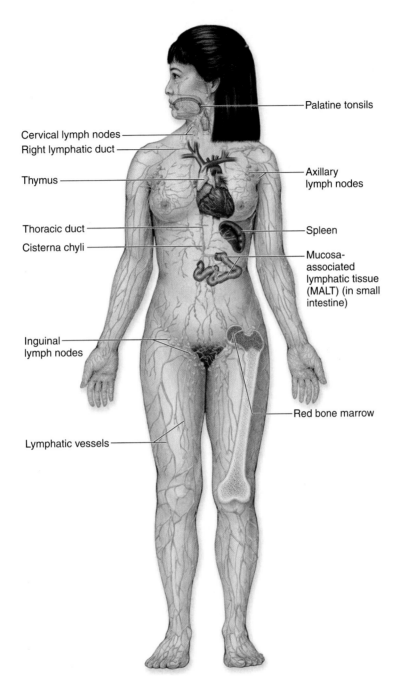

Cervical lymph nodes —
Right lymphatic duct —

Thymus —

Thoracic duct —
Cisterna chyli —

Inguinal —
lymph nodes

Lymphatic vessels —

— Palatine tonsils

— Axillary
    lymph nodes

— Spleen

— Mucosa-
    associated
    lymphatic tissue
    (MALT) (in small
    intestine)

— Red bone marrow

Lymphatic System
Figure 24.1

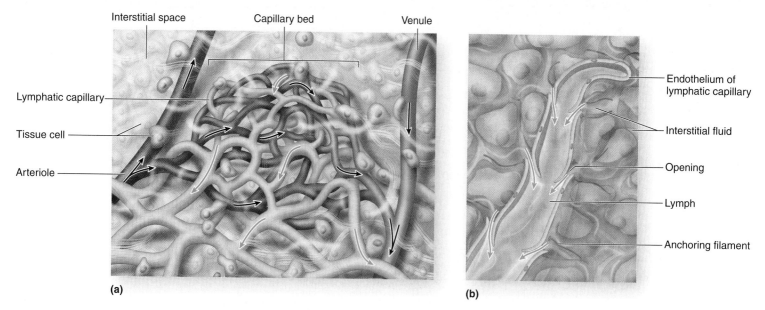

Interstitial space   Capillary bed   Venule

Lymphatic capillary—

Tissue cell—

Arteriole—

(a)

Endothelium of
lymphatic capillary

Interstitial fluid

Opening

Lymph

Anchoring filament

(b)

Lymphatic Capillaries
Figure 24.2

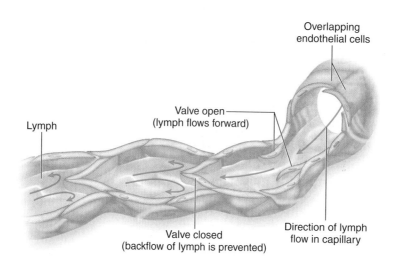

Overlapping
endothelial cells

Valve open
(lymph flows forward)

Lymph

Valve closed
(backflow of lymph is prevented)

Direction of lymph
flow in capillary

(a)

Lymphatic Vessels and Valves
Figure 24.3a

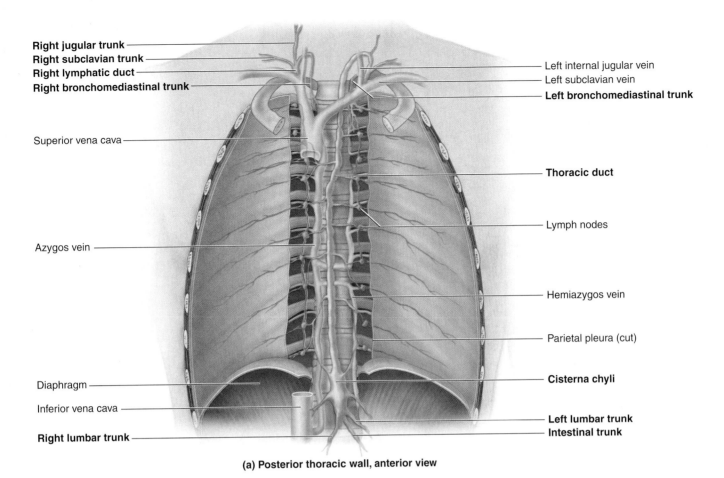

Right jugular trunk
Right subclavian trunk
Right lymphatic duct
Right bronchomediastinal trunk

Left internal jugular vein
Left subclavian vein
Left bronchomediastinal trunk

Superior vena cava

Thoracic duct

Lymph nodes

Azygos vein

Hemiazygos vein

Parietal pleura (cut)

Diaphragm

Cisterna chyli

Inferior vena cava

Left lumbar trunk
Intestinal trunk

Right lumbar trunk

(a) Posterior thoracic wall, anterior view

Lymphatic Trunks and Ducts
Figure 24.4a

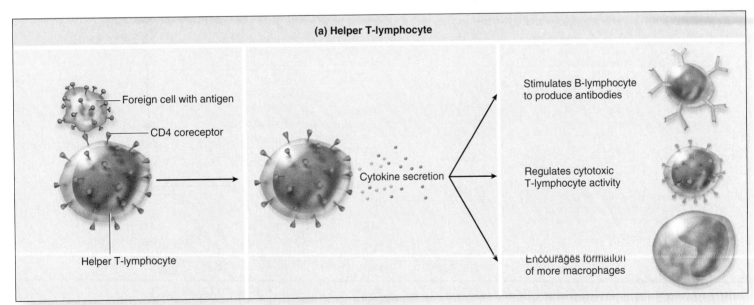

(a) Helper T-lymphocyte

Foreign cell with antigen

CD4 coreceptor

Cytokine secretion

Stimulates B-lymphocyte
to produce antibodies

Regulates cytotoxic
T-lymphocyte activity

Encourages formation
of more macrophages

Helper T-lymphocyte

① Helper T-lymphocyte recognizes antigen.

② Helper T-lymphocyte secretes cytokines and begins to undergo cell division to form more helper T-lymphocytes.

③ Cytokines secreted by helper T-lymphocytes initiate and control the immune response.

Helper T-lymphocyte
Figure 24.5a

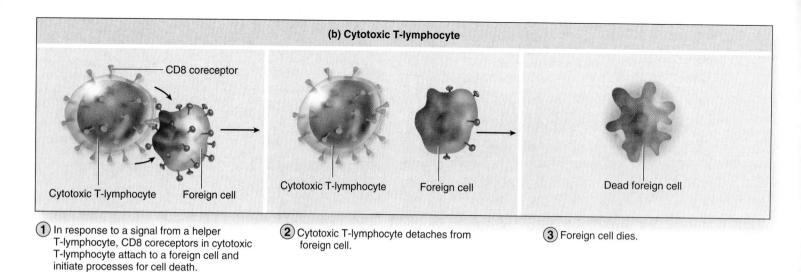

(b) Cytotoxic T-lymphocyte

CD8 coreceptor

Cytotoxic T-lymphocyte   Foreign cell

Cytotoxic T-lymphocyte   Foreign cell

Dead foreign cell

① In response to a signal from a helper T-lymphocyte, CD8 coreceptors in cytotoxic T-lymphocyte attach to a foreign cell and initiate processes for cell death.

② Cytotoxic T-lymphocyte detaches from foreign cell.

③ Foreign cell dies.

Cytotoxic T-lymphocyte
Figure 24.5b

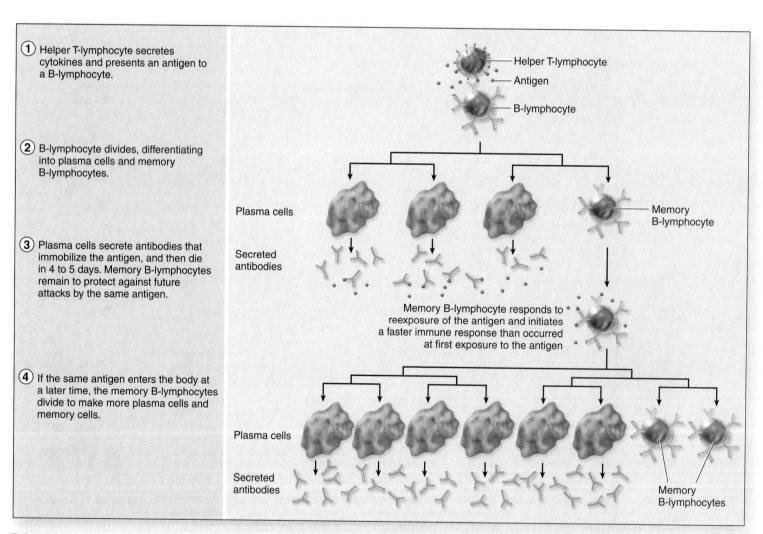

① Helper T-lymphocyte secretes cytokines and presents an antigen to a B-lymphocyte.

② B-lymphocyte divides, differentiating into plasma cells and memory B-lymphocytes.

③ Plasma cells secrete antibodies that immobilize the antigen, and then die in 4 to 5 days. Memory B-lymphocytes remain to protect against future attacks by the same antigen.

④ If the same antigen enters the body at a later time, the memory B-lymphocytes divide to make more plasma cells and memory cells.

Helper T-lymphocyte

Antigen

B-lymphocyte

Plasma cells

Secreted antibodies

Memory B-lymphocyte

Memory B-lymphocyte responds to reexposure of the antigen and initiates a faster immune response than occurred at first exposure to the antigen

Plasma cells

Secreted antibodies

Memory B-lymphocytes

B-lymphocytes and Their Role in the Immune Response
Figure 24.6

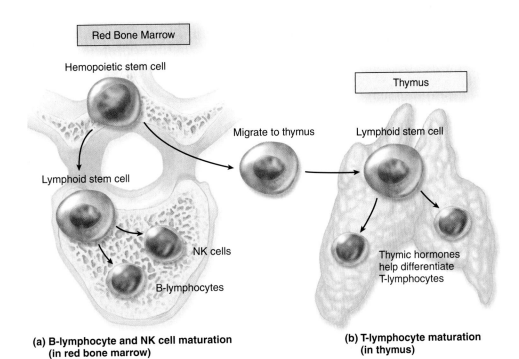

(a) B-lymphocyte and NK cell maturation
(in red bone marrow)

(b) T-lymphocyte maturation
(in thymus)

Lymphopoiesis
Figure 24.7

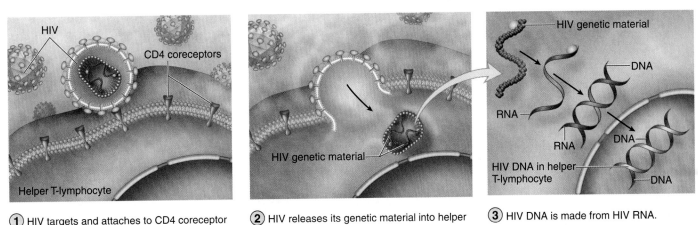

(1) HIV targets and attaches to CD4 coreceptor on helper T-lymphocyte.

(2) HIV releases its genetic material into helper T-lymphocyte.

(3) HIV DNA is made from HIV RNA.

(4) HIV DNA incorporates itself into the helper T-lymphocyte DNA.

(5) The helper T-lymphocyte becomes an "HIV-factory," producing HIV viruses that will be released from the helper T-lymphocyte and travel throughout the body.

HIV Targets Helper T-Lymphocytes in a Five-Stage Process
Clinical View p. 745

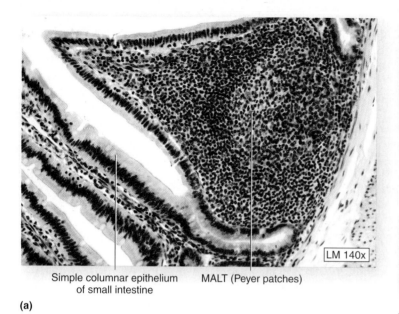

Simple columnar epithelium
of small intestine

MALT (Peyer patches)

LM 140x

(a)

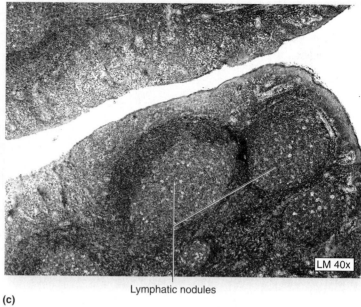

LM 40x

Lymphatic nodules

(c)

Tonsils and Lymphatic Nodules
Figure 24.8a,c

Figure 24.8a,c: © The McGraw-Hill Companies, Inc./Photo by Dr. Alvin Telser

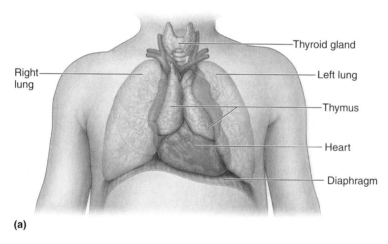

Thyroid gland

Right
lung

Left lung

Thymus

Heart

Diaphragm

(a)

Thymus
Figure 24.9a

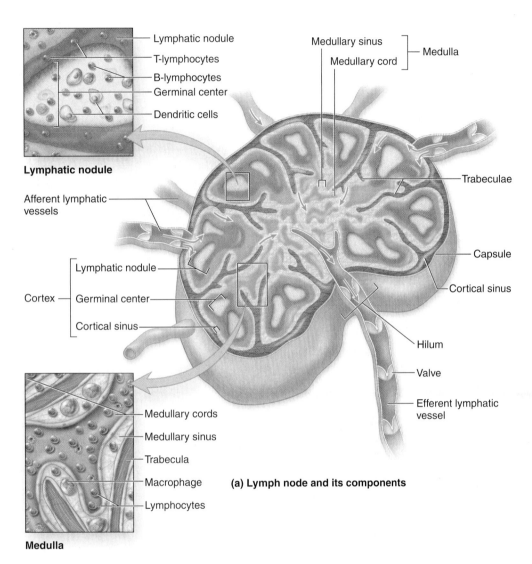

Lymphatic nodule
T-lymphocytes
B-lymphocytes
Germinal center
Dendritic cells

Lymphatic nodule

Afferent lymphatic vessels

Cortex
Lymphatic nodule
Germinal center
Cortical sinus

Medullary cords
Medullary sinus
Trabecula
Macrophage
Lymphocytes

Medulla

Medullary sinus
Medullary cord
Medulla

Trabeculae

Capsule

Cortical sinus

Hilum

Valve

Efferent lymphatic vessel

(a) Lymph node and its components

Lymph Nodes
Figure 24.10a

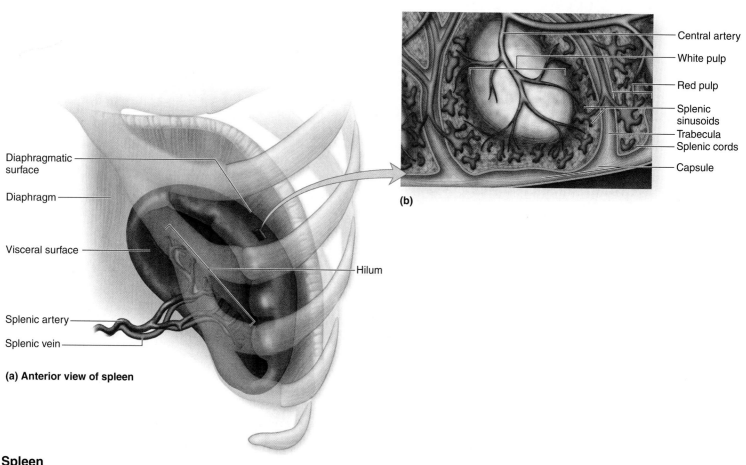

Diaphragmatic surface

Diaphragm

Visceral surface

Splenic artery

Splenic vein

Hilum

(a) Anterior view of spleen

(b)

Central artery

White pulp

Red pulp

Splenic sinusoids

Trabecula

Splenic cords

Capsule

Spleen
Figure 24.11a,b

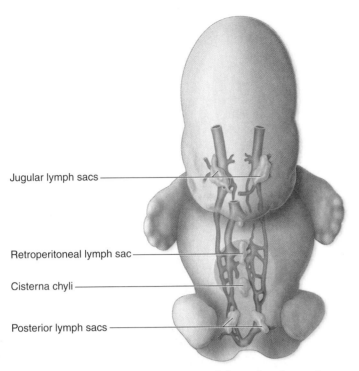

Jugular lymph sacs

Retroperitoneal lymph sac

Cisterna chyli

Posterior lymph sacs

(a) Week 6: Primary lymph sacs form

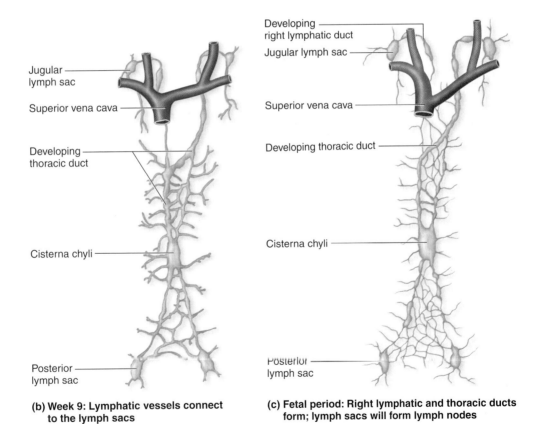

Jugular lymph sac

Superior vena cava

Developing thoracic duct

Cisterna chyli

Posterior lymph sac

(b) Week 9: Lymphatic vessels connect to the lymph sacs

Developing right lymphatic duct

Jugular lymph sac

Superior vena cava

Developing thoracic duct

Cisterna chyli

Posterior lymph sac

(c) Fetal period: Right lymphatic and thoracic ducts form; lymph sacs will form lymph nodes

Development of the Lymphatic System
Figure 24.12

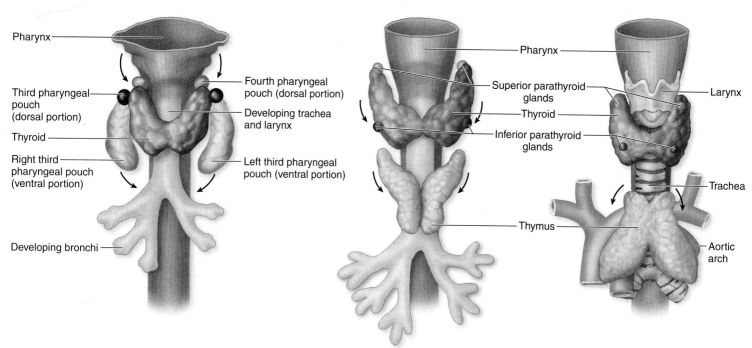

Pharynx

Third pharyngeal pouch (dorsal portion)

Thyroid

Right third pharyngeal pouch (ventral portion)

Developing bronchi

Fourth pharyngeal pouch (dorsal portion)

Developing trachea and larynx

Left third pharyngeal pouch (ventral portion)

(a) Week 5: Ventral portions of left and right third pharyngeal pouches migrate inferiorly

Pharynx

Superior parathyroid glands

Thyroid

Inferior parathyroid glands

Thymus

(b) Week 7: Left and right third pharyngeal pouches fuse to form the thymus

Pharynx

Larynx

Trachea

Aortic arch

(c) Fetal period: Bilobed thymus is positioned in mediastinum

Development of the Thymus
Figure 24.13

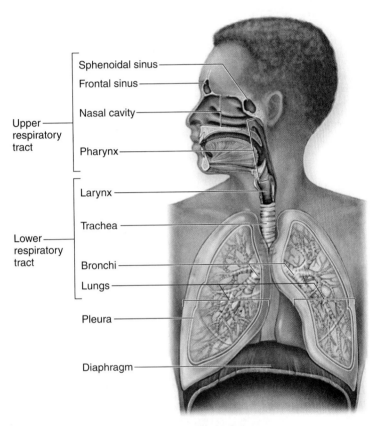

Sphenoidal sinus

Frontal sinus

Nasal cavity

Upper respiratory tract

Pharynx

Larynx

Trachea

Lower respiratory tract

Bronchi

Lungs

Pleura

Diaphragm

Gross Anatomy of the Respiratory System
Figure 25.1

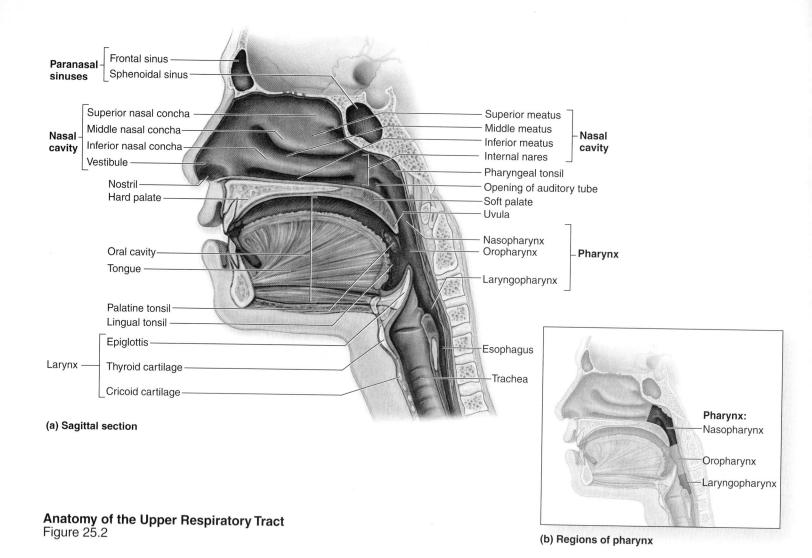

Paranasal sinuses
- Frontal sinus
- Sphenoidal sinus

Nasal cavity
- Superior nasal concha
- Middle nasal concha
- Inferior nasal concha
- Vestibule

- Nostril
- Hard palate

- Oral cavity
- Tongue

- Palatine tonsil
- Lingual tonsil

Larynx
- Epiglottis
- Thyroid cartilage
- Cricoid cartilage

- Superior meatus
- Middle meatus
- Inferior meatus
- Internal nares **Nasal cavity**

- Pharyngeal tonsil
- Opening of auditory tube
- Soft palate
- Uvula

- Nasopharynx
- Oropharynx **Pharynx**
- Laryngopharynx

- Esophagus
- Trachea

(a) Sagittal section

Anatomy of the Upper Respiratory Tract
Figure 25.2

Pharynx:
- Nasopharynx
- Oropharynx
- Laryngopharynx

(b) Regions of pharynx

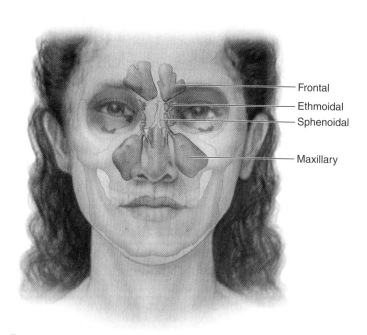

- Frontal
- Ethmoidal
- Sphenoidal
- Maxillary

Paranasal Sinuses
Figure 25.3

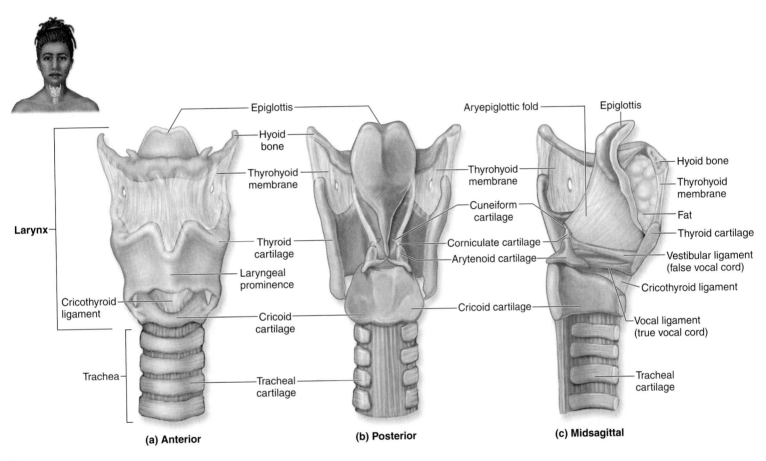

Epiglottis

Hyoid bone

Thyrohyoid membrane

Thyroid cartilage

Laryngeal prominence

Larynx

Cricothyroid ligament

Cricoid cartilage

Trachea

Tracheal cartilage

(a) Anterior

Aryepiglottic fold

Thyrohyoid membrane

Cuneiform cartilage

Corniculate cartilage

Arytenoid cartilage

Cricoid cartilage

Tracheal cartilage

(b) Posterior

Epiglottis

Hyoid bone

Thyrohyoid membrane

Fat

Thyroid cartilage

Vestibular ligament (false vocal cord)

Cricothyroid ligament

Vocal ligament (true vocal cord)

Tracheal cartilage

(c) Midsagittal

Larynx
Figure 25.4

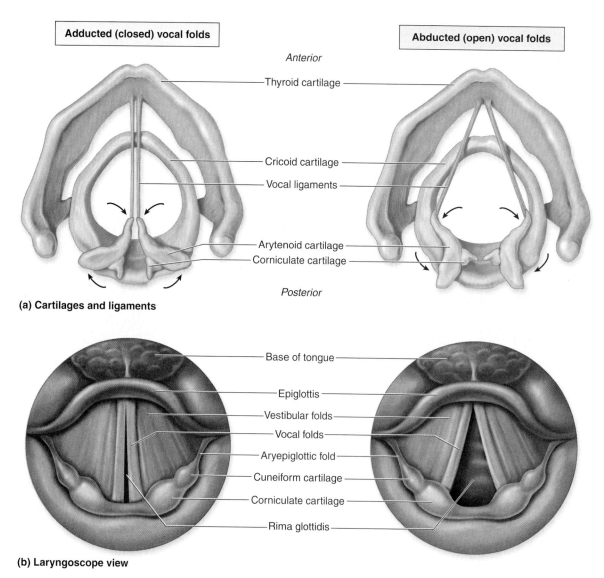

Adducted (closed) vocal folds

Abducted (open) vocal folds

Anterior

Thyroid cartilage

Cricoid cartilage

Vocal ligaments

Arytenoid cartilage

Corniculate cartilage

Posterior

(a) Cartilages and ligaments

Base of tongue

Epiglottis

Vestibular folds

Vocal folds

Aryepiglottic fold

Cuneiform cartilage

Corniculate cartilage

Rima glottidis

(b) Laryngoscope view

Vocal Folds
Figure 25.5

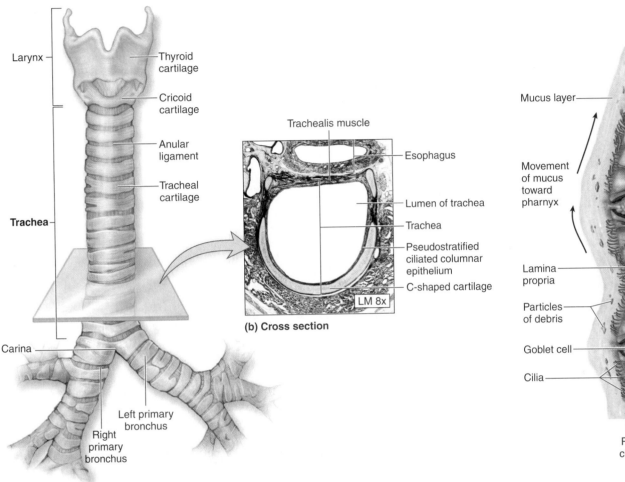

Larynx

Thyroid
cartilage

Cricoid
cartilage

Anular
ligament

Tracheal
cartilage

Trachea

Carina

Right
primary
bronchus

Left primary
bronchus

(a) Anterior view

Trachealis muscle

Esophagus

Lumen of trachea

Trachea

Pseudostratified
ciliated columnar
epithelium

C-shaped cartilage

LM 8x

(b) Cross section

Mucus layer

Movement
of mucus
toward
pharnyx

Lamina
propria

Particles
of debris

Goblet cell

Cilia

Pseudostratified
ciliated columnar
epithelium

(c) Microscopic view of tracheal lining

Trachea
Figure 25.7
Figure 25.7b: © Science VU/Visuals Unlimited

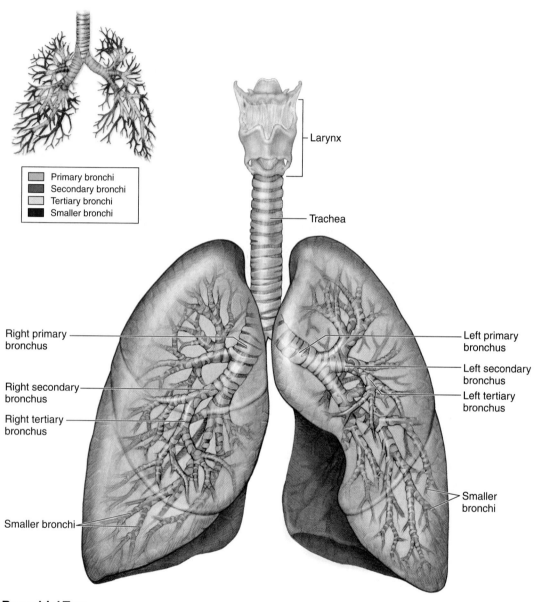

Primary bronchi
Secondary bronchi
Tertiary bronchi
Smaller bronchi

Larynx

Trachea

Right primary bronchus

Left primary bronchus

Left secondary bronchus

Right secondary bronchus

Left tertiary bronchus

Right tertiary bronchus

Smaller bronchi

Smaller bronchi

Bronchial Tree
Figure 25.8

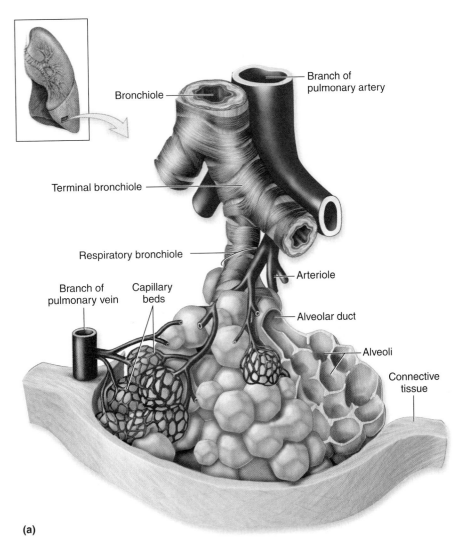

Bronchiole

Branch of
pulmonary artery

Terminal bronchiole

Respiratory bronchiole

Branch of
pulmonary vein

Capillary
beds

Arteriole

Alveolar duct

Alveoli

Connective
tissue

(a)

Bronchioles and Alveoli
Figure 25.9a

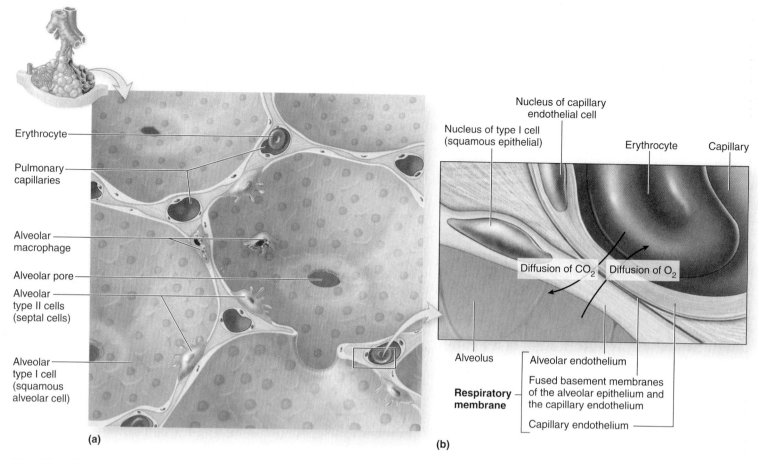

Erythrocyte

Pulmonary
capillaries

Alveolar
macrophage

Alveolar pore

Alveolar
type II cells
(septal cells)

Alveolar
type I cell
(squamous
alveolar cell)

(a)

Nucleus of capillary
endothelial cell

Nucleus of type I cell
(squamous epithelial)

Erythrocyte

Capillary

Diffusion of CO_2 Diffusion of O_2

Alveolus

Respiratory
membrane

Alveolar endothelium

Fused basement membranes
of the alveolar epithelium and
the capillary endothelium

Capillary endothelium

(b)

Alveoli and the Respiratory Membrane
Figure 25.10

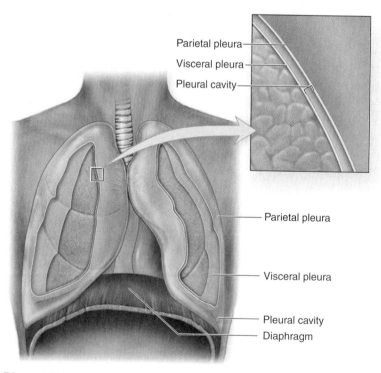

Parietal pleura
Visceral pleura
Pleural cavity

Parietal pleura

Visceral pleura

Pleural cavity
Diaphragm

Pleural Membranes
Figure 25.11

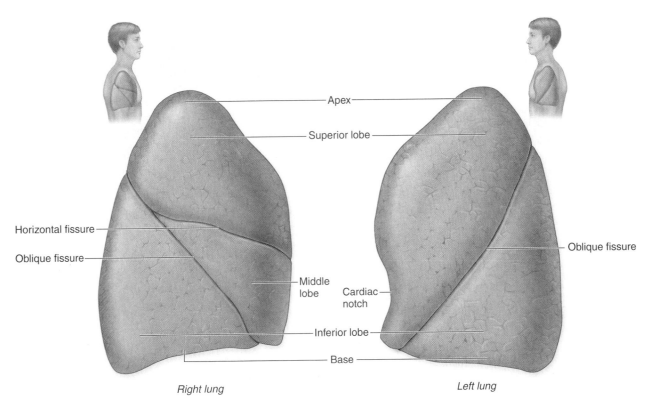

(a) Lateral views

Apex

Superior lobe

Horizontal fissure

Oblique fissure

Middle lobe

Cardiac notch

Inferior lobe

Base

Right lung

Oblique fissure

Left lung

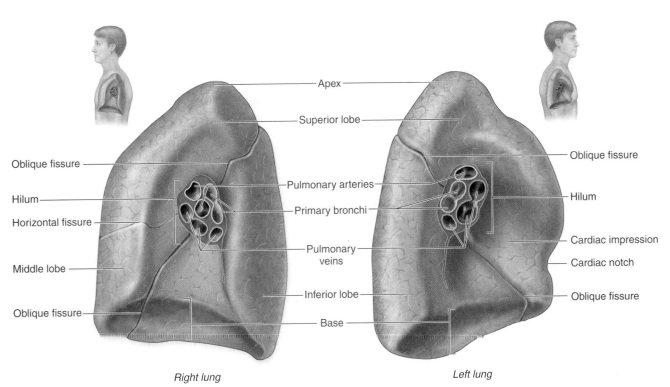

(b) Medial views

Apex

Superior lobe

Oblique fissure

Hilum

Horizontal fissure

Middle lobe

Oblique fissure

Pulmonary arteries

Primary bronchi

Pulmonary veins

Inferior lobe

Base

Oblique fissure

Hilum

Cardiac impression

Cardiac notch

Oblique fissure

Right lung

Left lung

Gross Anatomy of the Lungs
Figure 25.12

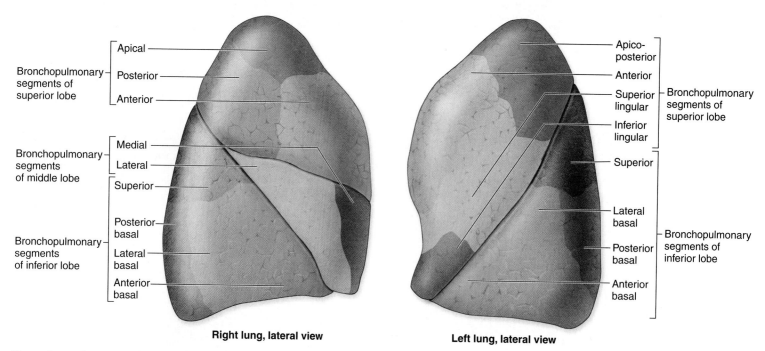

Bronchopulmonary segments of superior lobe
- Apical
- Posterior
- Anterior

Bronchopulmonary segments of middle lobe
- Medial
- Lateral

Bronchopulmonary segments of inferior lobe
- Superior
- Posterior basal
- Lateral basal
- Anterior basal

Right lung, lateral view

Bronchopulmonary segments of superior lobe
- Apico-posterior
- Anterior
- Superior lingular
- Inferior lingular

Bronchopulmonary segments of inferior lobe
- Superior
- Lateral basal
- Posterior basal
- Anterior basal

Left lung, lateral view

Bronchopulmonary Segments of the Lungs
Figure 25.13

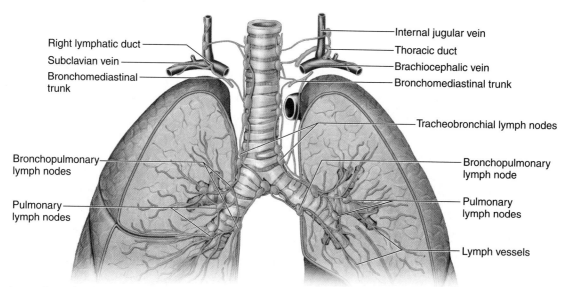

- Right lymphatic duct
- Subclavian vein
- Bronchomediastinal trunk
- Internal jugular vein
- Thoracic duct
- Brachiocephalic vein
- Bronchomediastinal trunk
- Tracheobronchial lymph nodes
- Bronchopulmonary lymph nodes
- Pulmonary lymph nodes
- Bronchopulmonary lymph node
- Pulmonary lymph nodes
- Lymph vessels

Lung Lymphatic Drainage
Figure 25.14

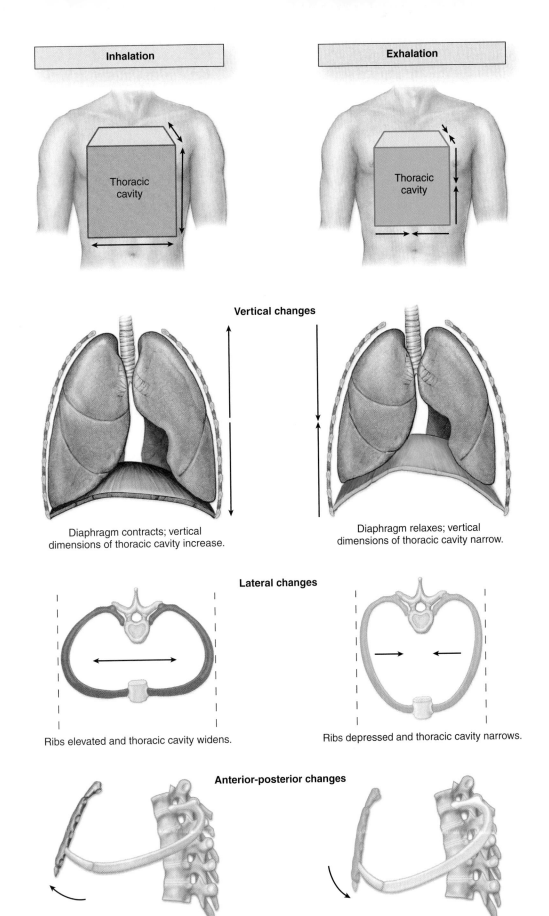

Inhalation	Exhalation

Vertical changes

Diaphragm contracts; vertical dimensions of thoracic cavity increase.

Diaphragm relaxes; vertical dimensions of thoracic cavity narrow.

Lateral changes

Ribs elevated and thoracic cavity widens.

Ribs depressed and thoracic cavity narrows.

Anterior-posterior changes

Inferior portion of sternum moves anteriorly.

Inferior portion of sternum moves posteriorly.

Thoracic Cavity Dimensional Changes Associated with Breathing
Figure 25.15

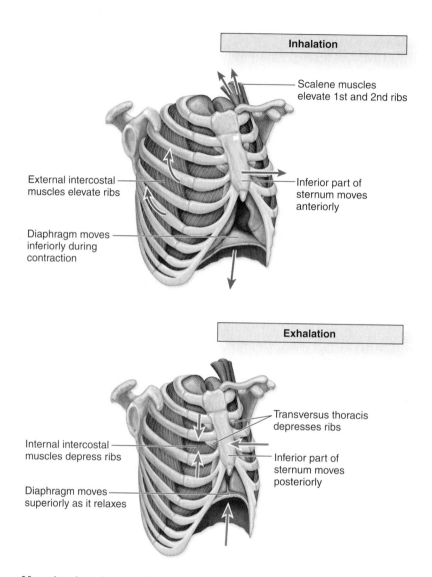

Inhalation

Scalene muscles
elevate 1st and 2nd ribs

External intercostal
muscles elevate ribs

Inferior part of
sternum moves
anteriorly

Diaphragm moves
inferiorly during
contraction

Exhalation

Transversus thoracis
depresses ribs

Internal intercostal
muscles depress ribs

Inferior part of
sternum moves
posteriorly

Diaphragm moves
superiorly as it relaxes

Muscles Involved in Respiration
Figure 25.16

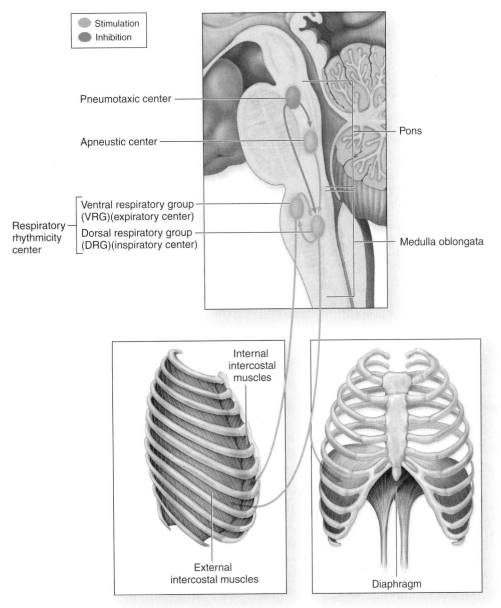

Stimulation
Inhibition

Pneumotaxic center

Apneustic center

Respiratory rhythmicity center

Ventral respiratory group (VRG)(expiratory center)

Dorsal respiratory group (DRG)(inspiratory center)

Pons

Medulla oblongata

Internal intercostal muscles

External intercostal muscles

Diaphragm

Respiratory Control Centers in the Brainstem
Figure 25.17

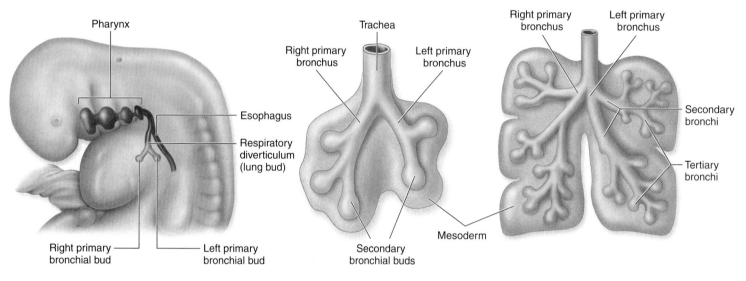

(a) Week 4: Respiratory diverticulum and primary bronchial buds form

Pharynx

Esophagus

Respiratory diverticulum (lung bud)

Right primary bronchial bud

Left primary bronchial bud

(b) Week 5: Secondary bronchial buds form

Trachea

Right primary bronchus

Left primary bronchus

Secondary bronchial buds

Mesoderm

(c) Week 6: Tertiary bronchi form

Right primary bronchus

Left primary bronchus

Secondary bronchi

Tertiary bronchi

Development of the Respiratory System
Figure 25.18

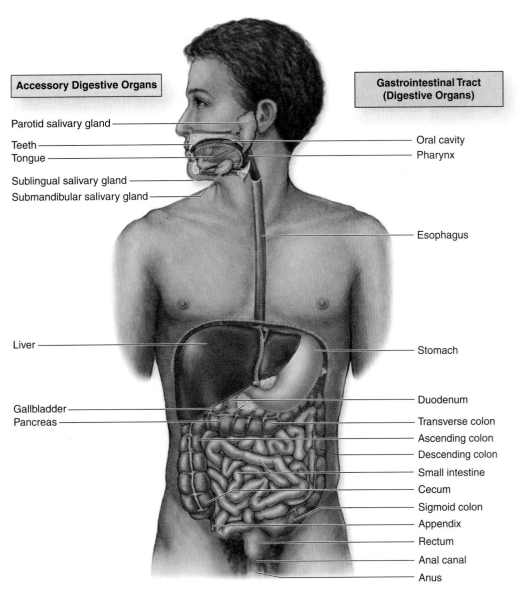

Parotid salivary gland

Teeth

Tongue

Sublingual salivary gland

Submandibular salivary gland

Oral cavity

Pharynx

Esophagus

Liver

Stomach

Gallbladder

Pancreas

Duodenum

Transverse colon

Ascending colon

Descending colon

Small intestine

Cecum

Sigmoid colon

Appendix

Rectum

Anal canal

Anus

Digestive System
Figure 26.1

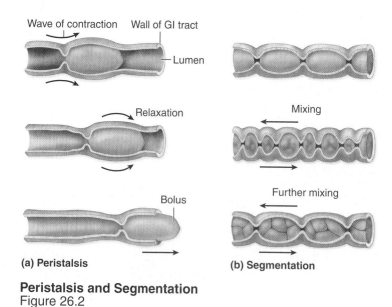

Wave of contraction   Wall of GI tract

Lumen

Relaxation

Mixing

Bolus

Further mixing

(a) Peristalsis

(b) Segmentation

Peristalsis and Segmentation
Figure 26.2

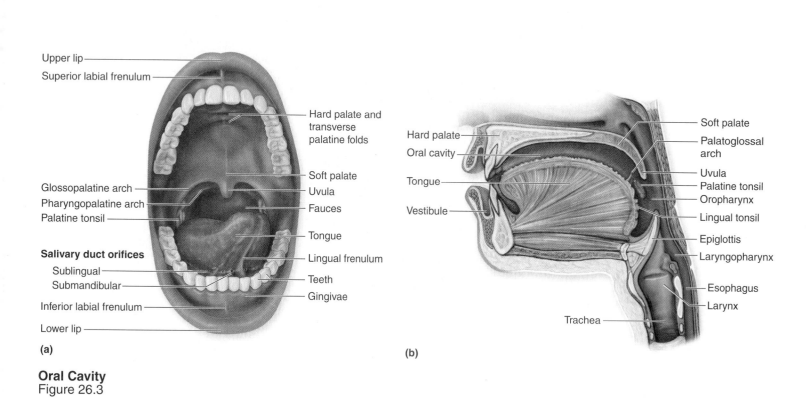

Upper lip

Superior labial frenulum

Hard palate and transverse palatine folds

Soft palate

Uvula

Glossopalatine arch

Pharyngopalatine arch

Palatine tonsil

Fauces

Tongue

Salivary duct orifices

Lingual frenulum

Sublingual

Teeth

Submandibular

Gingivae

Inferior labial frenulum

Lower lip

(a)

Hard palate

Oral cavity

Tongue

Vestibule

Soft palate

Palatoglossal arch

Uvula

Palatine tonsil

Oropharynx

Lingual tonsil

Epiglottis

Laryngopharynx

Esophagus

Larynx

Trachea

(b)

Oral Cavity
Figure 26.3

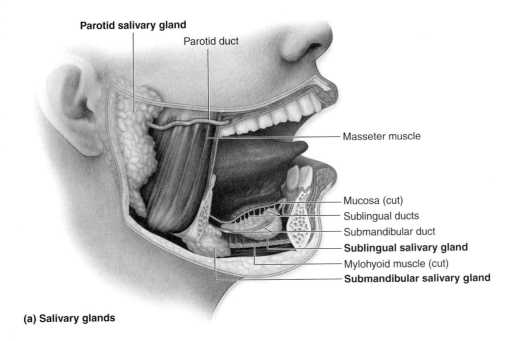

Parotid salivary gland

Parotid duct

Masseter muscle

Mucosa (cut)

Sublingual ducts

Submandibular duct

Sublingual salivary gland

Mylohyoid muscle (cut)

Submandibular salivary gland

(a) Salivary glands

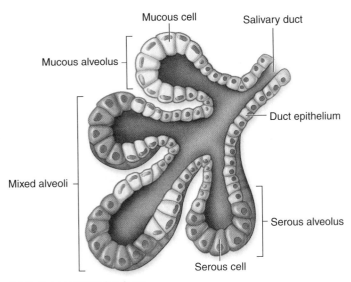

Mucous cell

Salivary duct

Mucous alveolus

Duct epithelium

Mixed alveoli

Serous alveolus

Serous cell

(b) Salivary gland histology

Salivary Glands
Figure 26.4a, b

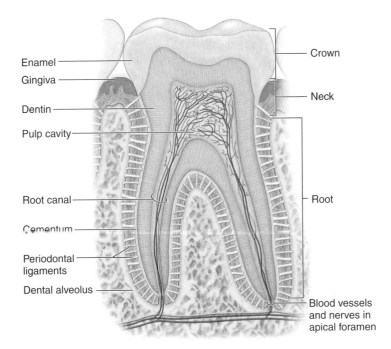

Enamel

Gingiva

Dentin

Pulp cavity

Root canal

Cementum

Periodontal ligaments

Dental alveolus

Crown

Neck

Root

Blood vessels and nerves in apical foramen

Anatomy of a Molar
Figure 26.5

371

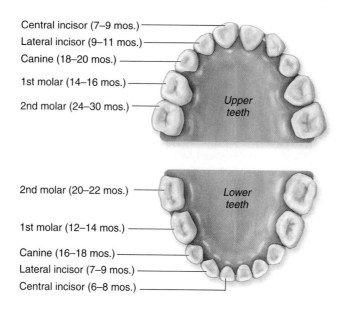

Central incisor (7–9 mos.)
Lateral incisor (9–11 mos.)
Canine (18–20 mos.)
1st molar (14–16 mos.)
2nd molar (24–30 mos.)

Upper teeth

2nd molar (20–22 mos.)
1st molar (12–14 mos.)
Canine (16–18 mos.)
Lateral incisor (7–9 mos.)
Central incisor (6–8 mos.)

Lower teeth

(b) Deciduous teeth

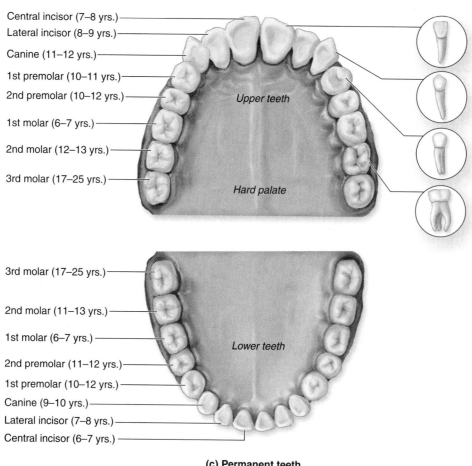

Central incisor (7–8 yrs.)
Lateral incisor (8–9 yrs.)
Canine (11–12 yrs.)
1st premolar (10–11 yrs.)
2nd premolar (10–12 yrs.)
1st molar (6–7 yrs.)
2nd molar (12–13 yrs.)
3rd molar (17–25 yrs.)

Upper teeth

Hard palate

3rd molar (17–25 yrs.)
2nd molar (11–13 yrs.)
1st molar (6–7 yrs.)
2nd premolar (11–12 yrs.)
1st premolar (10–12 yrs.)
Canine (9–10 yrs.)
Lateral incisor (7–8 yrs.)
Central incisor (6–7 yrs.)

Lower teeth

(c) Permanent teeth

Deciduous and Permanent Teeth
Figure 26.6b, c

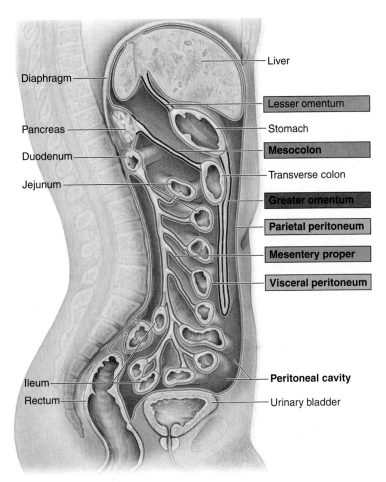

Diaphragm

Pancreas

Duodenum

Jejunum

Liver

Lesser omentum

Stomach

Mesocolon

Transverse colon

Greater omentum

Parietal peritoneum

Mesentery proper

Visceral peritoneum

Peritoneal cavity

Ileum

Rectum

Urinary bladder

Peritoneum and Mesenteries
Figure 26.7

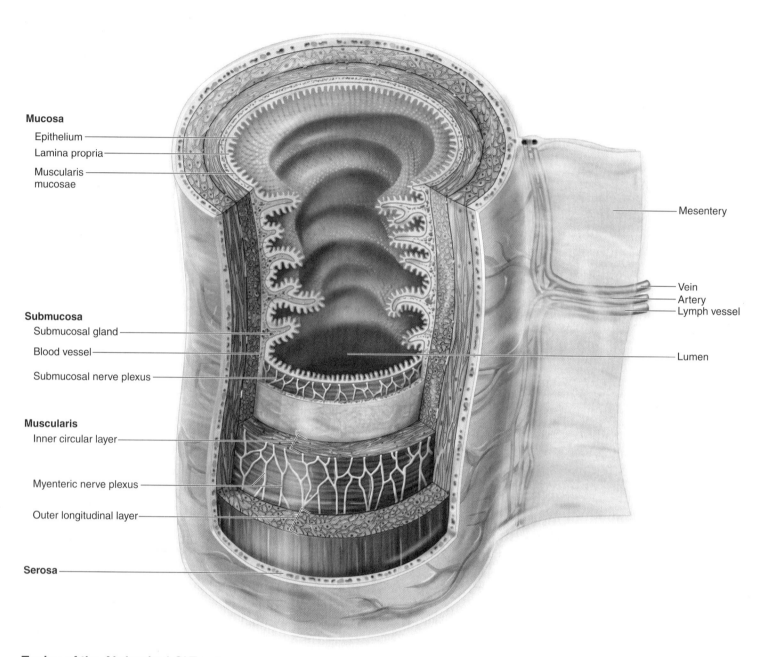

Mucosa
Epithelium
Lamina propria
Muscularis
mucosae

Submucosa
Submucosal gland
Blood vessel
Submucosal nerve plexus

Muscularis
Inner circular layer

Myenteric nerve plexus

Outer longitudinal layer

Serosa

Mesentery

Vein
Artery
Lymph vessel

Lumen

Tunics of the Abdominal GI Tract
Figure 26.9

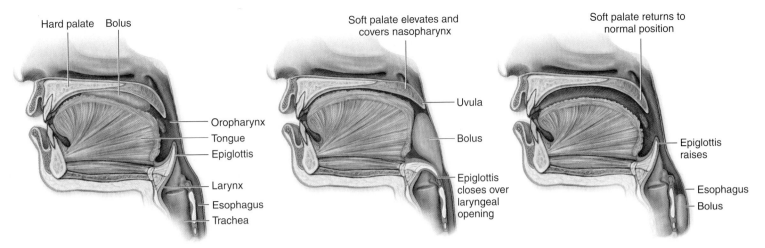

① **Voluntary phase:** Bolus of food is pushed by tongue against hard palate and then moves toward oropharynx.

② **Pharyngeal phase** (involuntary): As bolus moves into oropharynx, the soft palate closes off nasopharynx, the epiglottis closes over laryngeal opening.

③ **Esophageal phase** (involuntary): Esophageal muscle contractions push bolus toward stomach; soft palate and epiglottis return to their pre-swallowing positions.

Phases of Swallowing
Figure 26.11

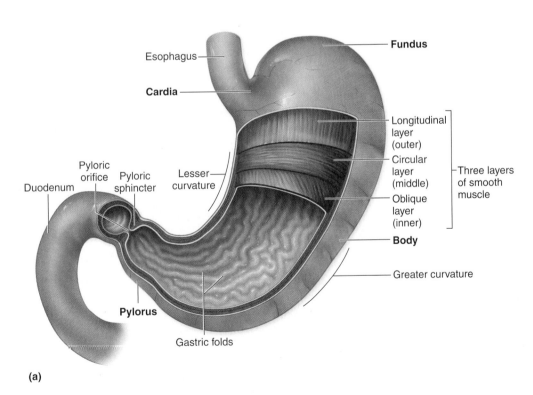

(a)

Gross Anatomy of the Stomach
Figure 26.12a

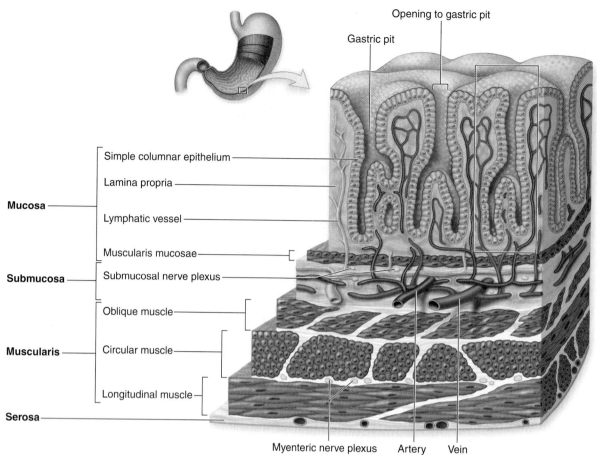

Opening to gastric pit

Gastric pit

Simple columnar epithelium

Lamina propria

Mucosa

Lymphatic vessel

Muscularis mucosae

Submucosa

Submucosal nerve plexus

Oblique muscle

Muscularis

Circular muscle

Longitudinal muscle

Serosa

Myenteric nerve plexus   Artery   Vein

(a) Sectional view

Histology of the Stomach Wall
Figure 26.13a

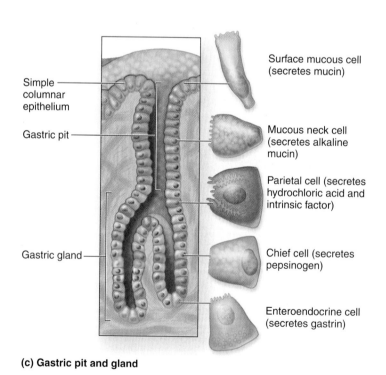

Surface mucous cell
(secretes mucin)

Simple
columnar
epithelium

Mucous neck cell
(secretes alkaline
mucin)

Gastric pit

Parietal cell (secretes
hydrochloric acid and
intrinsic factor)

Gastric gland

Chief cell (secretes
pepsinogen)

Enteroendocrine cell
(secretes gastrin)

(c) Gastric pit and gland

Gastric Pit and Gland
Figure 26.13c

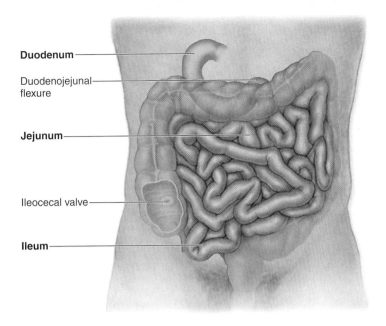

Gross Anatomy of the Small Intestine
Figure 26.14

Duodenum

Duodenojejunal flexure

Jejunum

Ileocecal valve

Ileum

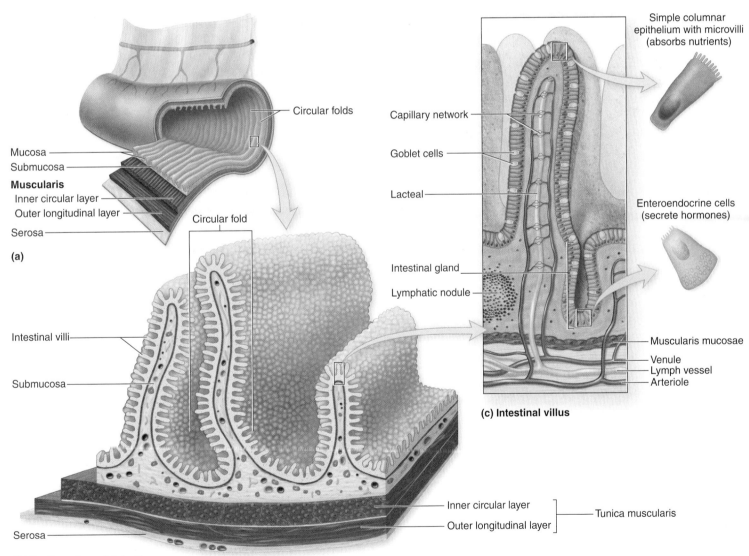

Circular folds

Mucosa
Submucosa
Muscularis
Inner circular layer
Outer longitudinal layer
Serosa

(a)

Circular fold

Intestinal villi

Submucosa

Serosa

(b) Section of small intestine

Capillary network

Goblet cells

Lacteal

Intestinal gland

Lymphatic nodule

Simple columnar epithelium with microvilli (absorbs nutrients)

Enteroendocrine cells (secrete hormones)

Muscularis mucosae
Venule
Lymph vessel
Arteriole

(c) Intestinal villus

Inner circular layer
Outer longitudinal layer
Tunica muscularis

Histology of the Small Intestine
Figure 26.15a,b,c

377

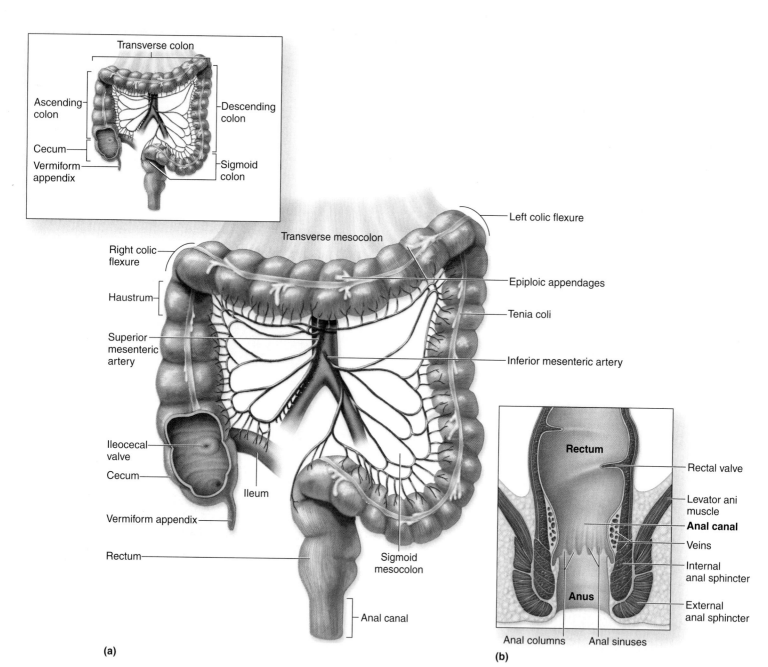

Transverse colon

Ascending colon

Descending colon

Cecum

Vermiform appendix

Sigmoid colon

Transverse mesocolon

Left colic flexure

Right colic flexure

Epiploic appendages

Haustrum

Tenia coli

Superior mesenteric artery

Inferior mesenteric artery

Ileocecal valve

Cecum

Vermiform appendix

Ileum

Rectum

Sigmoid mesocolon

Rectum

Rectal valve

Levator ani muscle

Anal canal

Veins

Internal anal sphincter

Anus

External anal sphincter

Anal columns

Anal sinuses

Anal canal

(a)

(b)

Gross Anatomy of the Large Intestine
Figure 26.16

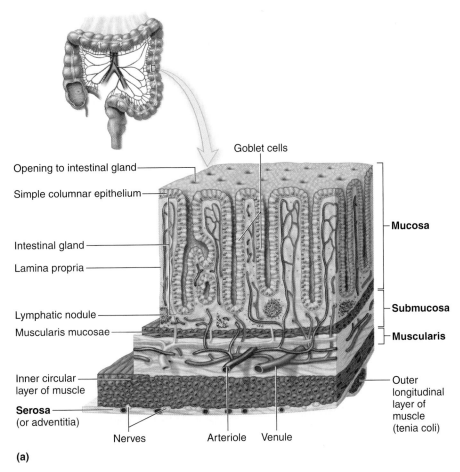

Opening to intestinal gland

Simple columnar epithelium

Intestinal gland

Lamina propria

Lymphatic nodule

Muscularis mucosae

Inner circular layer of muscle

Serosa (or adventitia)

Goblet cells

Mucosa

Submucosa

Muscularis

Outer longitudinal layer of muscle (tenia coli)

Nerves Arteriole Venule

(a)

Histology of the Large Intestine
Figure 26.17a

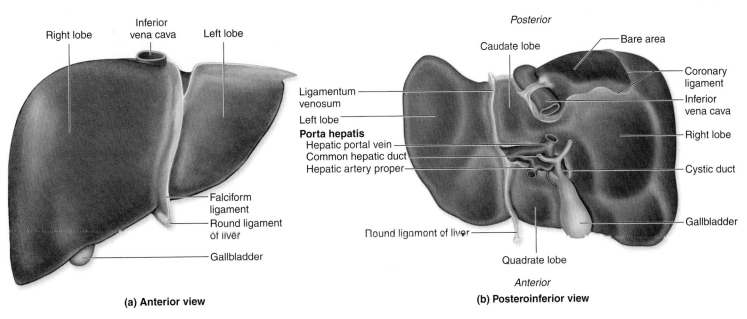

Right lobe

Inferior vena cava

Left lobe

Falciform ligament

Round ligament of liver

Gallbladder

(a) Anterior view

Posterior

Caudate lobe

Bare area

Ligamentum venosum

Left lobe

Porta hepatis
Hepatic portal vein
Common hepatic duct
Hepatic artery proper

Coronary ligament

Inferior vena cava

Right lobe

Cystic duct

Round ligament of liver

Quadrate lobe

Gallbladder

Anterior

(b) Posteroinferior view

Gross Anatomy of the Liver
Figure 26.18

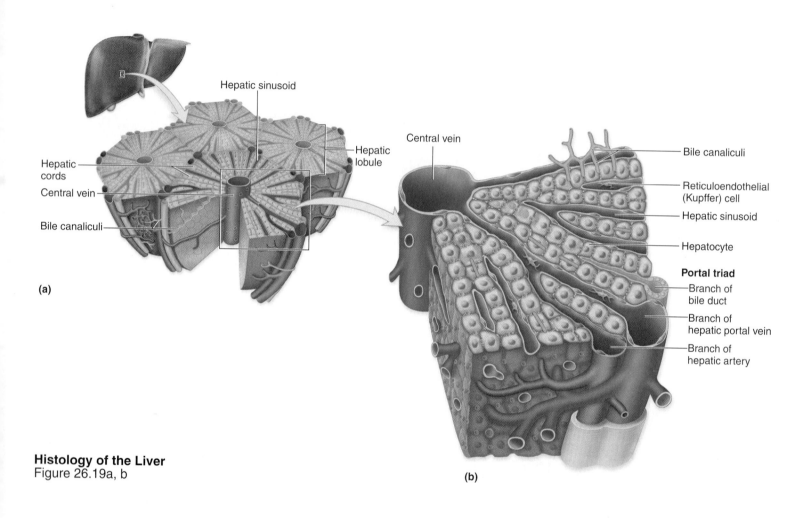

(a)

Central vein

Bile canaliculi

Reticuloendothelial (Kupffer) cell

Hepatic sinusoid

Hepatocyte

Portal triad
Branch of bile duct

Branch of hepatic portal vein

Branch of hepatic artery

(b)

Hepatic sinusoid

Hepatic lobule

Hepatic cords

Central vein

Bile canaliculi

Histology of the Liver
Figure 26.19a, b

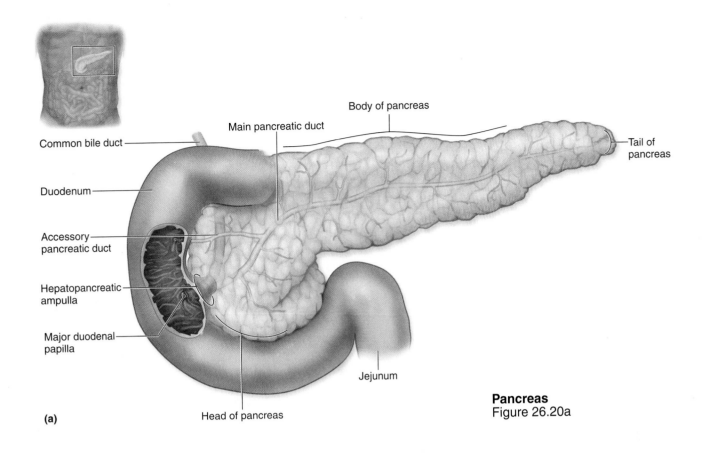

Common bile duct

Duodenum

Accessory pancreatic duct

Hepatopancreatic ampulla

Major duodenal papilla

Main pancreatic duct

Body of pancreas

Tail of pancreas

Jejunum

Head of pancreas

(a)

Pancreas
Figure 26.20a

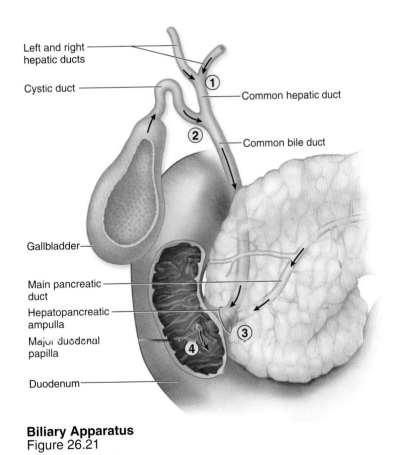

Left and right hepatic ducts

Cystic duct

Gallbladder

Main pancreatic duct

Hepatopancreatic ampulla

Major duodenal papilla

Duodenum

Common hepatic duct

Common bile duct

Biliary Apparatus
Figure 26.21

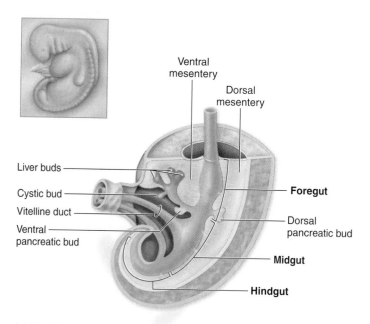

Ventral mesentery

Dorsal mesentery

Liver buds

Cystic bud

Vitelline duct

Ventral pancreatic bud

Foregut

Dorsal pancreatic bud

Midgut

Hindgut

(a) Week 4: Liver, gallbladder, and pancreatic buds develop

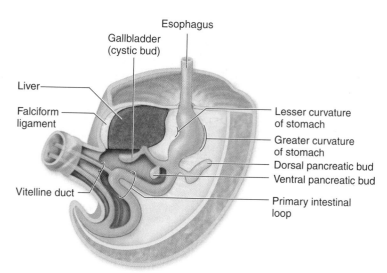

Esophagus

Gallbladder (cystic bud)

Liver

Falciform ligament

Lesser curvature of stomach

Greater curvature of stomach

Dorsal pancreatic bud

Ventral pancreatic bud

Vitelline duct

Primary intestinal loop

(b) Week 5: Greater and lesser curvatures of stomach form

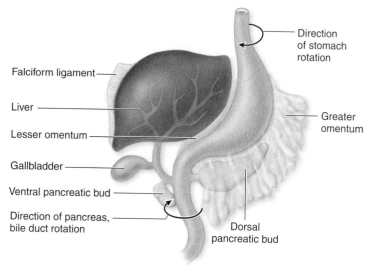

Direction of stomach rotation

Falciform ligament

Liver

Lesser omentum

Gallbladder

Ventral pancreatic bud

Direction of pancreas, bile duct rotation

Greater omentum

Dorsal pancreatic bud

(c) Weeks 6–7: Rotation of stomach, pancreatic buds

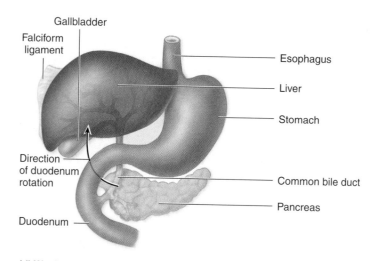

Gallbladder

Falciform ligament

Esophagus

Liver

Stomach

Direction of duodenum rotation

Common bile duct

Pancreas

Duodenum

(d) Week 8: Postnatal position of organs attained

Development of the Foregut
Figure 26.22

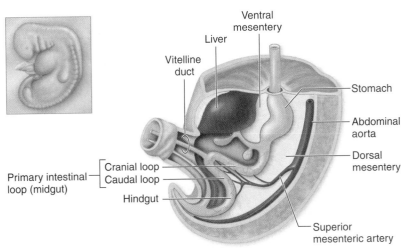

(a) Week 5: Primary intestinal loop forms

Ventral mesentery
Liver
Vitelline duct
Stomach
Abdominal aorta
Dorsal mesentery
Primary intestinal loop (midgut) — Cranial loop, Caudal loop
Hindgut
Superior mesenteric artery

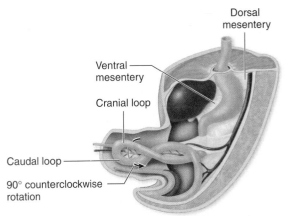

(b) Week 6: Herniation of loop; 90° counterclockwise rotation

Dorsal mesentery
Ventral mesentery
Cranial loop
Caudal loop
90° counterclockwise rotation

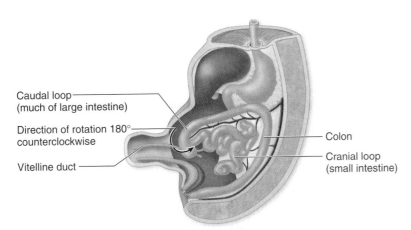

(c) Weeks 10–11: Retraction of intestines back into abdominal cavity; 180° counterclockwise rotation

Caudal loop (much of large intestine)
Direction of rotation 180° counterclockwise
Vitelline duct
Colon
Cranial loop (small intestine)

Development of the Midgut
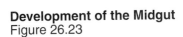
Figure 26.23

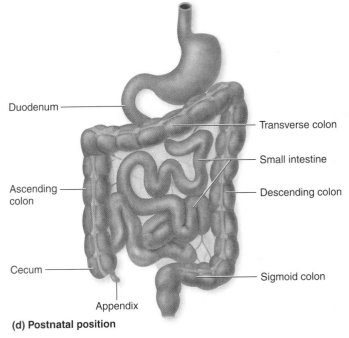

(d) Postnatal position

Duodenum
Transverse colon
Small intestine
Ascending colon
Descending colon
Cecum
Sigmoid colon
Appendix

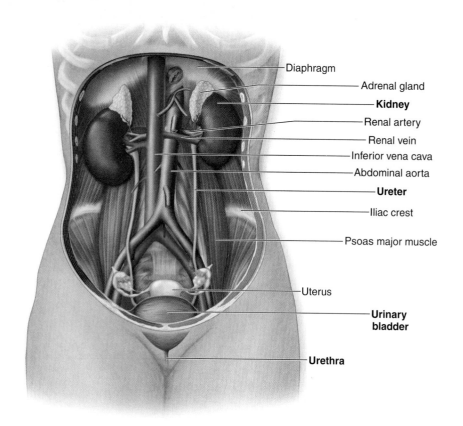

- Diaphragm
- Adrenal gland
- **Kidney**
- Renal artery
- Renal vein
- Inferior vena cava
- Abdominal aorta
- **Ureter**
- Iliac crest
- Psoas major muscle
- Uterus
- **Urinary bladder**
- **Urethra**

(a) Anterior view

Urinary System—Anterior View
Figure 27.1a

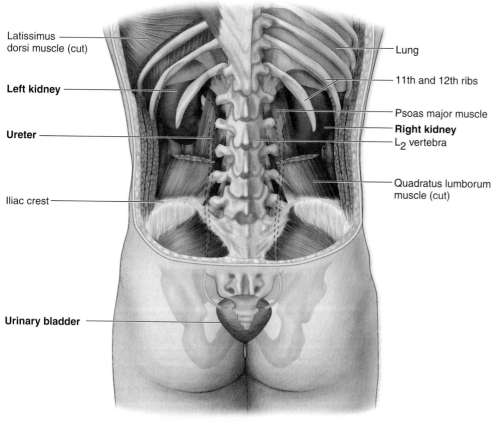

- Latissimus dorsi muscle (cut)
- **Left kidney**
- **Ureter**
- Iliac crest
- **Urinary bladder**
- Lung
- 11th and 12th ribs
- Psoas major muscle
- **Right kidney**
- L$_2$ vertebra
- Quadratus lumborum muscle (cut)

(b) Posterior view

Urinary System—Posterior View
Figure 27.1b

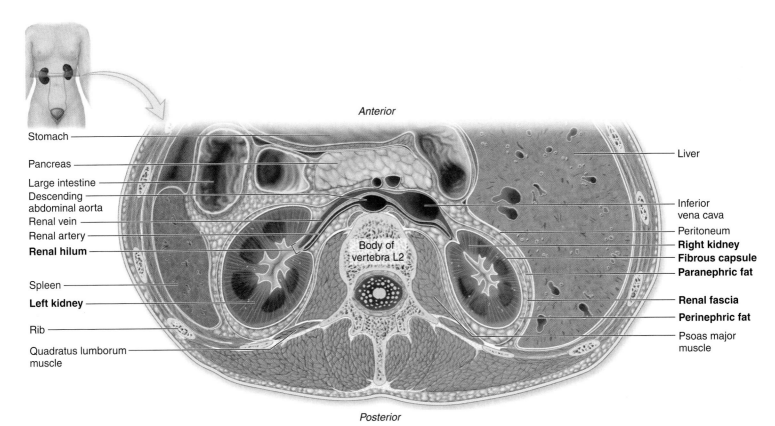

Anterior

Stomach

Pancreas

Large intestine

Descending abdominal aorta

Renal vein

Renal artery

Renal hilum

Spleen

Left kidney

Rib

Quadratus lumborum muscle

Liver

Inferior vena cava

Peritoneum

Right kidney

Fibrous capsule

Paranephric fat

Renal fascia

Perinephric fat

Psoas major muscle

Body of vertebra L2

Posterior

Position and Stabilization of the Kidneys
Figure 27.2

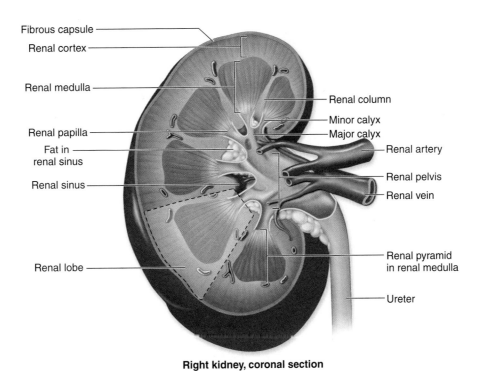

Fibrous capsule

Renal cortex

Renal medulla

Renal papilla

Fat in renal sinus

Renal sinus

Renal lobe

Renal column

Minor calyx

Major calyx

Renal artery

Renal pelvis

Renal vein

Renal pyramid in renal medulla

Ureter

Right kidney, coronal section

Gross Anatomy of the Kidney
Figure 27.3

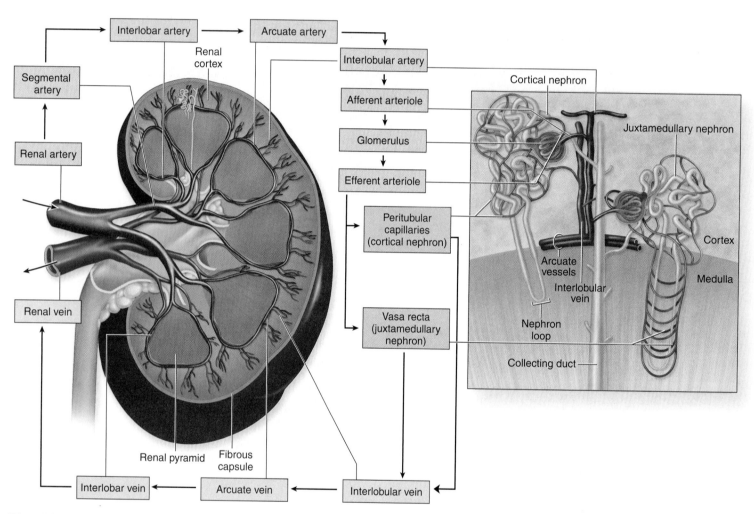

Blood Supply to the Kidneys
Figure 27.4

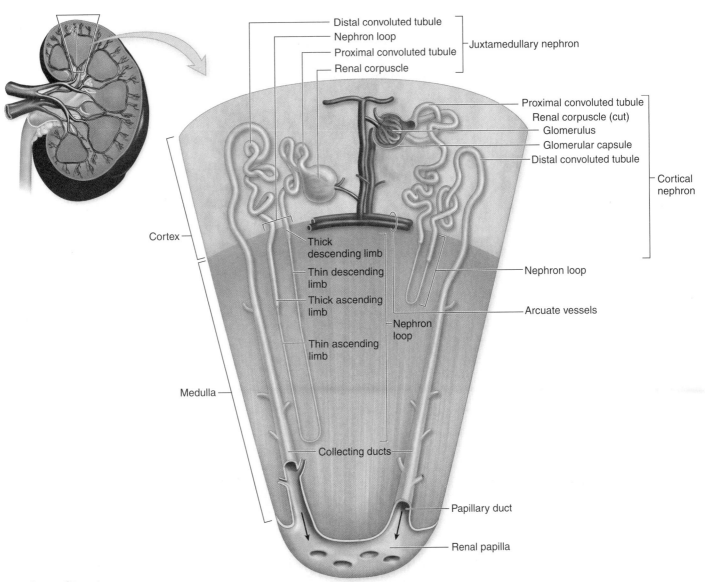

Distal convoluted tubule
Nephron loop
Proximal convoluted tubule
Renal corpuscle
Juxtamedullary nephron

Proximal convoluted tubule
Renal corpuscle (cut)
Glomerulus
Glomerular capsule
Distal convoluted tubule
Cortical nephron

Cortex

Thick descending limb
Thin descending limb
Thick ascending limb
Thin ascending limb
Nephron loop

Nephron loop

Arcuate vessels

Medulla

Collecting ducts

Papillary duct

Renal papilla

Nephron Structure
Figure 27.5

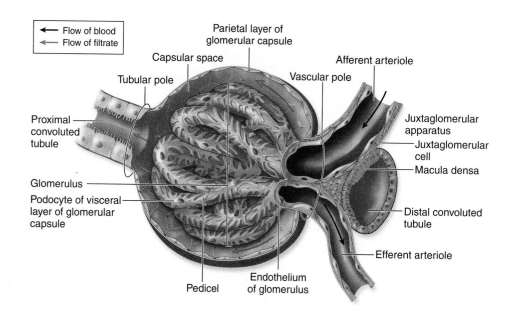

Flow of blood
Flow of filtrate

Parietal layer of glomerular capsule

Capsular space

Tubular pole

Afferent arteriole

Vascular pole

Proximal convoluted tubule

Juxtaglomerular apparatus

Juxtaglomerular cell

Macula densa

Glomerulus

Podocyte of visceral layer of glomerular capsule

Distal convoluted tubule

Efferent arteriole

Pedicel

Endothelium of glomerulus

(a) Renal corpuscle

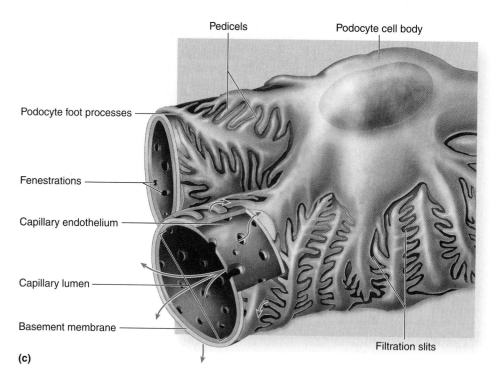

Pedicels

Podocyte cell body

Podocyte foot processes

Fenestrations

Capillary endothelium

Capillary lumen

Basement membrane

Filtration slits

(c)

Renal Corpuscle
Figure 27.6a,c

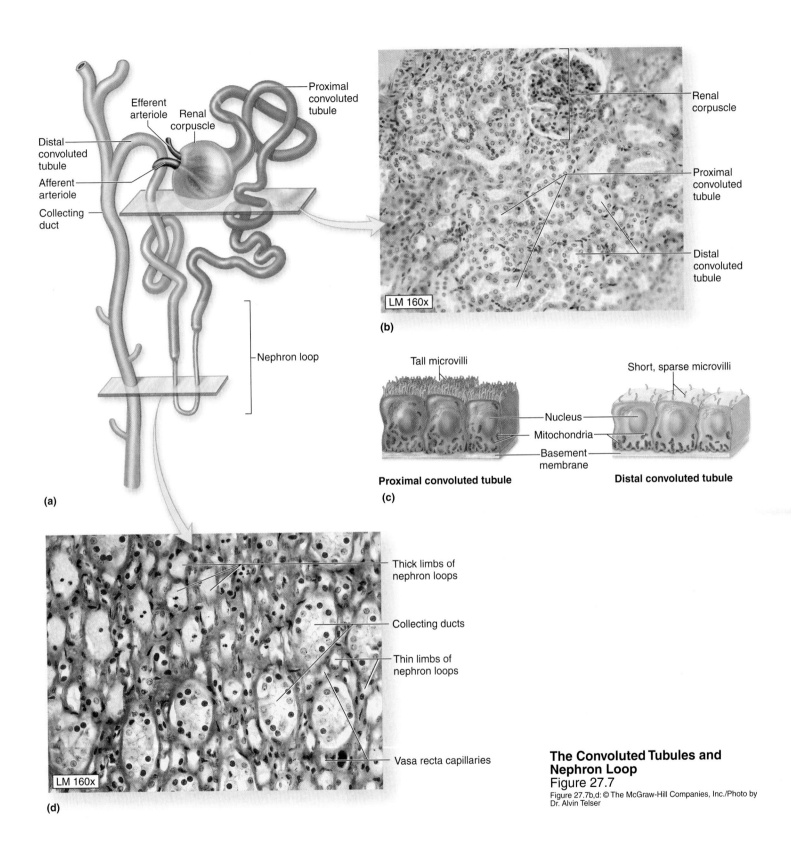

(a)

Efferent arteriole

Renal corpuscle

Proximal convoluted tubule

Distal convoluted tubule

Afferent arteriole

Collecting duct

Nephron loop

(b)

LM 160x

Renal corpuscle

Proximal convoluted tubule

Distal convoluted tubule

(c)

Tall microvilli

Short, sparse microvilli

Nucleus

Mitochondria

Basement membrane

Proximal convoluted tubule

Distal convoluted tubule

(d)

LM 160x

Thick limbs of nephron loops

Collecting ducts

Thin limbs of nephron loops

Vasa recta capillaries

The Convoluted Tubules and Nephron Loop

Figure 27.7

Figure 27.7b,d: © The McGraw-Hill Companies, Inc./Photo by Dr. Alvin Telser

389

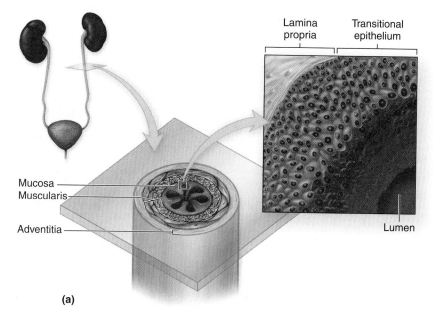

Ureters
Figure 27.8a

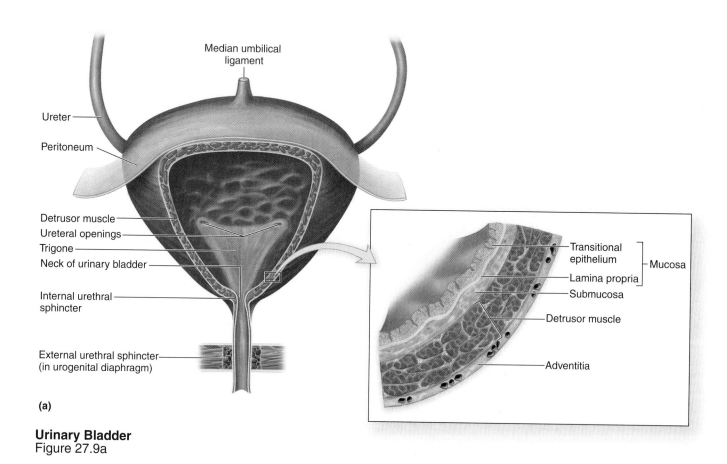

Urinary Bladder
Figure 27.9a

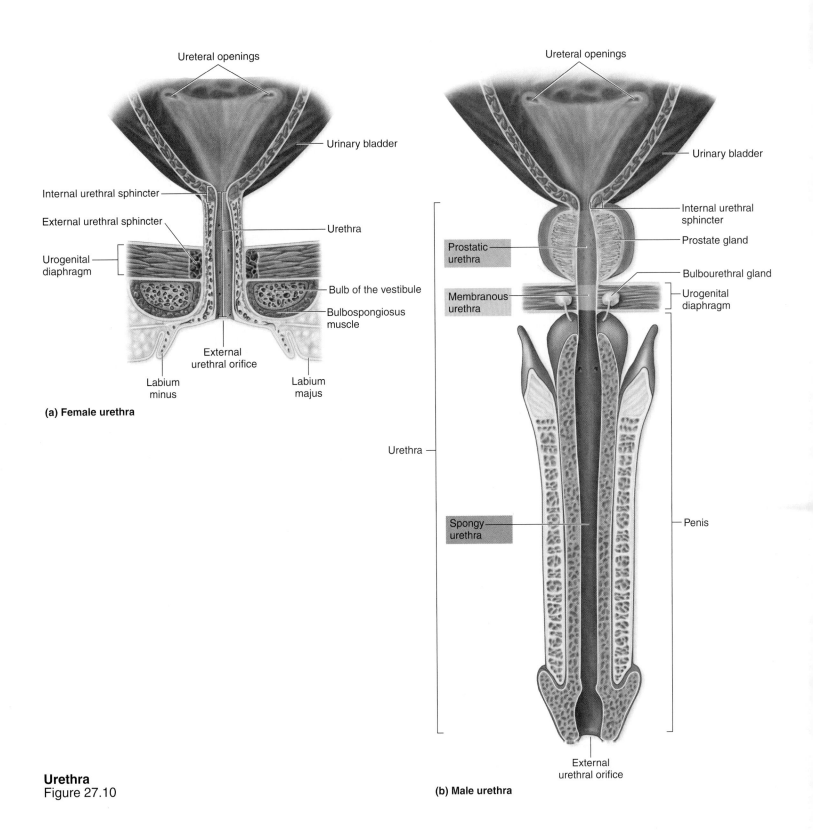

Ureteral openings

Urinary bladder

Internal urethral sphincter

External urethral sphincter

Urethra

Urogenital diaphragm

Bulb of the vestibule

Bulbospongiosus muscle

External urethral orifice

Labium minus

Labium majus

(a) Female urethra

Ureteral openings

Urinary bladder

Internal urethral sphincter

Prostatic urethra

Prostate gland

Bulbourethral gland

Membranous urethra

Urogenital diaphragm

Urethra

Spongy urethra

Penis

External urethral orifice

(b) Male urethra

Urethra
Figure 27.10

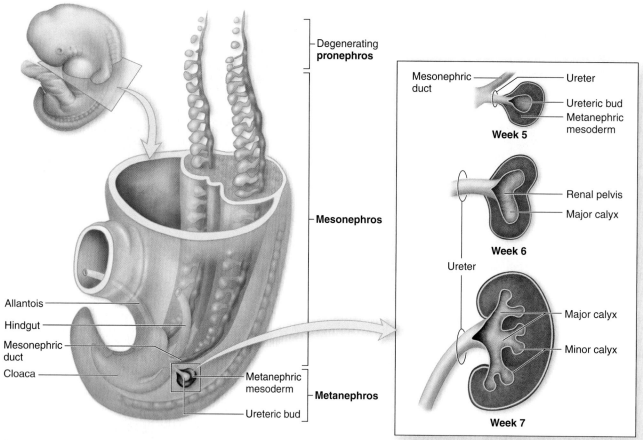

- Degenerating **pronephros**

- **Mesonephros**

Allantois
Hindgut
Mesonephric duct
Cloaca

Metanephric mesoderm

Ureteric bud

- **Metanephros**

(a) Week 5

Mesonephric duct — Ureter
— Ureteric bud
— Metanephric mesoderm
Week 5

Renal pelvis
Major calyx
Week 6

Ureter

Major calyx
Minor calyx
Week 7

(b) Metanephric kidney formation

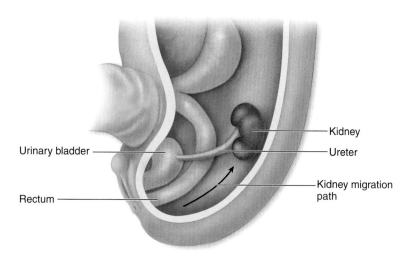

Urinary bladder — — Kidney
— Ureter
Rectum — — Kidney migration path

(c) Weeks 6–9: Kidney migrates from pelvis to lumbar region

Kidney Development
Figure 27.11

392

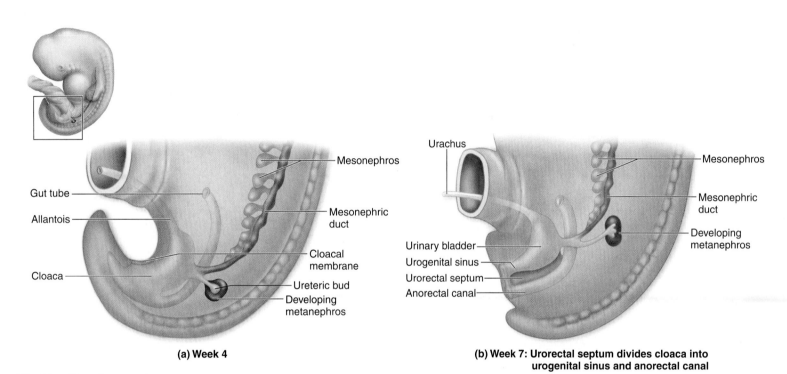

Gut tube

Allantois

Cloaca

Mesonephros

Mesonephric duct

Cloacal membrane

Ureteric bud

Developing metanephros

(a) Week 4

Urachus

Urinary bladder

Urogenital sinus

Urorectal septum

Anorectal canal

Mesonephros

Mesonephric duct

Developing metanephros

(b) Week 7: Urorectal septum divides cloaca into urogenital sinus and anorectal canal

Bladder Development
Figure 27.12

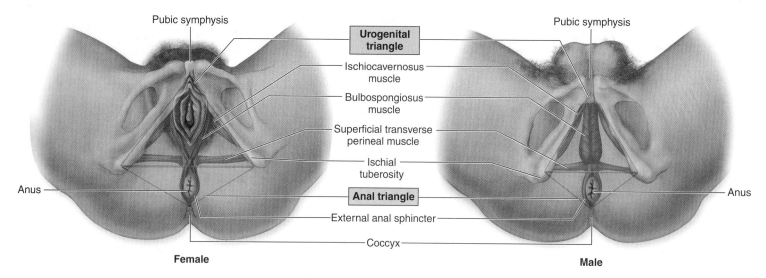

Pubic symphysis

Urogenital triangle

Ischiocavernosus muscle

Bulbospongiosus muscle

Superficial transverse perineal muscle

Ischial tuberosity

Anal triangle

Anus

External anal sphincter

Coccyx

Pubic symphysis

Anus

Female

Male

Perineum
Figure 28.1

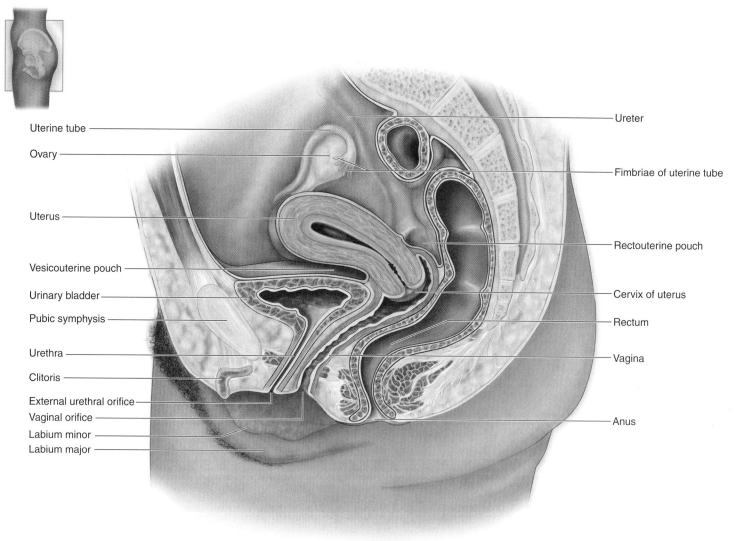

Uterine tube

Ovary

Uterus

Vesicouterine pouch

Urinary bladder

Pubic symphysis

Urethra

Clitoris

External urethral orifice

Vaginal orifice

Labium minor

Labium major

Ureter

Fimbriae of uterine tube

Rectouterine pouch

Cervix of uterus

Rectum

Vagina

Anus

Sagittal Section of the Female Pelvic Region
Figure 28.2

394

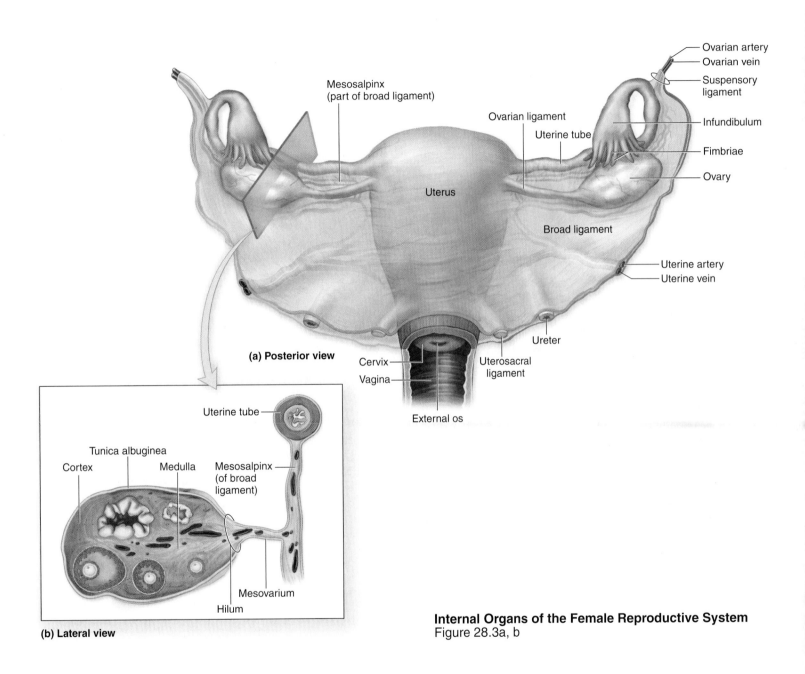

Mesosalpinx
(part of broad ligament)

Ovarian artery
Ovarian vein
Suspensory ligament

Ovarian ligament

Uterine tube

Infundibulum

Fimbriae

Ovary

Uterus

Broad ligament

Uterine artery
Uterine vein

Ureter

(a) Posterior view

Cervix

Vagina

Uterosacral ligament

External os

Uterine tube

Tunica albuginea

Cortex

Medulla

Mesosalpinx
(of broad ligament)

Mesovarium

Hilum

(b) Lateral view

Internal Organs of the Female Reproductive System
Figure 28.3a, b

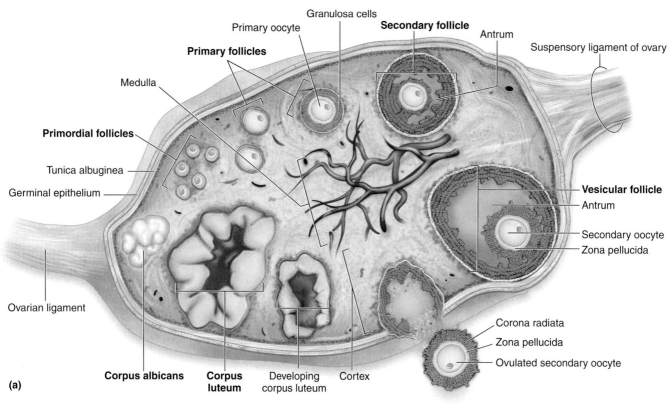

Granulosa cells

Primary oocyte

Secondary follicle

Antrum

Suspensory ligament of ovary

Primary follicles

Medulla

Primordial follicles

Tunica albuginea

Germinal epithelium

Vesicular follicle

Antrum

Secondary oocyte

Zona pellucida

Ovarian ligament

Corona radiata

Zona pellucida

Ovulated secondary oocyte

(a)

Corpus albicans

Corpus luteum

Developing corpus luteum

Cortex

Coronal Section of Ovary
Figure 28.4a

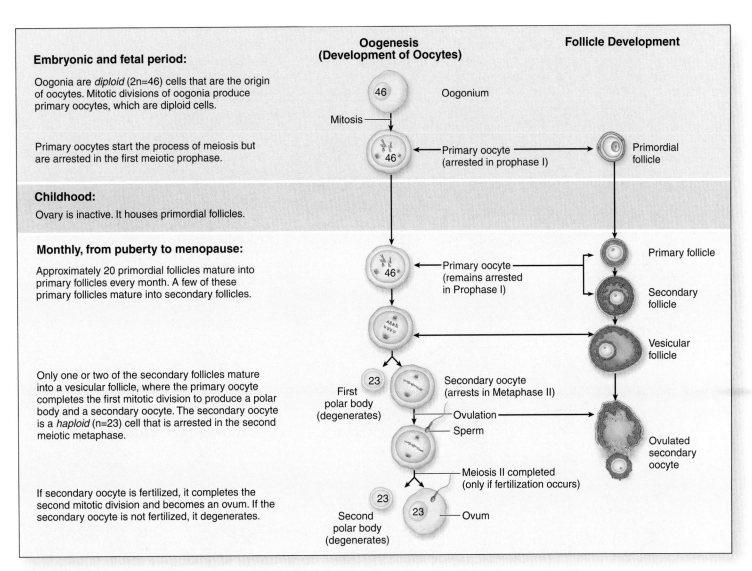

Embryonic and fetal period:

Oogonia are *diploid* (2n=46) cells that are the origin of oocytes. Mitotic divisions of oogonia produce primary oocytes, which are diploid cells.

Primary oocytes start the process of meiosis but are arrested in the first meiotic prophase.

Childhood:

Ovary is inactive. It houses primordial follicles.

Monthly, from puberty to menopause:

Approximately 20 primordial follicles mature into primary follicles every month. A few of these primary follicles mature into secondary follicles.

Only one or two of the secondary follicles mature into a vesicular follicle, where the primary oocyte completes the first mitotic division to produce a polar body and a secondary oocyte. The secondary oocyte is a *haploid* (n=23) cell that is arrested in the second meiotic metaphase.

If secondary oocyte is fertilized, it completes the second mitotic division and becomes an ovum. If the secondary oocyte is not fertilized, it degenerates.

**Oogenesis
(Development of Oocytes)**

Follicle Development

46 — Oogonium

Mitosis

46 — Primary oocyte (arrested in prophase I) — Primordial follicle

46 — Primary oocyte (remains arrested in Prophase I) — Primary follicle / Secondary follicle

Vesicular follicle

23 First polar body (degenerates) — Secondary oocyte (arrests in Metaphase II)

Ovulation — Sperm — Ovulated secondary oocyte

Meiosis II completed (only if fertilization occurs)

23 Second polar body (degenerates) — 23 Ovum

Oogenesis
Figure 28.5

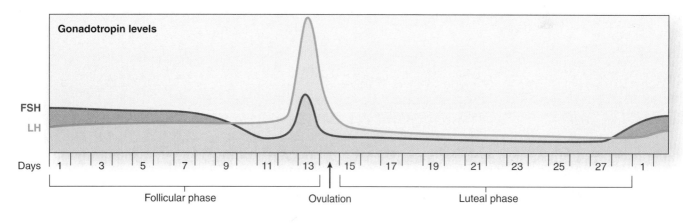

Gonadotropin levels

FSH
LH

Days 1 3 5 7 9 11 13 ↑ 15 17 19 21 23 25 27 1

Follicular phase · Ovulation · Luteal phase

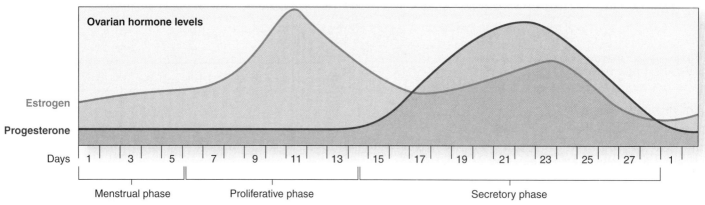

Ovarian hormone levels

Estrogen
Progesterone

Days 1 3 5 7 9 11 13 15 17 19 21 23 25 27 1

Menstrual phase · Proliferative phase · Secretory phase

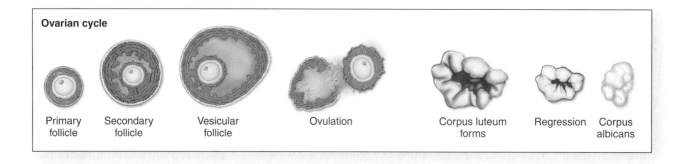

Ovarian cycle

Primary follicle · Secondary follicle · Vesicular follicle · Ovulation · Corpus luteum forms · Regression · Corpus albicans

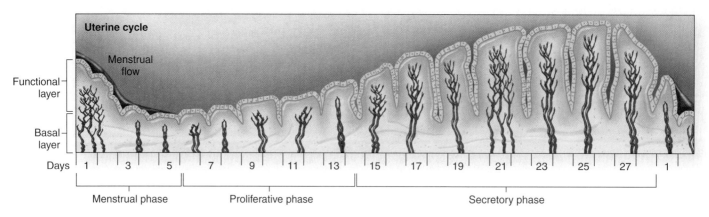

Uterine cycle

Menstrual flow

Functional layer

Basal layer

Days 1 3 5 7 9 11 13 15 17 19 21 23 25 27 1

Menstrual phase · Proliferative phase · Secretory phase

Hormonal Changes in Female Reproductive System
Figure 28.6

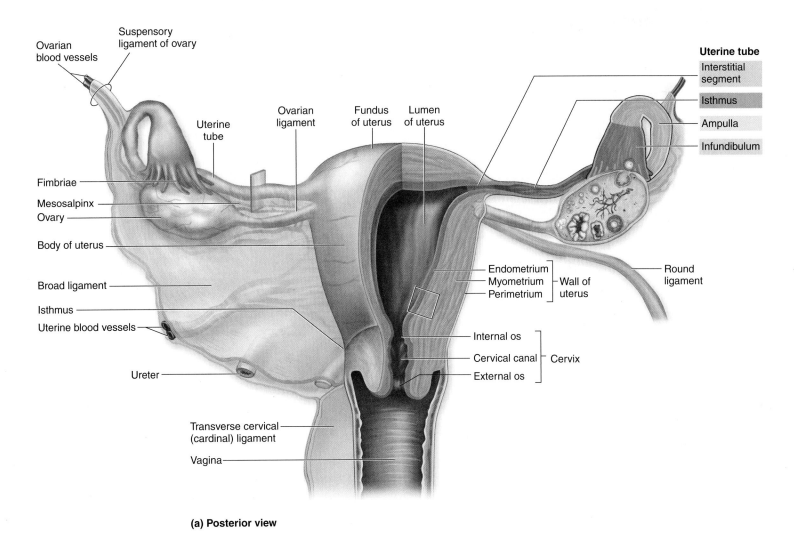

Uterine Tubes and the Uterus
Figure 28.7a

(a) Posterior view

Ovarian blood vessels

Suspensory ligament of ovary

Uterine tube

Ovarian ligament

Fundus of uterus

Lumen of uterus

Uterine tube
- Interstitial segment
- Isthmus
- Ampulla
- Infundibulum

Fimbriae

Mesosalpinx

Ovary

Body of uterus

Broad ligament

Isthmus

Uterine blood vessels

Ureter

Endometrium
Myometrium — Wall of uterus
Perimetrium

Round ligament

Internal os
Cervical canal — Cervix
External os

Transverse cervical (cardinal) ligament

Vagina

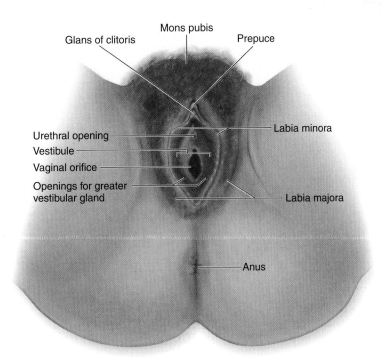

Female External Genitalia
Figure 28.9

Glans of clitoris

Mons pubis

Prepuce

Urethral opening

Vestibule

Vaginal orifice

Openings for greater vestibular gland

Labia minora

Labia majora

Anus

399

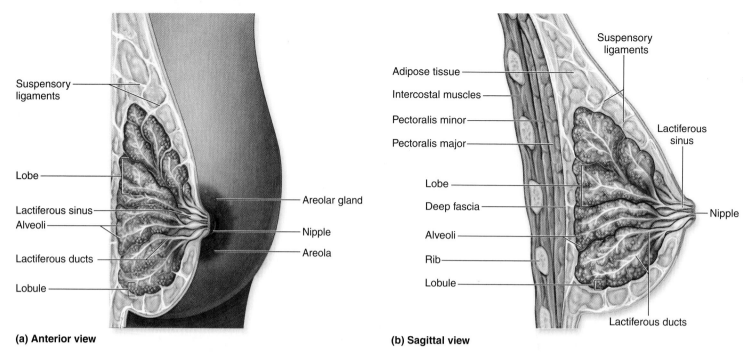

Suspensory
ligaments

Lobe

Lactiferous sinus

Alveoli

Lactiferous ducts

Lobule

Areolar gland

Nipple

Areola

(a) Anterior view

Adipose tissue

Intercostal muscles

Pectoralis minor

Pectoralis major

Lobe

Deep fascia

Alveoli

Rib

Lobule

Suspensory
ligaments

Lactiferous
sinus

Nipple

Lactiferous ducts

(b) Sagittal view

Mammary Glands
Figure 28.10

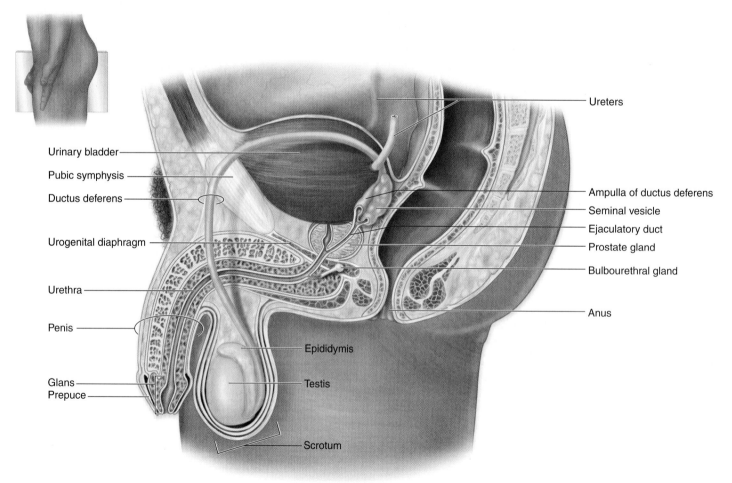

Urinary bladder

Pubic symphysis

Ductus deferens

Urogenital diaphragm

Urethra

Penis

Glans

Prepuce

Epididymis

Testis

Scrotum

Ureters

Ampulla of ductus deferens

Seminal vesicle

Ejaculatory duct

Prostate gland

Bulbourethral gland

Anus

Male Pelvic Region
Figure 28.11

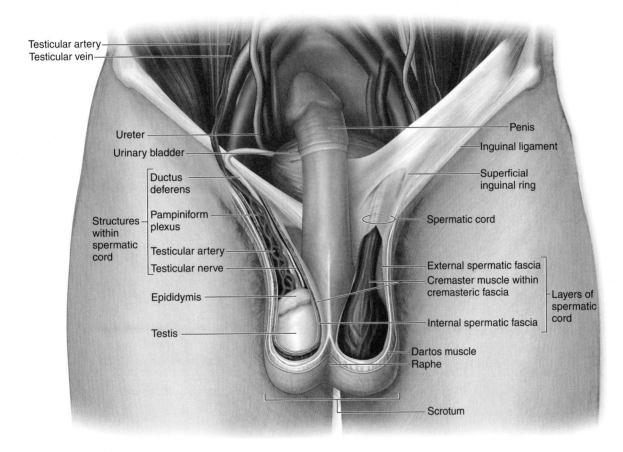

Testicular artery
Testicular vein

Ureter
Urinary bladder

Structures within spermatic cord
- Ductus deferens
- Pampiniform plexus
- Testicular artery
- Testicular nerve

Epididymis

Testis

Penis
Inguinal ligament

Superficial inguinal ring

Spermatic cord

External spermatic fascia
Cremaster muscle within cremasteric fascia

Internal spermatic fascia

Dartos muscle
Raphe

Layers of spermatic cord

Scrotum

Scrotum and Testes
Figure 28.12

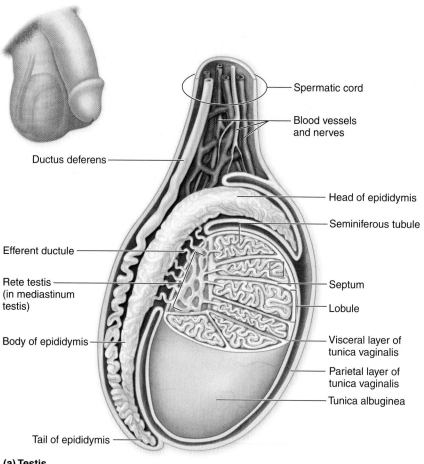

Spermatic cord

Blood vessels
and nerves

Ductus deferens

Head of epididymis

Seminiferous tubule

Efferent ductule

Rete testis
(in mediastinum
testis)

Septum

Lobule

Body of epididymis

Visceral layer of
tunica vaginalis

Parietal layer of
tunica vaginalis

Tunica albuginea

Tail of epididymis

(a) Testis

Testis
Figure 28.13a

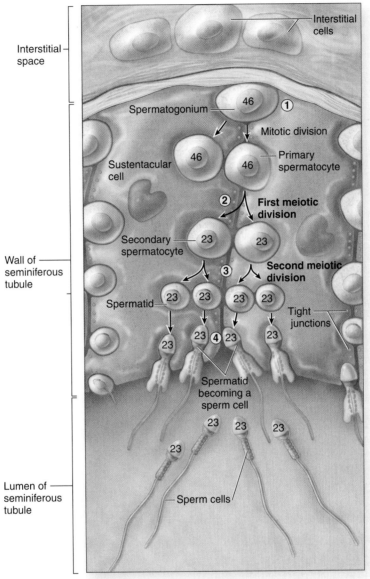

Interstitial cells

Interstitial space

Spermatogonium — 46 ①

Mitotic division

Sustentacular cell 46 46 — **Primary spermatocyte**

② **First meiotic division**

Secondary spermatocyte 23 23

③ **Second meiotic division**

Spermatid 23 23 23 23 — **Tight junctions**

23 ④ 23 23

Spermatid becoming a sperm cell

23 23 23

Lumen of seminiferous tubule

Wall of seminiferous tubule

23

Sperm cells

(a) Spermatogenesis

Spermatogenesis
Figure 28.14a

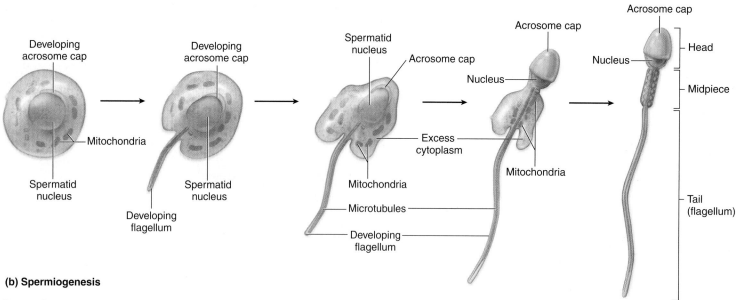

(b) Spermiogenesis

Spermiogenesis
Figure 28.14b

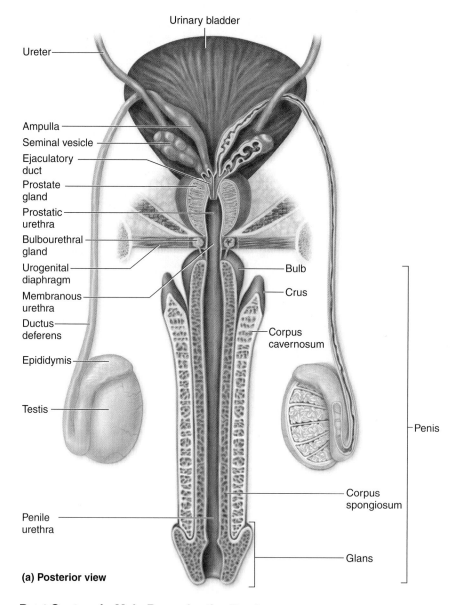

(a) Posterior view

Duct System in Male Reproductive Tract
Figure 28.15a

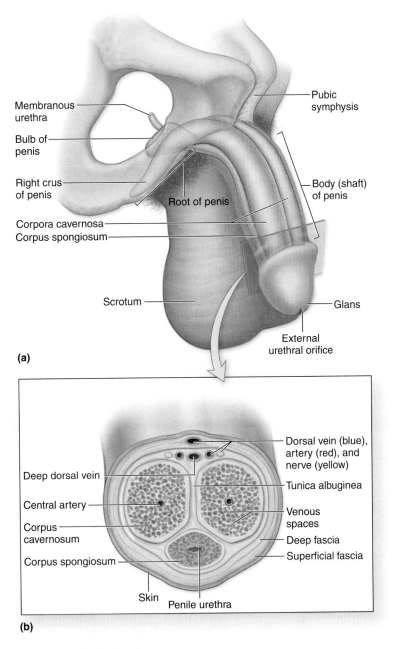

(a)

Membranous urethra

Bulb of penis

Right crus of penis

Root of penis

Corpora cavernosa

Corpus spongiosum

Scrotum

Pubic symphysis

Body (shaft) of penis

Glans

External urethral orifice

(b)

Deep dorsal vein

Central artery

Corpus cavernosum

Corpus spongiosum

Skin

Penile urethra

Dorsal vein (blue), artery (red), and nerve (yellow)

Tunica albuginea

Venous spaces

Deep fascia

Superficial fascia

Anatomy of the Penis
Figure 28.17

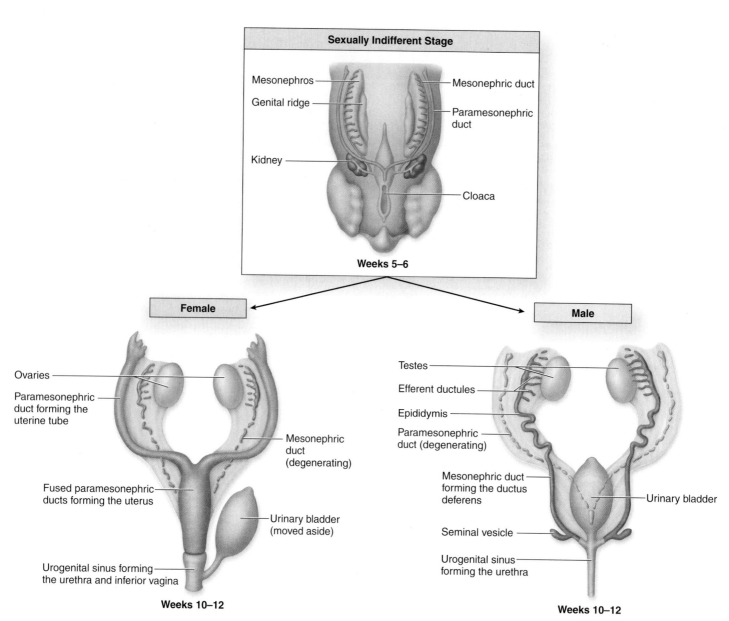

Embryonic Development of the Female and Male Reproductive Tracts—Indifferent Stage and Weeks 10-12

Figure 28.18 top

406

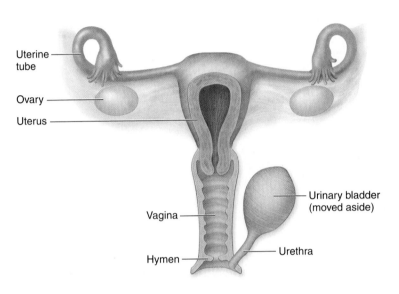

Uterine tube

Ovary

Uterus

Vagina

Hymen

Urinary bladder (moved aside)

Urethra

At birth

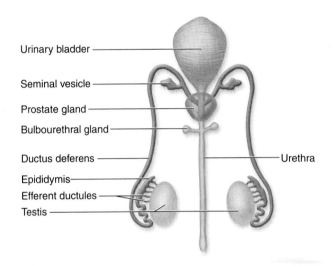

Urinary bladder

Seminal vesicle

Prostate gland

Bulbourethral gland

Ductus deferens

Epididymis

Efferent ductules

Testis

Urethra

At birth

Embryonic Development of the Female and Male Reproductive Tracts—At Birth
Figure 28.18 bottom

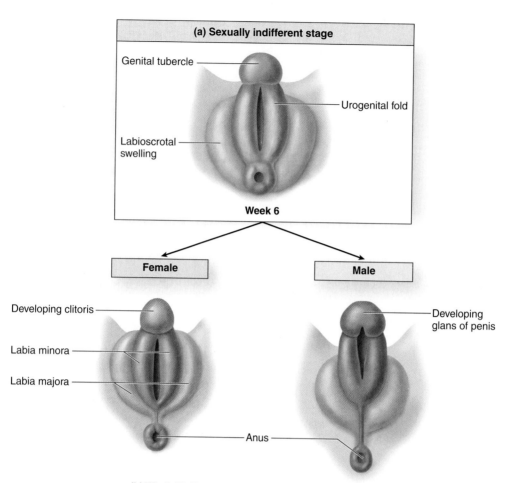

(a) Sexually indifferent stage

Genital tubercle

Urogenital fold

Labioscrotal swelling

Week 6

Female

Male

Developing clitoris

Developing glans of penis

Labia minora

Labia majora

Anus

(b) Week 12: Urogenital folds begin to fuse in the male

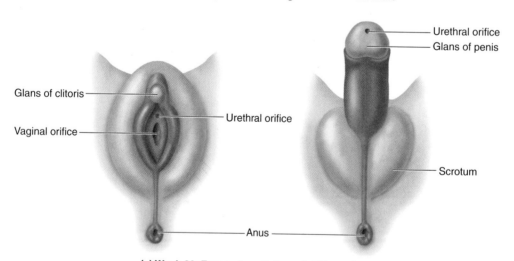

Urethral orifice

Glans of penis

Glans of clitoris

Urethral orifice

Vaginal orifice

Scrotum

Anus

(c) Week 20: External genitalia well differentiated

Development of External Genitalia
Figure 28.19

NOTES

NOTES

NOTES

NOTES

NOTES

NOTES

NOTES

NOTES

NOTES

NOTES

NOTES

NOTES

NOTES

NOTES